CW00506428

THE HISTORY OF THE WELSH PONY

The History of the Welsh Pony

Published by
Medina Publishing Ltd
9 St John's Place
Newport
Isle of Wight
PO30 1LH
www.medinapublishing.com

ISBN 978-1-909339-02-6

Designed by Kitty Carruthers
Printed and bound by Toppan Leefung Printing Ltd, China

CIP Data: A catalogue record for this book is available from the British Library.

THE HISTORY OF THE WELSH PONY

Tom Best

Acknowledgements

I would like to thank all the many people who have provided information, photographs and interesting material to allow the completion of this book.

In particular, I would like to thank David Blair and Dr Wynne Davies for their encouragement and for proof reading the text.

Photographic Credits

Every effort has been made to trace the copyright owners of the photographs used in this book but if there are any omissions or inaccuracies the author and publishers would be pleased to be informed.

I Aeron 115; Aldridge Family 33,41,47; B Astor 203; Bassettlaw Museum 7; D Blair 105; T ten Brinke 247, 267, 272, 284, 302, 307, 325; M Butterworth 263; K Cheetham 145, 253; S Coles 295; A Corbishley Collection15; J Couttie 193; M Daley 227; De Walden Estates 34,39; Dr E W Davies Archive 10, 13, 28, 31, 32, 50, 52, 64-66, 69, 86, 132, 138, 142, 147, 174; D Davies 115; W Davies 148, 174; S Deane 90; I Delaitre 229,291,304,310,324; J Edwards 311; E Ingram Collection 9,10; The Event Photographer 233, 276, 277, 338, 339, 341; F Fennell 312; C Gilson 166, 228, 244, 249, 261, 263, 274, 288, 318; M Grundie 184; J Hainke 187; Horse & Hound 72; Horse Power Photography 329; D Hounsham 297; A James 229; J Jones Collection 75, 76, 82, 84, 186; P Jones 162; R Jones 325; V Karlestrand 181; M Keijzen 279; Lord Kenyon 131; A Knowles 212; F Leadbitter 254; L Lane 217; E Mansfield-Parnell 144, 180; L Mayall 146, 148, 164, 194, 201, 221, 206; L Mayes 127, 249; R Miller 223; NPS Stud Book 11, 27, 53, 70; Monty 189; Noble Photos 315; PPS Stud Book 8; R Parker 303; G Parsons 31; Photonews 191, 193, 232, 242 Pleasure Prints 213, 240, 284; J Pringle 218; RTI 269, 273, 292, 318, 331; R Rees 114, 115, 136; A Reynolds 118,137, 214, 219, 225, 230, 247, 250, 251, 257, 287, 308, 323, 328, 330, 337; L Reynolds 68, 198; Riding magazine 17, 111; N Robertson 301; H Sloane 183; T Smalley 200, 208; Sport & General 121; J Teewen 182; The Field 104; University of Reading 26; P Upton 18, 25, 29, 30, 31; Valentine Postcards 150; W Rouche & Co 1, 54,91, 102, 127

Contents

Dr Wynne Davies MBE FRAgS receives the Sir Bryner Jones Award in 2002 from The Princess Royal

Foreword

Ever since I wrote *One Hundred Glorious Years* to commemorate the centenary of the Welsh Pony and Cob Society, there have been no publications in relation to section B of the Welsh Stud Book. In the book, I was restricted in space available and also to a maximum of 25 section B Studs and sadly several Studs worthy of mention had to be omitted, but this situation has now been rectified substantially in this book.

Tom Best has researched minutely the families of the ponies and the persons involved and has discovered several interesting facts, which I certainly did not previously know. This must have taken him an enormous amount of time and effort but his enthusiasm for the breed would have spurred him on.

There is no one better to write a book on Welsh ponies section B. The successes of Waxwing Stud owned by Tom Best and his business partner David Blair have been phenomenal. They head the list of the most successful breeders with regards to numbers of Horse of the Year Show ridden qualifiers since the Mountain and Moorland classes began with wins in sections A, B and C. At the 2007 HOYS they had eight qualifiers, with Waxwing Thumbs Up ending up supreme riding pony of the Show.

They have won more than twenty section B first prizes at the Royal Welsh Show, culminating with the section B championship with Cwrtycadno Perlen in 2013. It is fitting that this was the year the Queen's Cup was offered to section B, for only the second time since it was first presented in 1958.

Dr Wynne Davies, MBE, FRAgS
President of the WPCS in 1984

The author receives the Queen's Cup from the judge, Meirion Davies, following victory with Cwrtycadno Perlen in the Section B championship at the Royal Welsh in 2013.

Introduction

The Welsh breeds, renowned throughout the world for their beauty, hardiness and versatility, were first registered in their own stud book in 1902 and within which there is a specific section for the Welsh Section B pony. It has been developed over the past eighty years to meet the demand for ponies suitable for children to ride. Admittedly, there were always ponies of a riding type bred in Wales but they tended to be smaller and failed to meet the demand for larger ponies more suited to older children, especially as riding became more popular throughout Britain during the 20th century.

Dedicated breeders and enthusiasts set about breeding ponies with Welsh characteristics which satisfied the demand for a larger child's pony up to 13.2 hands (138cms). They made use of the traditional Welsh Mountain Pony as a foundation for the new Welsh Pony of Riding Type, which, judiciously crossed with both the Thoroughbred and Arab, over time produced a pony suitable for children to ride while still retaining the characteristics for which the Welsh breeds are famed. As demand grew, so too did the popularity of the Welsh Section B and soon they were to be found not only in many areas of competition but also in the hunting field and at the Pony Club. Slowly but surely numbers of Welsh Section B ponies rose, as did the number of those interested in breeding them both at home and abroad. Their popularity as a competition pony has similarly risen over the years and during the early part of the 21st century they are to be found at the very top of the line at prestigious events such as the Horse of the Year Show.

At times, the development of the Welsh Section B followed a rocky path; it had its critics and there were many dissenters along the way. Nevertheless, a dedicated band of knowledgeable breeders persisted and their foresight and fortitude managed to establish a recognisable breed within the Welsh Stud Book, the Welsh Section B. A key factor was the establishment of a Foundation Register which was used to build up numbers after two World Wars, when the number of Welsh ponies and cobs being bred had hit a worrying low. By using registered sires on 'Foundation' fillies, within three generations following inspection a consistent type of riding pony emerged showing Welsh characteristics. By giving the gene pool both breadth and depth, the time was right for the Welsh Section B to flourish within the Stud Book.

It is because the history of the development of the Welsh Section B is relatively recent, combined with the fact that its breeding details have been recorded within the Welsh Stud Book, that little has been written about the people and ponies who have contributed towards its history. It is a fascinating story – and, above all, it is current. By recording this story in more detail than previously afforded to the Welsh Section B, it is hoped that breeders and enthusiasts alike will gain a deeper knowledge of the personalities involved in this Welsh breed and how their vision of breeding a Welsh Pony of Riding Type came to fruition, flourished, and became what it is today.

Tom Best
March 2013

A herd of Welsh ponies.

Chapter I

Clumber/Hardwick, Dyoll

Mrs Greene's Grove ponies take the Mountain & Moorland Group award at Islington in 1922.

Early stud books in which Welsh ponies and cobs were originally registered, showing the change of name from Polo Pony Stud book to National Pony Stud Book.

Centuries before ponies were used for leisure purposes they had been bred in the mountain and moorland areas of Britain, primarily for agricultural use. Famously, Wales provided the ancestral home for two diverse types which can loosely be described as ponies and cobs. They were used for transport – either ridden or driven; for pulling power – ploughs and later coal tubs; and as a conveyance for shepherding. Depending on the use for which they were bred, just as elsewhere in the country, they were crossed with other breeds by way of improvement and in order to increase their value. Hackneys, Thoroughbreds and Arabs had proved popular 'improvers' for almost two centuries prior to the establishment of stud books, which for the first time formally recorded breed specifications which were moulded by 'societies' of like-minded breeders.

Although the current crop of ponies and cobs can be directly traced through the Welsh Stud Book for just over 100 years, it is important to recognise the role played by breeders prior to this time. It was their judicious selection of breeding stock that would determine how useful a pony or cob would become, how beautiful and striking its appearance and, most importantly, how sound it would prove to be when put to work. Through these efforts they brought together a gene pool that enabled subsequent generations of breeders to build upon their success. Before the official stud books were established breeders-cum-farmers at best kept their own written records and at worst passed down oral records that over time became unreliable or lost altogether. Folklore and hearsay played their part in distorting the facts.

The Victorian age of breeding was characterised by its need to compartmentalise and to strive for purity, so the emergence of specific breeds as opposed to types was inevitable. As a consequence, an increasing number of breeders of native ponies established their own breed societies, starting with the Shetland Pony Stud Book Society, formed in 1890 and its first stud book published the following year. Other breeds followed, including the Welsh in 1901, New Forest in 1906, Dales in 1917, Dartmoor in 1920, Exmoor in 1921, Highland in 1925, Fell in 1927 and the English Connemara in 1946. However, an early start was made in the late 1800s by breeders who had become active in breeding ponies for the game of polo and who formed their own stud book in 1893. While the farmers and pony breeders in Wales had fashioned the main breed characteristics of the ponies and cobs that set them apart from the other breeds of the British Isles, history shows that the polo pony breeders played a pivotal role in the determination of the riding pony type within the Welsh Stud Book that would become the Welsh Section B.

John Hill of Church Stretton, Shropshire, a founder member of the Welsh Pony and Cob Society (WPCS), was coincidentally the prime mover in the formation of the Polo Pony Stud Book in 1894 (which became the National Pony Society Stud Book in 1913). Towards the end of the 19th century, as well as breeding Welsh ponies, he also crossed them with Arabs, then with small Thoroughbreds, to achieve the perfect polo pony. Hill's vision was borne out in his statements in the preface to Volume I of the Polo Pony Stud Book. He writes:

> In the first place this Stud Book was formed for the registration of all Ponies suitable for riding purposes, or for the production of such; the 'weight-carrying' blood Polo Pony being the type which the Society considers the one to be aimed at. ... Although Welsh ponies have of late been terribly neglected, and the original beauty and distinctive character in many districts spoilt by crosses of the Hackney for the purposes of gaining size, still there are evident signs of the awakening appreciation of the pure old blood, which has all the time been carefully guarded by men who value it as much as the eastern breeders do the high caste Arab. Living on the boundary of Wales and being intimate with its ponies all my life, I have no hesitation in saying that there is no safer or more appropriate foundation for the object which the Society has in view and which I have endeavoured to describe, than this pure Welsh pony blood. ...
>
> The indomitable pluck, endurance, and good temper of the Welsh pony, together with his substance and dash, will be found an invaluable cross for the thoroughbred and eastern bred ponies.

Significantly, he states that the Polo Pony Stud Book should be for 'all' ponies and that they should be for 'riding purposes'. Most important was his vision of 'fixing the type', a vision which Welsh breeders would subsequently achieve with the development of the Welsh Section B. Of the 57 stallions entered in the first Polo Pony Stud Book, 24 were Arab or Barb (mainly imported from the Middle East and India) and 11 were small Thoroughbreds. Among the other pony stallions there features two cobs, Trotting Briton and Trotting Comet, and three ostensibly mountain ponies, Lord Brockhurst (No 26), of unknown breeding, entered by a Mr Campbell-Hyslop from Church Stretton; Taffy (No 47), again of unknown breeding and entered by the Lords Cecil of Tonbridge, Kent; and Tyrant (No 54) 'Sire: A Welsh Stallion. Dam: A Longmynd Mare by A Longmynd Stallion', bred by G Robinson and registered by W J Roberts, both of Church Stretton.

The section 'Mares and their Produce' painted a different picture as a very wide range of ponies of varying heights was entered, including a good number of Welsh-bred mares. Messrs Hill, Roberts and Campbell-Hyslop took advantage of the new stud book, as did a few other breeders of Welsh ponies from Wales. For example, Welsh Fashion (No 305), owned by Messrs J Jones and Sons from Llandudno, appears in one of the few photographic illustrations. Fashion, a 12.2-hands chestnut foaled in 1887, had wins to her credit in shows at Brecon, Rhyl, Llanrwst and Llangefni.

Polo aside, during this time the breeding of ponies for the army was of importance and here too the Welsh breeds played a significant role. Another well-known breeder, Sir Richard D Green Price, stresses the importance of breeding ponies of riding type in the 1901 *Live Stock Journal Almanac*:

> No doubt we have to thank the game of polo, an aristocratic national game, for this boom in ponies. And yet this is not all. It has been discovered that ponies are an essential factor in modern warfare. In addition to this I have many personal assurances from officers in the army, and men fresh from South Africa itself that, without their ponies, the Boers would long ago have been conquered and that those in our Colonial and Lovat Scouts' ranks have been of the greatest value to our service. It was only a few weeks since that I had testimony of my son, an Imperial Yeoman who has been through the whole campaign, and who, writing home said, 'You are doing good work in strongly advocating the use of ponies in our army service; they have fairly outlasted the bigger horses in this campaign, and without the aid of our mounted men, the infantry would have been unable to tackle the Boers.'
>
> The idea of raising what I may call the nucleus of pony battalions in Great Britain may sound a crude one, and no doubt it is so in the present moment, yet there are many far-seeing men who would fain agree with me, and have already begun to pave the way by breeding to a type of improved riding pony. Here the Thoroughbred or the best type of Arab, on the sire's side, is one of the most likely to succeed, and the height to be aimed at is from 14 hands to 14 hands two inches. This may seem difficult at first, but it is proverbial of the Welsh pony at least that it breeds animals bigger than itself, if they are not absolutely starved in their youth, and no doubt the same can be said of other notable pony breeds.
>
> Let me beg breeders on a large scale on hilly countries to eschew the Hackney, or mere half-bred trotting cob, if they intend to build up a herd of ponies fitted for riding purposes. By the use of Hackney sires they will be defeating some of the finest attributes that Nature has molded in the real hill pony, viz. stamina, tractability, and sure-footedness, which can never be attained in rough hilly country by an animal that tosses its legs about, and tires after going a few miles.

Such was the success of the Polo Pony Stud Book that an increasing number of native ponies were being registered in it, since at the time there was no other means of recording the breeding of these ponies. In 1899, mountain and moorland pony sections were opened in Volumes V and VI of the Polo Pony Stud Book for Dartmoor, Exmoor, Fell, New Forest, Welsh, Connemara and Highland. The Welsh entrants were divided into two different sections, for 'South Wales Ponies' and 'North Wales Ponies'. The overall provision for mountain and moorland ponies within the Stud Book had two principal objectives:

1. To save from extinction and offer endeavour to the Native Breeds
2. To establish a foundation stock of approved mares easily accessible to the breeders of Polo Ponies – through the Stud Book.

So there was method in the madness. Not only was it in the interest of the polo pony breed to have a good supply of native pony mares suitable for crossing with the bigger stallions, but it was also in the breeders' interests to maintain a strong base within each breed to ensure that the supply of the native stock should neither dry up nor deteriorate.

Modern breeders of native ponies owe a great debt of gratitude to the polo pony breeders and their new society; without their interest and dedication, who knows what might have happened to our natives breeds? Nevertheless, the breeding policies that would be adopted by many of the Welsh enthusiasts did little to advance the cause of the riding pony type that would soon prevail, particularly through the use of the hackney horse and pony so openly criticised by Sir Richard D Green Price. An early member of the Polo Pony Society and a great enthusiast of the Welsh breeds was John Marshall Dugdale, one of the founder members of the WPCS and its Chairman from 1901 until his death in 1918. Together with his sons, he bred a large number of ponies and cobs with the Llwyn prefix at Llanfyllin near Oswestry, Shropshire, at the turn of the 19th century.

Prior to the establishment of the Welsh Stud Book his entries in the Polo Pony Stud Book show that he used Hackney stallions extensively within his herd. For example, Eiddwen Flyer, a 14.2-hh chestnut Hackney, was used to breed mares like Llwyn Hazel which, at only 11.1½-hh, was consequently registered in Section A of the Welsh Stud Book. Likewise, he registered Llwyn Prince of Wales (12.1½ hh) in Section A of Volume II of the Welsh Stud Book; his sire was described as a Dartmoor Hackney, while his dam was a bay Hackney by a Mountain Pony. With over 200 mares and 40 stallions registered in volumes I to XVII of the Welsh Stud Book, Dugdale made a huge contribution to the gene pool of the breed – and wasn't alone in that respect.

Further south, in Brecon, another co-founder of the WPCS, William S Miller, had established his Forest Stud at around the same time and he too registered large numbers of ponies and cobs in the early stud books – as many as 500 mares and stallions by the outbreak of World War I. One of his entries, Forest Morning Star, born in 1873, was one of the most elderly ponies to be registered. The registrations could be somewhat misleading as some of the Forest-prefixed ponies had been bought in. One such pony which sired a number of "Section A"s was Forest Adbolton Sir Horace, a small Hackney pony stallion by the famous stallion, Sir Horace, purchased from Roy Charlton of the Linnels Stud near Hexham. It is noticeable that, like many of the early breeders, the majority of his mountain ponies were of solid colours, with few of them grey. This colour was about to enter the Stud Book with a flourish, on the strength of the popularity of a brilliant show stallion, Dyoll Starlight, and his sons (although it is noteworthy that only two of the Forest stallions were by Dyoll Starlight and only two by his son, Greylight).

Outside Wales, there were other breeders who had taken an interest in the Welsh Mountain Pony – possibly inspired by ponies they saw at the London shows. One such person was the Duchess of Newcastle, a highly-acclaimed dog breeder whose kennel of Borzois, Fox Terriers, Whippets, Spaniels and Scottish Deerhounds housed many show champions. She also had Clumber Spaniels – indeed, she lived at the 3,500-acre Clumber Park, Nottinghamshire, which had been the home of the Dukes of Newcastle-under-Lyne for more than 300 years. The huge mansion house in which the Duchess lived was demolished in 1938 and the Park is now owned by the National Trust, which also owns another of their former homes, Hardwick Hall. It was from these residences that the Duchess took the names for her pony stud – Hardwick for the stallions and Clumber for the mares.

She owned, bred and showed some of the very best dogs in the country and her aims for her pony stud were no less ambitious. In 1897, she purchased from John Jones & Son of Dinarth Hall, Llandudno, one of the leading show mares of the day, Lady White – a winner of no fewer than 50 prizes. Jones had a great reputation as a breeder and exhibitor so it was no wonder that the Duchess selected six mares from Dinarth Hall as a foundation for her own stud, later purchasing mares from Mrs Greene's Grove Stud. She bred a large number of ponies and showed them very successfully, registering a few with the Polo Pony Society and the majority with the WPCS.

Her breeding policy was an interesting one – she married the best show animals of the day to achieve her desired effect but these were not restricted to Welsh ponies. Having selected a band of the best mares she could find, she crossed them with a pony called Linnel Don, a small dark brown stallion by Little Wonder II, one of top ponies of his day for Sir Walter Gilbey and one of the first entries in the newly-established Polo Pony Stud Book. Little Wonder had been bred by Christopher 'Kit' Wilson of Kirby Lonsdale, whose strict breeding programme resulted in a quality pony which laid the foundations of the modern Hackney pony.

Clumber House photographed in 1907.

Linnel Don (foaled in 1899) was bred by Roy Charlton. In his book *A Lifetime with Ponies,* Charlton recalls seeing Linnel Don in a driving class at the International Horse Show in the Great Hall at Olympia in 1907. Since being sold as a yearling, the little Hackney stallion had become the top driving pony of his height in the country for the Duchess of Newcastle. Charlton was amused to learn that Linnel Don's daughter, Clumber Blacky, and granddaughter, Clumber Starlight, were themselves great winners in Welsh pony classes.

LINNEL DON 1061.

Brown. *Foaled* 1899. *Height* 12-0.

Owner and Breeder, ROY B. CHARLTON, The Linnels, Hexham.
Sire, LITTLE WONDER II. 25, dark brown, 13-3.
Dam, PRIDE, dark brown, 11-3, by AUCHENDENNAN by MARS.
G. dam, POLLY, by SIR GARNET.

Prizes :

1906 : 1st, R.A.S.E. at Willesden. 1914 : 1st, Carnarvon.
1907 : 2nd, Olympia.

Entry for Linnel Don in the Polo Pony Stud Book

Clumber Blacky (f. 1907) Section A by Linnel Don out of Lady Jones II. Champion Doncaster 1912.

In breeding Starlight, the Duchess had used the very fashionable Dyoll Starlight on Blacky. However, the dam side was completely inbred to her Hackney stallion, Linnel Don – Blacky was by him and her dam, Clumber Lady Jones II, was by him out of a famous show mare, Clumber Lady Janet III, who in turn was by Hardwick Sensation, a son of Linnel Don. The Stud Book entries of this stud show that the combination of relatives of Linnel Don was commonplace. This level of inbreeding (I am confident she would have preferred to call it 'line breeding') was unrivalled among Welsh pony breeders and was probably modelled on the breeding of dogs for which the Duchess was famed. Of the 32 Clumber mares registered in the Welsh Stud Book, 15 were by Linnel Don, while another 11 were by his son, Hardwick Sensation.

The Duchess of Newcastle was one of many who took advantage of having her ponies registered with the Polo Pony Society. In addition to its Stud Book, the Society promoted the exhibition of ponies and held its first show for polo and riding ponies at Ranelagh in 1895. The Society also provided prize money at the Royal Show in 1897, 1898 and 1899 and initiated the awarding of Gold and Silver Medals at nominated shows throughout the country. In 1900 it held its first Spring Show at the Agricultural Hall, Islington, where there were classes for native ponies. John Hill – writing in 1899 and unaware how much controversy this stallion and his dam would create in the future – reported in the *Live Stock Journal Almanac* of 1900:

> Ponies in 1899: The exhibition of Welsh ponies at the Royal and Crystal Palace, although small in number, was exceptionally meritorious. The grey stallion shown by Mr Meuric Lloyd is undoubtedly one of the best which has ever been seen; he took the first prize at the Royal last year and this, and also at the Palace. It is not often a winner can boast of a dam which is of the highest show quality; but

Dyoll Moonlight, dam of Dyoll Starlight, grazing at Glanrannell Park.

> those who had the pleasure of seeing Mr Lloyd's beautiful grey mare at the Palace Show will endorse my opinion that mother and son, as a pair, would be hard to beat in any breed competition.

The mare was Moonlight and the stallion her son, Dyoll Starlight (simply registered as Starlight in Volume VI of the Polo Pony Stud Book in 1901), to whom Dr Wynne Davies referred in Welsh Ponies and Cobs:

> It can safely be said that Dyoll Starlight revolutionised the Welsh Mountain Pony Breed. Dyoll Starlight was in his prime the most beautiful in the whole world, and he was progenitor of a whole new dynasty of beautiful ponies.

There were questions about the origins of many of the Welsh ponies and cobs bred prior to the formation of the WPCS in 1901, and despite every effort to seek the true breeding of the ponies entered in the first Stud Book at the start of the 20th century, it was simply not always possible to separate fact from fiction. A good example of this (and arguably the most contentious of this time) was indeed Dyoll Starlight (Vol I, registration number 4). He was recognised as one of the most beautiful and influential stallions of the Stud Book, whose children and grandchildren not only took the show ring by storm but exerted a very strong influence on generations to come, including the Welsh Section B. Such was his influence that the type of pony established through his blood was referred to as the 'Starlight Strain.' But the authenticity of his pedigree was publicly disputed some 20 years after the death of Meuric Lloyd, Starlight's breeder.

Howard Meuric Lloyd was born in 1853 in the West Indies, where his father was a magistrate in the Colonial Service. He studied Law at Oxford, gained an MA and later became High Sheriff of Carmarthenshire, where the Lloyd Family owned the Danyrallt and Cynghordy Estates. Lloyd's interest in Welsh mountain ponies grew during the 1880s and it was in the following decade that he started his stud in earnest, leading to the birth of Dyoll Starlight in 1894. (The stud name Dyoll came from his family name of Lloyd spelled backwards.) He and his wife lived at the large house of Glanyrannell, where a century later former WPCS President David Davies (affectionately known as Dai Glanyrannell) successfully ran a country house hotel which was famed for its Welsh Cob Sale Week.

Until 1919, Starlight remained with his breeder. As

Handwritten notes by Meuric Lloyd, pictured alongside, on the breeding of Dyoll Starlight, with his Polo Pony Stud Book registration number.

Lloyd's health deteriorated, the stallion was sold to Lady Wentworth, the world-renowned breeder of Arab horses who was keen to establish a herd of mountain ponies at her Crabbet Stud. His breeding prowess continued in his new home and, despite his age, he produced many champions. It was Lady Wentworth herself who publicly cast doubt on her stallion's pedigree by writing an article for 'Riding' in 1944 in which she stated:

Llwyn Nell (f. 1890) Section A by Eiddwen Flyer (Hackney Stud Book) out of Tibbie.

Rumour says that owing to an error his dam was mated, without his owner's knowledge, to a very beautiful Arab called Apricot, the property of Mr Wynn of Rhug, near Corwen, instead of to his reputed sire. Be that as it may, she certainly produced such a foal as she did not produce again.

The motives for spreading this 'rumour' are debatable. Some say that she did not want to associate Starlight and his progeny at Crabbet with his sire, Glasallt – which by all accounts was pretty plain and was gelded soon after Starlight's dam was mated with him. For obvious reasons, the association with Apricot, the Arab cross Welsh stallion (not purebred as she stated) not only helped explain the tendency of Starlight to throw ponies showing Arab characteristics but also added credence to the role of the Arab horse which Lady Wentworth was only too keen to promote. On both accounts, it was to her advantage financially.

Others might believe that it was part of a bitter rivalry between Lady Wentworth and Lloyd's daughter Lorna, who had spent many years overseas with her husband Lt Col Raleigh Blandy, who served with the Gurkha Rifles in India and North Africa (for which he was awarded the Military Cross). Meuric Lloyd had died in 1922 and it was on her return home in 1925 that Lorna contested the sale of Starlight to Spain (where he died, aged 35) on the basis that it was not part of the written agreement between her father and Lady Wentworth. There was no love lost between the two women, so it is little wonder that Mrs Blandy sprang to the defence of her father, whose competence (rumour or not) and even integrity had been brought into question by Lady Wentworth. However forcefully Lorna presented the evidence supporting her father's record of Starlight's breeding, Lady Wentworth was determined to have the last word – she compounded the original 'rumour' with yet another unsubstantiated claim:

> If Mrs Blandy really wishes to insist on trying to prove that Starlight was by a common ugly pony without pedigree, not good enough to be retained as a stallion, we must reluctantly conclude he was not so well bred as we thought, and she is, of course, quite entitled to her opinion. At the same time his appearance and prepotency in transmitting a strong Arabian type seems to contradict this theory. Personally, I am a practical breeder unwilling to believe that a silk purse can ever be made out of a pig's ear, and Apricot or no Apricot, I feel pretty certain that Starlight was one of those happy accidents which breeders can welcome without shame or regret and, indeed, be thankful in Providence. May there be many more such!

In the event, the last word went some years later to doyenne pony breeder and respected Welsh pony judge Mrs Nell Pennell, who recorded for posterity the 'Story of Dyoll Starlight' in the April 1960 edition of 'Riding'. With the help of Mrs Blandy she was able to put the ghosts of Starlight's breeding to rest calling on stud records, letters and photographs belonging to Meuric Lloyd. She attributes the 'Arab' attributes passed on by Starlight to his dam, Moonlight, which she describes thus:

> It appears that Moonlight had passed through several hands before that day [when she was purchased by Meuric Lloyd] at Llanddeusant, for she was bred by Mr Thomas, The Pentre, under the Brecon Beacons, and descended beyond doubt from the Crawshay Bailey Arab turned out on these hills about 1850. She was foaled in 1886, and is described by Mr Lloyd as a 'Miniature Arab' full of quality with a lovely head and good shoulder. Her sire was a 12 hh white pony, but no record of her dam exists. It is true that Moonlight and her

Tom Jones Evans at the 1936 Royal Welsh Show at Abergele with his champion Mountain pony stallion Grove Sprightly (left), and class second and reserve champion Grove Will O'The Wisp. It is interesting to not that the discs of card on the bridle denote the prizes won at a time when rosettes were not commonplace.

> son, Starlight, inherited the Arab characteristics that he undoubtedly possessed.

Interestingly, Dr Davies makes mention of Apricot and Dyoll Starlight when he writes:

> These famous names of the nineteenth Century belonging to Cobs, the proliferation of Comets, Flyers and Expresses, so what of the ponies? As already mentioned, the Williams-Wynn family turned out Merlin (Thoroughbred) on the Ruabon Hills in the mid-seventeen hundreds. It seems about 50 years later Col. Vaughan of Nannau and his brother Sir Robert Vaughan of Rug, near Corwen imported some 'Barb Arabs' one of whom sired 'Apricot' from a 'pure Welsh mare'.

Apricot was in great demand as a sire and very soon Col Vaughan's stud became the 'best of its kind'. In addition, Apricot set up records at racecourses all over North Wales, at Harlech, Ruabon and Mold – indeed, at Mold it is recorded that he won four races the same day! John Hill records in the introduction to the Welsh Stud Book, Volume I that:

> When staying at Rug as a boy about 1850, I noticed two distinct types of Welsh Ponies around 13 hands in height, the one 'old' type, thick set, well made carrying the shooting parties with wonderful ease and safety. The other type showed the influence of Apricot, lighter with more quality, they make excellent children's ponies; the mares of this sort are exactly the ones to mate with Thoroughbred or Arab stallions for breeding Polo Ponies.

With both Davies and Hill dating Apricot's influence to the early part of the 19th century, doubt is cast on Lady Wentworth's original theory regarding the breeding of Moonlight, who was foaled in 1886. Lloyd's records clearly showed that Dyoll Starlight was by Dyoll Glasallt, a colt he had purchased and named after the farm near Carmarthen where he was bred. Glasallt's sire was the grey Mountain pony Flower of Wales, while his dam had been bred by Mary Storey-Maskelyne, who lived near Sennybridge in the Brecon Beacons. Her father was the well-respected politician, Sir John Talbot Dillwyn-Llewelyn, a Justice of the Peace for Glamorgan, Carmarthenshire and Breconshire and later a Member of Parliament for Swansea between 1895 and 1900. As the recording of breeding had little significance then, as for many other ponies of the day this side of Starlight's pedigree has been lost in the mists of time.

The story of Dyoll Starlight and his influence on the Welsh breeds reflects development from two distinct sources within Wales. Firstly, there were the farmers whose families had bred ponies for generations – their efforts would provide the foundations on which the breeds would be built during the late 19th and early 20th centuries. Secondly, there was a growing number of men and women of means who lived mainly on large estates and who would follow a trend within their social élite to breed and exhibit the best of equines – be they shire or Welsh pony. Like Hill, they would help steer the course of the Welsh pony during the ensuing years. Motivated by

a desire to breed a pony that was fit for riding, the Welsh breed would be pushed in that direction, sometimes by the vagaries of fashion, at other times by the prospect of financial gain identified by a few astute breeders, or by enthusiasts with a desire to breed a pony true to Welsh

Total (f. 1904) Section B by Klondyke. Although registered as a Section B, Total is very much of cob type. Note the side and bearing reins and the docked tail.

type and character, with size and action to accommodate the modern child.

By recording the ancestry of the recognised Welsh breeds, the formation of a Stud Book was an essential first step towards developing them. Over time the breeds would witness change and the development of discrete sections which would include the Welsh Section B as we know it today. The policymakers played an important role in guiding the development, but it was the breeders who saw that the policies were put into practice.

Greylight and his pedigree as it appears in the personal diary of Charles Coltman Rogers, one of the leading figures in the formation of the Welsh Stud Book in 1901 and Chairman of the Welsh Pony & Cob Society from 1918 to 1927.

Chapter II

Skowronek, King Cyrus, Grove, Craven Cyrus, Incoronax

252 RIDING March, 1941

SKOWRONEK

Champion Arab (white) 14.2 h.h. Sire, IBRAHIM; Dam, YASKOULKA (Crabbet Park Stud).

Skowronek appeared on the cover of 'Riding' magazine in March 1941.

King Cyrus by Skowronek was used on a few Welsh mares before his return to Crabbet. (Drawing by Peter Upton.)

Looking at the Welsh pony today, the close resemblance between it and the Arab breed is obvious – the beautifully chiselled head, the high tail set carried well and the free movement with floating action are all distinguishing features. These traits belonged to no other recognisable breed at the time, so it was no wonder that mares like Moonlight were described as 'miniature Arabs'.

Well over a century ago, the connection between the Arab horse and the Welsh pony was widely discussed and recorded, not only among intellectuals and historians but also among Welsh pony enthusiasts keen to know more about its ancestry. One such person was the highly successful American businessman Charles A Stone who, from humble beginnings in Boston, reached the pinnacle of the financial world of East Coast America during the early part of the 20th century, having amassed a fortune along with his business partner Edwin Webster. Between 1919 and 1923, Stone was a director of a variety of financial institutions on Wall Street, including the influential American Federal Bank.

Like others in that milieu, his lifestyle was akin to that of an English gentleman. He owned a smart townhouse on New York's Fifth Avenue as well as an estate in Virginia, where he successfully bred Thoroughbreds, and another estate in New Hampshire where he bred the best of Morgan horses and Welsh ponies. On a trip to Wales in 1912 he was impressed by ponies there. In an attempt to increase his own knowledge of the Welsh pony, which he described as 'superficial', as well as to highlight its qualities to his fellow Americans, he enlisted the help of his cousin's friend, Mrs Olive Tilford Dargan, to make a study of the breed while she toured England and Wales during the early 1900s. This she undertook with great enthusiasm. Her research took her to the leading academics of the day, including Professor Ewart of the British Museum and Professor Ridgeway of Cambridge University. She also attended leading horse shows such as those at Olympia and Crystal Palace where she discussed pony breeding with prominent polo pony breeders such as Lord Arthur Cecil and Sir Walter Gilbey, and explored the hills of Wales where she spoke with breeders and Welsh pony enthusiasts and historians.

All of this she related in two letters to Stone's cousin, Miss Anne Whitney, which Stone published in the privately-financed book *The Welsh Pony, Described in Two Letters to a Friend* in 1913, the same year that Stone imported the Welsh Mountain Pony stallion, Grey Light A1 (f. 1920), a grandson of Dyoll Starlight. Of his 1912 visit, Stone wrote in his introduction to the book:

> On this trip I found the Welsh country so charming and the ponies so attractive and so different from any ponies I had known before, that I spent several weeks in Wales and the border counties selecting a herd which finally amounted to about twenty-five of the best of the true mountain type that I could obtain.

Mrs Dargan's academic studies took her back to Roman times, when the military horses brought to these shores from the Mediterranean were mated with the indigenous mountain ponies of Wales. She wrote:

> So the Romans left to Wales not only a heritage of legendary stone, such as the old camp, Y Caer Bannau, which is shown you in Breconshire, but a far more valued legacy which is yet animate in the veins of the Welsh pony. The invaders were busy in Wales for four hundred years, during which time the packhorse became a domestic type, and gradually the acclimated Arabian blood crept up the hills and among the wildest herds – a slow infusion that left the hill pony, retaining all the hardiness that made life possible on the scanty-herbaged peaks.

Mrs Dargan refers in her letters to the influence of other breeds brought into Britain to enhance the native stock, such as draught horses from continental Europe, and concedes that they in turn would have exerted some influence on the Welsh pony. However, she directs the reader to more recent times when she returns to the influence of the Arab blood by stating:

> We must also remember that two centuries ago, when adventures in breeding began, the English had commenced those prudent experiments with the Arab cross which has fixed the Thoroughbred in his sovereign place. There had been occasional importations of the Arab since Roman days, but the English horses were of such numerous and diffused types, and so unlike the Eastern horse in build and nature, that such spasmodic introductions had no permanent effect. The great improvement came with the determined enthusiasm and patience of the eighteenth century breeders; and it seems providential again that as the ways of breeding between England and Wales became promiscuously open, the Eastern blood was becoming prevalent in England.
>
> From this source the Welsh breeders started renewing the beneficent strain in the slow, best manner. Merlin, a descendant of the Brierly Turk, after his brilliant years on the turf, was brought to Wales and turned out with the ponies on the Ruabon hills to become the foundation of a famous and prolific line. Mr Richard Crashaw [Crawshay] secured for his county the Arab sire of Cymro Llwyd; and in Merionethshire, the half-Arab, Apricot, of multiple progeny, became an imperishable tradition. Seventy or eighty years ago, Mr Morgan Williams put Arabian sires with his droves on the hills behind Aberpergwm; Moonlight was discovered, roving and unshod, by Mr Meuric Lloyd, and this dam of certain Arabian descent gave Wales Dyoll Starlight.

Mrs Dargan makes reference to the 18th century horse breeders and their preference for Eastern-bred stallions. Well documented and mentioned in her letters was a descendent of Richard Crawshay, who was one of the most prominent industrialists of South Wales. By 1810, the year of his death, Richard Crawshay had developed six blast furnaces for the production of iron at the Cyfarthfa Ironworks at Merthyr Tydfil and amassed an estate estimated at £1.5 million, an enormous sum by today's standards. His son, William, inherited the bulk of the estate and built for the family a castellated mansion which he named Cyfarthfa Castle and which stood in 156 acres of grounds and now belongs to Merthyr Tydfil Council. Meanwhile, Richard Crawshay's nephew, Crawshay Bailey, joined his uncle in the iron business and soon made his own fortune from foundaries in Ebbw Vale and coal mines in the Rhondda Valley.

Bearing in mind that huge numbers of ponies were required in industry as well in the coal mines of the South Wales valleys, it was little wonder that the Victorians took an interest in them. Moreover, as we have seen, the improvements involved crossing them with a variety of breeds. One member of the family – most likely William or his son Richard Crawshay – introduced Eastern-bred sires for the herds of ponies in this area around 1840. Several stallions bear the Crawshay prefix, such as Crawshay Dance with Me, reputedly the sire of the Crawshay Bailey Arabian – the most famous of them all. It would be this pony who would appear in many famous Welsh cob pedigrees through his son, Cymro Llwyd (f. about 1850) and through his likely granddaughter, Moonlight (f. 1886), the dam of the highly-acclaimed mountain pony, Dyoll Starlight, of whom more later.

Far from the industrial heart of South Wales, another example of early improvements in breeding stock could be found in Scotland where, unlike the Crawshay Bailey family, written records had been maintained. Indeed, one of the earliest records of Arabian importation to Britain is housed at the Broomhall Estate in Fife. The estate is owned by the Earl of Elgin, whose ancestor the 7th Earl is most famously associated with bringing the Parthenon Marbles to Britain in 1802–1812. In a correspondence with the author, the present Earl of Elgin wrote:

> It is sometimes not appreciated but the 7th Earl was a great lover of horses and he arranged for a stud of Arabian horses to be sent back overland from Constantinople, through Vienna to Bremen and then by sea to Gravesend in 1802/1803. The groom, Mr Walker, came on to Broomhall with four stallions after leaving the main stud in the south and, on arrival at Charlestown, he was paid £25, which was his annual emolument, and was then given some £19 to make his journey, with his wife, back to London. The stallions went out all over this part of Scotland and we have records of payments for services.

Loosely referred to as Arabs in this text, the horses bred around the Mediterranean were much admired by expatriates – particularly those in the Colonial Service and the British Army. The Victorians set great store by their beauty, speed, tractability and hardiness, using them for all sorts of pursuits such as polo, pig-sticking, tent-pegging and racing. Their aristocratic appearance also added a certain gravitas on the parade ground and

Lady Wentworth with her favourite Arab stallion Skowronek.

officers valued them highly. As 'working' horses, many were brought back to Britain following campaigns in India and North Africa and, as we have seen, they outnumbered any other breed in the section for males in the newly-formed Polo Pony Stud Book.

By 1918, the Arab Horse Society had been formed by a few notable enthusiasts. The Rev D B Montefiore, a former Polo and Riding Pony Society Chairman, was appointed the new society's Secretary. The first elected President was Wilfrid Scawen Blunt, with Sidney Hough as his Vice President. Prince Faisal, the Grand Sharif of Makkah (and later King of the Arabian Kingdom of Syria) accepted an invitation to be Patron to the new society.

Wilfrid and his wife Lady Anne Blunt – the granddaughter of Lord Byron and a woman of immense talent and energy – had in 1878 founded the Crabbet Stud, the most influential of the British-based Arab studs. They visited all the great horse-breeding tribes of the Northern deserts of Africa and purchased mares and stallions of their best breeds. In 1879 they returned to the Nejd, and in 1881 visited the Hamad, or Great Desert, south of Palmyra. It was becoming increasingly difficult to import horses from Arabia, but their most significant purchases were made in 1890 when Wilfrid acquired the remainder of the once-celebrated stud of Ali Pasha Sherif, who in turn had bought them from the estate of Abbas I Pasha, the Governor of Egypt and Sudan and a passionate breeder of desert Arabians.

Altogether the Blunts imported 16 stallions and 32 mares, including the very influential Mesaoud, which became a foundation sire at Crabbet before being exported in 1903 to Count Branicki's expanding Arabian stud at Kiev in Russia. Some of the Abbas Pasha horses were kept in Egypt at the Blunt's Sheikh Obeyd stud.

As the supply of good stock diminished in Egypt itself, the Crabbet Arabs were in great demand both at home and in Europe. To this day, the Crabbet name is associated with a type held in high regard throughout the world, and it is said that at least 90% of all Arabian horses alive today trace their pedigrees in one or more lines to Crabbet horses. The Blunts separated in 1906 and the stud was divided: Lady Anne kept Crabbet Park and Wilfrid established himself at Newbuildings Place, both in Sussex. It had been a tempestuous marriage but, although they still disagreed about breeding methods, there was a constant exchange of horses between the two studs. In 1915, Lady Anne Blunt went back to Sheikh Obeyd and she died there in 1917. She had bequeathed the Crabbet Stud to the granddaughters of the Blunts' only child Judith (Lady Wentworth), and this sparked a long legal battle between Wilfrid and his daughter and granddaughters, which he ultimately lost – though not before he had had sold or destroyed many of the horses.

Having regained control of the stud and bought back many of the horses from their new owners, Lady Wentworth began looking for an outcross for the Crabbet Arabs, which she considered were becoming somewhat inbred. She made the astute purchase of a small white stallion, Skowronek. Jane Llewellyn Ott took up his story in the August 1964 issue of the American magazine *Your Pony:*

> Skowronek was a completely new type; he has founded a new breed in his own image ... Nominally a Polish Arabian, Skowronek was foaled in 1908 at the Antinomy Stud farm of

Count Joseph Potocki in the Ukraine. Most of the Antinomy stock was destroyed in 1919, in the Bolshevik uprising, but by that time Skowronek was safe in England, having been imported there by Walter Winans in 1913.[1] He attracted no great attention until after his purchase by Mr HB Musgrave-Clark, a long-standing friend of Lady Anne Blunt who owned some of the best of the Egyptian bloodlines that had made the reputation of the Blunts' Crabbet Park Arabian Stud before Lady Wentworth inherited it. In 1920, not long after Clark began using him, Lady Wentworth acquired Skowronek for the Crabbet Stud itself, and it was there that his reputation was made.

Lady Wentworth's daughter, Lady Anne Lytton, said of Skowronek: 'He was like a horse from fairyland, of dazzling pure white with a long curved neck.' Lady Wentworth had secured in Skowronek the type of Arab for which her mother had searched in vain during the latter part of her life. According to his owner, Skowronek founded 'a new dynasty at Crabbet'. Like many of the Eastern-bred stallions being used at this time, he had the superb riding qualities and temperament required in the child's riding pony. He was held in such esteem by his owner that the Crabbet stud card in 1924 shows that the stud fee for Skowronek was set at the astronomical figure of 500 guineas, while that of his son, King Cyrus, was 15 guineas. It was no surprise that his name would appear in the pedigrees of future generations of ponies, including the Welsh. King Cyrus, as well as his grandson, Incoronax, would in turn play an important role when the time came.

King Cyrus entered the Welsh Stud Book through Tom Jones Evans, whose large family lived in Mid Wales. His father, Dafydd Evans, had a great knowledge of Welsh cobs, which proved to be invaluable when the Welsh Stud Book was first being compiled. Tom Jones Evans himself was a member of the Welsh Pony and Cob Society Council for over 30 years and its President from 1931 to 1932. He rented a farm near Craven Arms on the Grove Estate of Mrs Harriet Greene, a member of the WPCS since 1906. She had inherited the 4,000-acre Grove Estate in Shropshire upon her father's death in 1885. She and her husband, a Conservative Member of Parliament and son of a Director of the Bank of England, were extremely wealthy and she herself was keen on fox-hunting. It is highly likely that she had attended the Spring Show at Islington, which had become one of London's 'Society' events of the season, and had admired the ponies which were causing quite a stir among the polo pony breeders.

Like many countrywomen of means, she needed an interest to occupy her time (and that of her grooms outside the hunting season) so she decided upon the breeding and showing of Welsh ponies which carried her Grove prefix.

From her first purchases at auction in 1906, when she bought several ponies from Meuric Lloyd (Dyoll

---

[1] Walter Winans had a remarkable eye for a horse. Not only did he purchase Skowronek but he was also a horse sculptor of some note. He won an Olympic gold medal in 1912 for his sculpture of an American trotter, as well as an Olympic silver medal for shooting. His achievements in shooting included an Olympic gold medal in 1908 and he reputedly held the shooting rights to 250,000 acres of land in the Scottish glens.

Stud), to her final dispersal sale in June 1927, Mrs Greene proved highly successful both in the show ring and as a breeder. During this period she bought the best bloodlines and bred the best. Arguably the most famous was Sir Walter Gilbey's outstanding stallion, Bleddfa Shooting Star, by Dyoll Starlight, purchased at auction in 1915 for 240 guineas, which was an enormous sum for a pony in those days. A report in the 1913 *Live Stock Journal Almanac* comments:

> Mrs H D Greene has a coterie of Welsh ponies it would be difficult to match. All over the kingdom she sends her 'Groves' to win in the best of company. In 1912, forty-one first prizes and seven championships were won at the principal shows by fifteen ponies from this Stud.

Ponies carrying the Grove prefix were popular throughout the world and many were sold to the United States at a time when the market there was buoyant. In the show ring, her string of Welsh mountain ponies was almost unbeatable during a seven-year period after World War I – a time when Lady Wentworth was also a regular exhibitor in Welsh classes with ponies carrying her name as a prefix. It was little wonder that at the Grove dispersal sale Lady Wentworth secured some of the highest-priced lots, including the multiple champion Grove King Cole II (130 guineas) and the top price of the day, Grove Moonstone, who changed hands for 141 guineas.

Held in high esteem among Welsh pony breeders, Tom Jones Evans was closely associated with the Grove ponies as it was he who advised Mrs Greene on many of her purchases and her breeding policy. At the Grove dispersal in 1928, he and Lady Wentworth jointly bought the highly-acclaimed stallion Grove Sprightly for 126 guineas. Three years later he bought her share and the pony went on to win him seven consecutive Royal Welsh championships from 1930 to 1936 and another in 1939. (Tom Jones Evans judged in 1937, and there was no Royal Welsh Show in 1938 due to the fact that the Royal Agricultural Society Show was held in Cardiff that year.) In 1923, prior to their joint ownership of Grove Sprightly, Lady Wentworth had

Kibla, the Arab mare loaned to Tom Jones Evans.

Grove Peep O'Day, photographed at the Royal Welsh Show, 1917.

loaned to Evans an aged white Arab mare called Kibla, a daughter of Mesaoud and Makbula – two of Lady Anne Blunt's original 1896 purchases from Ali Pasha Sherif. The foal at her side was King Cyrus, a grey by Skowronek, which eventually returned to his breeder, Lady Wentworth, before being sold to South Africa. Before leaving Britain he was used by Tom Jones Evans on some Welsh mares to produce a small number of progeny which would be registered in the Welsh Stud Book. These included the mares Craven Kibla, Craven Araby and Craven Jessie. It was in 1927 that he produced a son to the Dyoll Starlight daughter Irfon Lady Twilight, a pony named Craven Cyrus who would have a major impact on future generations of Welsh ponies.

As there was an obvious need to respond to the increasing demand for quality children's riding ponies during the 1920s, Craven Cyrus, despite his half-Arab breeding, was allowed directly into the Welsh Stud Book. He was first registered as a Section A but later transferred to Section B as he grew to around 13 hh. One suspects that his acceptance into the Stud Book may have been partly due to Evans's influence at the time as well as the fine reputation of the Crabbet Arabians – especially Skowronek. Craven Cyrus was a good pony in his own right and recorded a win over the famous Tanybwlch Berwyn in the class for 'riding-type' Welsh stallions at the Royal Welsh Show in 1939. According to Dr Wynne Davies:

> They were a varied lot, one half Arab, one half Barb, one Section A, one pony of cob-type and one cob!

It was generally felt that the quality and refinement Craven Cyrus brought to the breed would be of use to future generations. History bears this out, since he features significantly in the pedigrees of the Welsh ponies in both Section A and B. According to Dr Wynne Davies in his *Welsh Ponies and Cobs*:

> It can safely be said that the Welsh Riding pony started with Tanybwlch Berwyn (f. 1924) and Craven Cyrus (f. 1927); these were the only two

164 WELSH PONIES. VOL. XXI.

Welsh Pony Section.

The description and characteristics of the Welsh Pony will be found on pages 4 and 5 of Volume 5.

WELSH STALLIONS.

CRAVEN CYRUS 1854 (**1441 W.S.B.**).

Roan or grey roan, blaze, near fore leg white to knee, off fore white to above fetlock, both hind white almost to hocks. Foaled 1927. *Height* 12.2. *Measured September* 11*th*, 1933.

Owner, Tom Jones Evans, Dinchope Farm, Craven Arms, Salop.
Breeder, Allan C. Lyell, Mayville, Neston, Cheshire.
Sire, King Cyrus (Arab), grey, 14.2.
Dam, (5478 W.S.B.) Irfon Lady Twilight, grey, 11.0, by Dyoll Starlight (4 W.S.B.), grey, 11.2½.
G. dam, (4525 W.S.B.) Seren Eppynt, chestnut roan, 11.2, by a Mountain pony.

Craven Cyrus (f. 1927) registered in Volume XXI of the NPS Stud Book and Volume XXVIII of the Welsh Stud Book.

strains until Criban Victor came along in 1944, Reeves Golden Lustre in 1945 and Solway Master Bronze in 1959.

What is not generally appreciated is the big effect which Craven Cyrus has had on the Welsh Mountain Ponies mainly through his daughter, Wentworth Grey Dapples. Considering Royal Welsh Show Champions, the following are direct descendants of his: Clan Pip, Ankwerwycke Clan Snowdon and Ready Token Glen Bride and of course, Clan Pip is responsible for the other Champions: Revel Cassino, Bengad Day Lily and Bengad Love-in-the-Mist. Craven Cyrus himself had been Champion at the 1939 Royal Welsh Show.

His influence on the Welsh Riding Pony will be considered at a later stage but suffice to say that the famous Section B stallion, Downland Chevalier, is directly descended from Craven Cyrus through his dam, Downland Misty Morning. As Peter Upton states in *Out of the Desert*:

> One of the best-known ponies tracing in two lines to Cyrus is Downland Chevalier, a chestnut stallion foaled in 1962 of extreme Arab type.

Although Craven Cyrus was registered by Tom Jones Evans, he was bred by Allen Lyell who, with his brother George, sought the advice of Evans when purchasing high-class Welsh ponies for the show ring. The brothers only bred a small number of ponies themselves, preferring to buy and register them with their own prefix, but they showed Welsh ponies and cobs successfully during the 1920s, winning at all the major shows. One of Lyell's cobs, Pontfaen Lady Model, shown by him in 1924, had already taken the George, Prince of Wales Cup at the RWS in 1922 and 1923 and was the grand-dam of Hendy Brenin, whose bloodlines are among the most sought-after in the Welsh Stud Book. Allen and George Lyell were partners in the Ness Stud which was based on the Wirral in Cheshire, only a short rail journey or ferry ride across the Mersey to Liverpool – an ideal location to cater to the prominent commission merchants buying and selling for clients plying their trade across the Atlantic to America. They were two of six brothers who were all traders like their father and grandfather, whose family home was Kinnordy Castle in Angus, Scotland. Their father, Sir Charles Lyell, a close friend of Charles Darwin, was best known for his pioneering work in geology for which he was knighted by Queen Victoria. Although Allen Lyell was President of the WPCS from 1927 to 1928, the family's commitment to the Welsh breeds was short-lived and shortly after the end of his Presidency he gave up all the ponies and cobs and retired.

Looking at the entries in the Welsh Stud Book, there was much Arab breeding to be found in the Appendix, which was started in 1930 at a time when registrations were extremely low – averaging only 60 annually over a four-year period. It was not surprising that the country had little appetite for breeding ponies during this period – the New York Stock Market crash in 1929 had heralded the beginning of the Great Depression. Nevertheless, a few stalwarts made every effort to maintain the numbers of Welsh ponies and cobs. A scheme to introduce more mares into the Stud Book was made possible by the inclusion of an appendix, thus creating a supply of 'Foundation Stock' for all sections. Following an initial inspection to allow mares of unknown or little-known breeding to enter as foundation stock (FS) by crossing them with purebred stallions, fillies could firstly enter as FS1; these crossed again to a registered Welsh-stallion-produced offspring that would gain entry to FS2. All colts from these crossings had to be castrated. It was only from FS2 mares that all the progeny from registered Welsh sires would be eligible for full registration. As we shall see, of all the sections it would be the Welsh Riding Pony (soon to be called Welsh Section B – but not quite

The Arab stallion Incoronax, ridden by Moses Griffith.

yet) which would gain most from the Foundation Stock scheme which proved to be a springboard for its success.

In the entries of the Foundation Stock scheme a variety of breeds of stallions can be found, including polo pony stallions, Thoroughbreds and, of course, Arabs. It was the latter that found most favour – probably because they had a look associated with the Welsh pony and possibly because it was fashionable at the time. Arab stallions were in abundance and readily available for purchase or lease, as we have seen in the case of King Cyrus. Arab sires in the Stud Book include Indian Grey, Iran, Irex, Jordan (bred by Miss Brodrick), Jotham, Riffayal, Rudan and Sahban as well as their part-bred offspring, Prince of Orange (Beryl

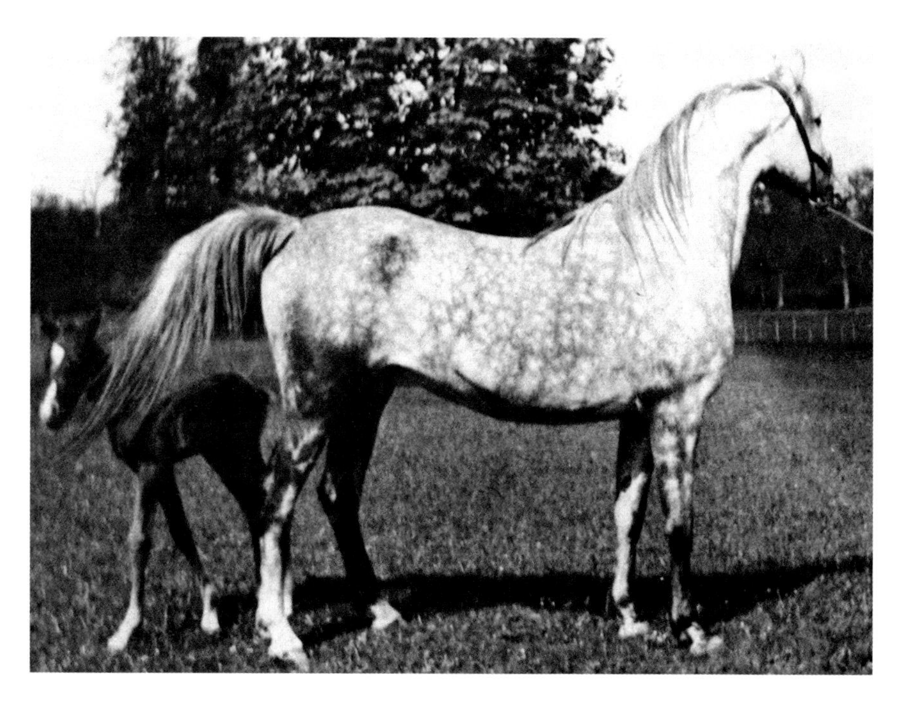

Incoronata, the dam of Incoronax.

Prior) and Munis (Emrys Griffiths). The Crabbet-bred sires were prolific in breeding circles throughout Britain at the time and the majority carried Skowronek within three generations in their pedigrees. One of Lady Wentworth's stallions which stands out for his contribution to the developing Welsh Riding Pony was the grey, Incoronax, by the highly-acclaimed Naseem. On both sides of his pedigree he carried a high degree of the original Arab blood imported by the Blunts but, most importantly, his dam, Incoronata, was by Skowronek.

Incoronax was purchased by Moses Griffith, grandson of the famous Victorian landscape artist of the same name. He was a well-known and well-respected agriculturalist, with great knowledge of grassland and its management. He was appointed Director of the Cahn Hill Improvement Scheme for the University College of Wales at Aberystwyth and moved nearby to Capel Bangor from his leased farm of Egryn where he had specialised in the breeding of Welsh Black cattle. In 1946, he was appointed Grassland Advisor for Wales. He was also co-director of a newly-formed chain of cafés (National Milk Bars) along with his brother William, who farmed extensively in mid- and north Wales, centred at Welshpool where the family still farm at Woodlands, Forden. The company was founded in 1933, the same year that Moses Griffith started his Egryn Stud where he bred Welsh ponies and cobs.

He obviously had a good eye for a pony, as he was responsible for bringing back from Ireland the highly influential Welsh cob stallion, Cahn Dafydd, which in later years became a prominent name in the pedigrees of many champion cobs. There is no record of how he came to own Incoronax but, having served in a cavalry regiment during World War I, he rode well and had an appreciation of the Arab horse as a breed and mount. A well-read man, he would have been aware of the increased demand of children's ponies so, like many of this era, he selected the Crabbet-bred stallion for this very purpose and not directly for the breeding of Welsh ponies. We know of his

Sahban, one of the Arab stallions found in the pedigrees in the early Welsh Stud Books. (Drawing by Peter Upton.)

appreciation of the Arab from comments he made during a talk given at a Welsh Cob Conference held at Lampeter on 25 November 1955. A paper on the talk was first published in a Welsh Stud Book (Volume XXXVIII) and later reprinted in the 1993 WPCS *Journal*. Although basically addressing the breeding of cobs, he states:

> It is claimed that the Welsh mountain pony also owes some of its superb qualities to Arab and Arab bred sires. There is no doubt that light horses throughout the whole world owe more to the introduction of Arab blood than to any other factor, and the great skill in breeding (maybe for over 2000 years) the ideal type of horse: one that can be cherished, fostered, and highly valued as a good horse – has come from the Arab.

Griffith was well known within the Welsh pony and cob community and was President of the WPCS from 1967 to 1968. Although better known for his cobs, his Arab stallion, Incoronax, would make his mark on the Welsh riding pony when he covered the mare Ceulan Silverleaf

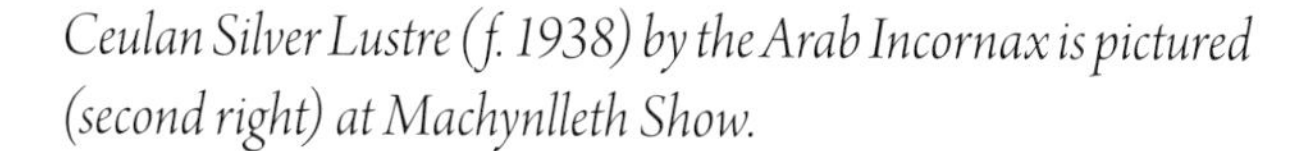

Ceulan Silver Lustre (f. 1938) by the Arab Incornax is pictured (second right) at Machynlleth Show.

E S Davies with his foundation mare Seren Ceulan (f. 1910).

in 1937 to produce the filly Ceulan Silver Lustre the following year. She was owned by ardent Welsh enthusiast and breeder, Evan Samuel Davies (always referred to as E S Davies), whose address of Ceulan Stores, Talybont, was a legend among Welsh breeders of all sections during the 1920s and 1930s. Like his son (Dr Wynne Davies), E S was often described as 'a walking stud book', such was his knowledge of the breed, its evolution and its development.

Silverleaf was a daughter of Davies' first purchase in 1915, Seren Ceulan (f. 1910), whose bloodlines still flow through the Ceulan ponies more than a hundred years later, making it one of the oldest studs in the breed's history. Among the many medals and championships to

Irex, one of the Arab stallions found in the pedigrees of FS mares in the Welsh Stud Book.

her credit, the championship at the 1928 Royal Welsh stands out. Her son, Ceulan Comet, was four times champion at the Royal Welsh before export to Australia in 1934 and her daughter, Ceulan Silverleaf, was reserve champion in 1932. The family tradition would continue for generations, starting with Silver Lustre's famous daughter, Reeves Fairy Lustre, which was Royal Welsh champion in 1973. We will return to this family later (see, e.g. chapter 10).

Incoronax may have gone to North Wales later in his life, since progeny by him were offered on some of the Coed Coch sales. It is interesting to note that when the Egryn Stud was finally dispersed at Devil's Bridge Market on 7 November 1951, as well as Incoronax breeding, other Arabs appeared in the pedigrees of the stock for sale. Youngstock were described in the catalogue as 'Riding Ponies Arab-Welsh' by Grey Star and brood mares 'Arab and Polo Pony Crosses' by Incoronax in foal to an Arab stallion by Refari.

Without doubt, while the Arab was proving to be commercially attractive to several of the pony breeders of Wales to produce children's ponies, the native Welsh pony itself played an extremely important part in establishing the type within the developing Welsh Section B. This trend was reflected over the border in England and it was from roots set down in the hunting field that an Arab cross Welsh stallion would appear which would have a remarkable impact on the development of this riding type of Welsh. Blessed with the most evocative of Welsh names, Tanybwlch Berwyn would stamp his mark on the Welsh Stud Book as indelibly as the hallmarks on sterling silver.

Chapter III

Chirk (Howard de Walden), Sahara, Tanybwylch

Sahara with his owner Denis Aldridge.

A painting by celebrated Irish artist John Lavery, commissioned by Lord Howard de Walden, showing the de Walden family in the saloon at Chirk Castle.

There is no question that Welsh farmers and breeders played a huge part in the success of the mountain ponies and cobs found in Britain by the end of the 19th century. Using all their skills and experience, they had brought the Welsh breeds to prominence over all the other native breeds, keeping their function current as they met the challenges of the Industrial Revolution. Mechanisation and the development of the railway engine and motor car posed particular threats to both Welsh ponies and cobs, as outlined in an article published in 'Riding' magazine in March 1937. The National Pony Society President and celebrated breeder of the Silverdale-prefixed polo ponies, Herbert Bright, made the following point:

> The breeders of native ponies are mostly farmers of very moderate means who cannot spare anything towards the expenses of their breed society, which has to provide prize money at local shows and to keep a register of brood mares and young ponies. These farmers used to breed ponies in hundreds which were sold at local fairs and bought by dealers who supplied the coal mines. The demand for the ponies has decreased very much in recent years, though in *The Times* of 6 January, 1937, a report of the Pit Ponies Protection Society, it is stated that last year 33,136 ponies were in the mines. In 1929 the number was 50,823. You see it will require a lot of people to buy riding ponies to make it worthwhile for the farmers to continue breeding.
>
> The National Pony Society had two riding classes for registered ponies at Islington, last March, and this year we have four classes. All ponies must prove their breeding by being registered by their breed society. This is the way the N.P.S. is encouraging the pure native breeds, and I hope that parents and children riders will try to help by buying native ponies and insisting on a registration certificate from the breed society.

Naturally, Bright used the platform to sing the praises of the National Pony Society. He was one of many who were acutely aware of the decline of horses and ponies used for work and the pony bred for polo as well as the rise in interest of ponies bred specifically for children. Little did he know that his own Silverdale stallions would have an influence on the Welsh Riding Pony later in its development.

Equally conscious of the changing pattern was the Past President of the Arab Horse Society, Lt-Col Patrick Stewart DSO, DL, an officer in the 1st Gordon Highlanders who fought in the Boer War and in India.

During his distinguished military career he was mentioned in despatches for his gallantry at Ypres with the 3rd (Prince of Wales's) Dragoon Guards. Colonel Stewart first lived at Moreton Hall, a mansion house near Moreton Morrell commissioned in 1906 by a wealthy American banker's son, Charles Garland. Completed in 1909, Moreton Hall became a privately-funded military training centre before Garland took British nationality in 1914 and joined the Household Cavalry. It is likely that the military connection brought Garland and Stewart together, resulting in the latter taking a teaching job at Moreton Hall's polo school and equestrian centre unaware that by the turn of the following century it would become well known as a prominent equestrian educational centre as part of Warwickshire College.

Stewart later moved to Chadshunt, near Kineton, Warwickshire, home of Lord Willoughby de Broke. The latter was Joint Master of the Warwickshire Hunt (1929–1935); we know that Stewart rode in point-to-points during this time and can safely assume that he had a hand in Willoughby de Broke's racehorse enterprise, which was considerable at the time. Lord Willoughby also held the posts of Chairman of Wolverhampton Race Course (1947–1971) and was President of the Hunter's Improvement Society (1957–1958). Indeed, Stewart made use of his own reputation as a highly regarded horseman to pen the highly successful books *Handling Horses* and *Training Race Horses.*

In the Foreword to Stewart's book *Handling Horses,* first printed in 1943, General Sir Beauvoir de Lisle wrote:

> This book, to which Colonel P. D. Stewart has given the title Handling Horses, is the result of some fifty years of experience; years spent in close association with horses of all breeds and sizes. In two campaigns he served with mounted troops; in both he was wounded and mentioned in despatches, and was awarded the DSO when serving with his regiment, the 3rd Dragoon Guards. In addition to his military training as a Cavalry Officer, Colonel Stewart has taken a prominent part in all kinds of sport.

Besides the two books mentioned, Stewart also wrote under the noms de plume 'Dandy Brush' and, more significantly, 'Crosbie'. His topics included 'Stable Hints' (as Dandy Brush) and 'Hints on Riding' (as Crosbie) in 'The Horse', the quarterly magazine of the Institute of the Horse and Pony Club (a precursor to the British Horse Society). 'Crosbie' is significant because this was also the name of a stallion of Arab origin reputedly bred by the Dursi tribe in Syria in 1908 and a race winner of the Municipal Cup in Alexandria prior to his export from Cairo to England by Stewart in 1914.

Crosbie was registered with the National Pony Society in Volume IV of its Stud Book and Volume I of the Arab Society Stud Book. He was exhibited by Stewart, who was by this time a member of the National Pony Society. Crosbie won extensively both in hand and under saddle. In 1916, he won the class for 'Eastern Sires' at the National Pony Society Spring Show at Islington and again in 1918 when the show transferred to Newmarket during World War I. In a report of the class, the judges, Mr Faudel-Phillips and the Rev D B Montefiore, wrote of him, 'The winner, Crosbie, is a nicely balanced little horse with plenty of substance.' He was later champion there in

1920, having stood reserve to the famous Skowronek at Ranelagh in 1919.

It is little wonder that Stewart had chosen to bring Crosbie back to England since, like many of the cavalrymen stationed overseas at that time, he had first-hand experience of the Arab breed when stationed in India. In *Handling Horses* he writes:

> The Arab, which was shipped to Bombay and Karachi in large numbers, predominated on the race course, the polo ground, and in the ranks of the cavalry; much the easier to feed, keep and ride, and with greater endurance, a jack-of-all-trades, he lacked the speed and weight of the other breeds (i.e. English Thoroughbred and Australian Waler).
>
> Unquestionably the Arab pony [1] was the best value for the British subaltern – he could ride him on parade in the morning, play polo on him in the afternoon; he would go hunting the next day and pig-sticking the following, and perhaps win a gymkhana race and forty-nine rupees for you on Saturday.

Crosbie was a popular sire of his day and in 1921 produced for his owner the grey colt Gray Cross, which was out of the 15 hh 'hunter cob' mare, Dolly Gray. This colt was shown only twice, taking a second and first place in the Arab Bred colt classes at the National Pony Show in London in 1922 and 1923 respectively. He was used extensively by the 8th Baron Howard de Walden on registered Welsh Section A mares which he kept at the mediaeval Chirk Castle near Wrexham in North Wales. Originally leased for 25 years in 1911, Howard de Walden loved his stay there so much that he extended the lease by ten years, leaving Chirk in 1945 for another of his properties, Dean Castle near Kilmarnock, Ayrshire, where he died in 1946.

In the Preface to a National Trust booklet, Lord Howard de Walden's grandson Thomas Seymour sums up his grandfather as, 'An Englishman turned Welshman ... polymath, poet, playwright, soldier, artist and Olympic sportsman'. Born in 1880, Thomas (Tommy) Evelyn Scott-Ellis, 8th Lord Howard de Walden, was all of these and more, a remarkable man who inherited a vast fortune on his father's death in 1899 and further estates in 1901. The title was created by Queen Elizabeth I in 1597 in recognition of Admiral Lord Thomas Howard's role in the defeat of the Spanish Armada in 1588. The title

An advertisement in 'Riding' magazine 29 September 1933 for Lord Howard de Walden's homebred four-in-hand team of ponies with their harness and coach.

CHIRK CASTLE PONY STUD.
The Property of Lord Howard de Walden.
FOR SALE, a typical FOUR-IN-HAND TEAM, home-bred PONY GELDINGS, 13 hands, first quality, with courage, powerful, clean limbs, true, fast, free action, used to all traffic; driven seven weeks in London. Also COACH and silver-mounted HARNESS.—Details and photographs on application to HENRY C. WEBB, Castle Office, Chirk. Telephone: Chirk 316 and 360.

1 According to Stewart, Arabs were referred to as ponies in India and not as horses, as in England.

moved out of the Howard family, first to Frederick Hervey, 4th Earl of Bristol and Bishop of Derry, then through the marriage of Elizabeth Hervey, the Earl-Bishop's great-granddaughter, to Charles Augustus Ellis. The Ellises originally came from Denbighshire and had become rich through the sugar plantations started by their ancestors in Jamaica. But the real wealth came from the marriage of the 6th Baron Howard de Walden to Lady Lucy Cavendish-Scott-Bentinck, daughter of the 4th Duke of Portland whose property portfolio included large parts of central London. Lady Lucy brought with her over 130 acres of fashionable Marylebone and extensive estates in Scotland. This led Tommy to change his surname first to Ellis-Scott (1901), then to Scott-Ellis (1917). The de Walden Family remains one of the wealthiest in Britain with assets valued at over one billion pounds.

Tommy Scott-Ellis had extraordinarily wide interests, most prominent among which were sport (a fencer, he represented Great Britain in the men's individual foil at the 1906 Olympics), an appreciation of the arts, a love of horses and a keen interest in all things Welsh. Two centuries after his ancestor Captain Ellis left for Jamaica, de Walden had a romantic idea of a 'home-coming' to Wales which led him to move from his fashionable London address to Chirk Castle, which underwent a complete refurbishment under his stewardship; this included reinstating the Great Hall to its medieval origins as well as setting out wonderful gardens and parkland. His embracing of all things Welsh included serving in the 10th Royal (Prince of Wales's Own) Hussars during the Boer War and the Royal Welch Fusiliers during World War I (acting as second-in-command of the 9th Battalion on the Western Front in 1916).

All things cultural interested him and he surrounded himself with many leading artists of the day. He wrote poems and plays under the pseudonym T E Ellis, the name of a well-known political leader of the Welsh nationalist movement. He learned the Welsh language and was a great supporter of the National Eisteddfod, where his first play was staged at Holyhead in 1933. He was also a great patron of the arts, commissioning works by notable artists, such as a bronze portrait of himself by Auguste Rodin and an oil painting by Augustus John. One of his most impressive purchases was a large painting of himself and his family in the sitting room at Chirk Castle, completed in 1933, by the celebrated Irish artist Sir John Lavery (see page 34).

Although de Walden's interests and pursuits were big and bold, he was by nature both retiring and reserved. He frequently argued against the cult of personality and had little time for the social élite, politicians and celebrities, but he took greater pleasure in the company of his employees whom he rated among his friends. One of these was his groom, Sam Fennell. Dr Wynne Davies recalls his own father's stories about staying in London with Fennell and driving round the city in Lord Howard de Walden's pony coach at weekends. Fennell divided his time between London and Chirk, where stables were built to accommodate de Walden's string of successful racehorses and in due course his Welsh ponies. De Walden was well known around the racecourses of England and was successful as an owner – although not as successful as his son, for whom horse racing was a passion. Breeding horses was de Walden's first love and Welsh ponies were an obvious choice for his new Welsh home, where he used the native Mountain Pony mare

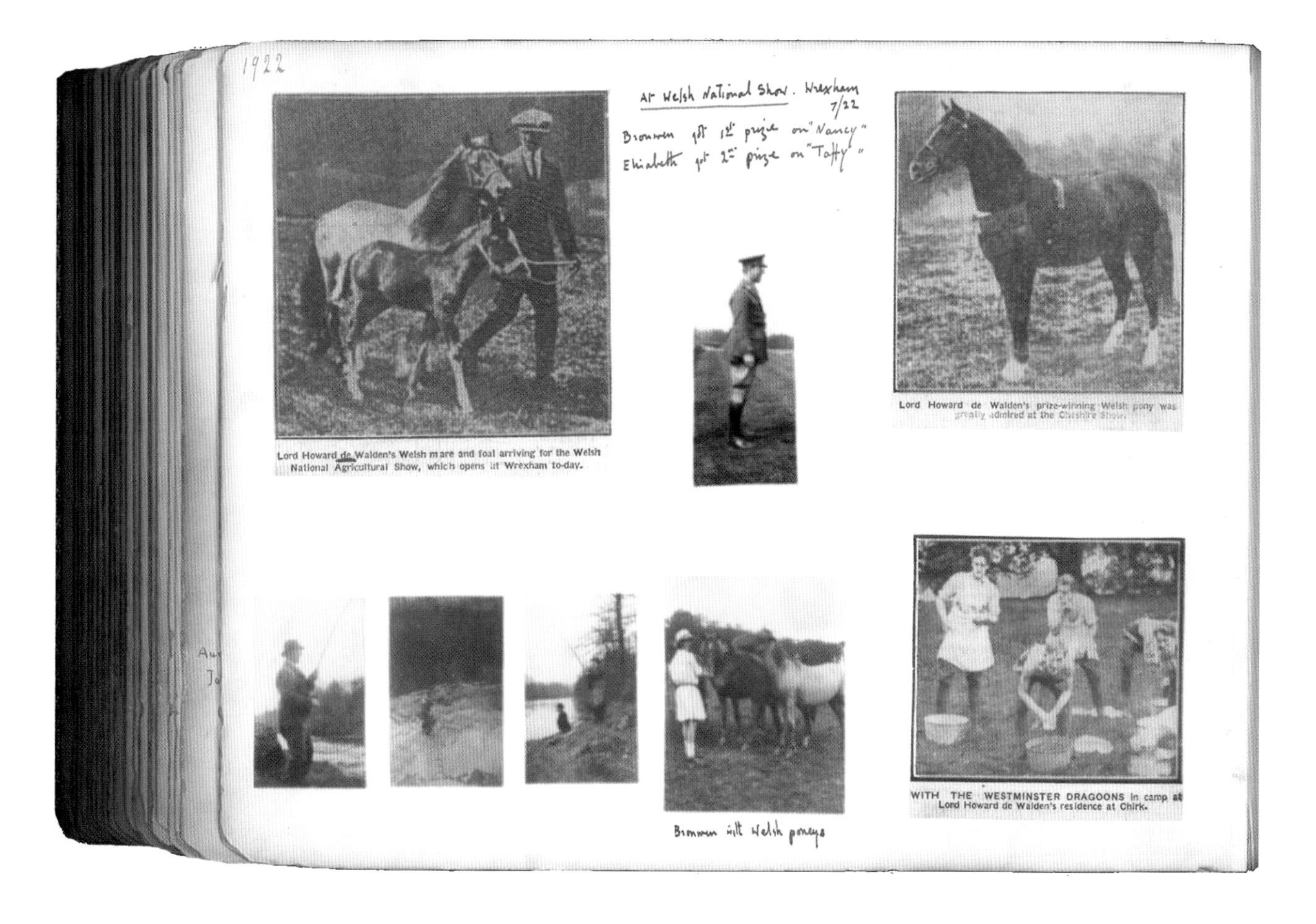

A page from Lord Howard de Walden's photograph album, July 1922, showing paper clippings and images of his ponies and a note of his daughters' placings at the Royal Welsh Show held at Wrexham.

to cross with Arab blood to breed the children's pony in demand at the time.

It was most likely that de Walden was acquainted with Lt-Col Stewart through his racing friend, Lord Willoughby de Broke, or he may have seen Gray Cross at Islington, so the sale of the colt to Chirk Castle was completed. Indeed, Lord Howard de Walden adopted Gray Cross as one of two prefixes for his newly-formed stud, the other being Chirk Castle. It was perhaps a sign of his idiosyncratic nature that he reserved the Chirk prefix for the purebred ponies but gave them numbers instead of names – Chirk Castle No 1 etc.; he registered

21 mares in Section A in Volume XV of the Welsh Stud Book. The Arab crosses were given names with the Gray Cross prefix. Taking advantage of the initial Foundation Stock Scheme initiated by the WPCS, Lord Howard de Walden registered 17 offspring of Gray Cross in the Foundation Stock register between 1928 and 1933. The auctioneers Frank Lloyd and Sons had the job of selling de Walden's ponies through the sale ring at Wrexham. The sale catalogues of 1926 and 1930 make interesting reading as they feature a selection of hunters, hacks and registered Welsh Mountain Ponies. By 1935, as well as registered Welsh, the ponies offered featured Welsh-Arab crosses and Arab-cross riding ponies.

In 1921, the year after Crosbie triumphed in the championship at Islington, another 'Eastern-bred' stallion came to the fore. This was the grey, Sahara, which had been admired by Denis Aldridge at the Royal Show staged at Cardiff in 1919 when shown by Enid Betty from West Wales. The daughter of Judge John Bishop, Deputy Lieutenant for Carmarthenshire and a well-respected figure in the area, Mrs Betty lived at her family's home of Dolgarreg near Llandovery. Her mother, Caroline, came from a prominent local family, the Pryse Lloyds, who lived at the Glansevin Estate in Carmarthenshire and whose name was to become synonymous with the best of Welsh Section Bs in the Stud Book much later in its history. It was with the privately-owned local pack of Harriers hunted by Dame Margaret Pryse-Price that Mrs Betty would regularly hunt her grey stallion Sahara.

Mrs Betty's husband, Major Paget Kemmis Betty CMG, DSO, was Secretary to the Governor of Gibraltar from 1886 to 1887 and a member of the Royal Engineers Expeditionary Force in South Africa. His wife travelled

Opposite: Denis Aldridge (on the left) with his great friend and fellow artist, Lionel Edwards.

overseas with him and it was in Gibraltar around 1913 that she first saw Sahara among a load of ponies from Morocco. She takes up the account of him in an article first published in the 1966 WPCS *Journal*:

> [Sahara] played polo, rather indifferently, for a season in Spain. He was over-grown and soft, and not a very good doer. However in 1914 war broke out and he was shipped home with other ponies as a charger, but being entire he was not accepted as such and was sent down to me in Wales. I then hunted him for two seasons and a very good hunter he was. Eventually the local farmers induced me to use him for stud and he travelled Carmarthenshire and Brecon districts for two seasons. The war then ended and I could not keep him or any of my other ponies. I took him to the Royal (that year at Cardiff) and did very well with him shown as a polo pony stallion. There Mr D Aldridge bought him and took some of his young stock and what mares I had in foal to him. He was the kindest and quietest horse I have ever owned and all his stock that I had dealings with were the same. I think he must have been a six year old when he came over to this country and it was most interesting to observe what a gawky, ugly pony could eventually become.

Denis Aldridge was an amateur painter whose lifelong interest in fox-hunting was reflected in his work. He

Above: Mr and Mrs Aldridge (foreground) hunting with the Atherstone Hunt, captured by Lionel Edwards.

Opposite: Mrs Inge's daughters with one of their mother's Welsh Mountain Pony mares (c.1910, artist unknown).

came from a hunting family whose wealth was based on manufacturing machinery for the hosiery trade. They lived at Sketchley Hall, near Hinckley in Staffordshire, a large house which his father, Charles Aldridge, extended to create a very impressive mansion reflecting its owner's success in business. Aldridge Senior kept the best horses for himself and his family to hunt and was always immaculately dressed – with violets in the buttonhole in his hunting coat. Dennis, too, was always smartly dressed, well mounted and immensely popular among the hunting fraternity. During his life he was Secretary to both the Atherstone and Quorn Hunts.

When Aldridge married Kathleen Crosfield in 1923, he became a farmer. His father built him a house and farm buildings adjacent to the Hall and he bred very good sheep and the Dairy Shorthorn cattle for which he became renowned. In 1922, 1923 and 1924 he took the Thornton Cup at the Dairy Show in London with homebred cows. The family tradition continued when his unmarried daughter, Rosemary, became a well-respected breeder of Dairy Shorthorns carrying the Eaves prefix, taken from the village of Woodhouse Eaves whence the Aldridge family moved in 1940 following the outbreak of war. It was at this time that Denis Aldridge sold his farm to take up an appointment with the War Office to collect horses for the army. He later moved to The Priory in Wymondham and was Quorn Hunt Secretary for 13 seasons.

His interest in hunting also brought Denis Aldridge into contact with the leading artists of the day. Among them were Cecil Aldin and Lionel Edwards, both household names whose paintings and prints of country life and the hunting field adorn many a wall. Both artists stayed with him during commissions in the 'Shires' (as this hunting area was called) and his hospitality was often reciprocated. Edwards was very close to Aldridge and encouraged him to paint, praising him particularly for his portrayal of deer. He also captured Aldridge and his wife in a well-known watercolour of the Atherstone Hunt, which became one of a popular series of hunting prints. Both Denis and Kathleen Aldridge were also acquainted with many of the great and the good associated with their hunts, including Siegfried Sassoon, whose *Memoirs of a Hunting Man* reflected much of Sassoon's own life within this hunting fraternity.

A doctor's daughter, Kathleen Aldridge came from an accomplished family and she herself penned an autobiography of her life up to the time the family moved to Woodhouse Eaves. Entitled *The Most of the Bubble*, part of her account tells us about the stallion, Sahara, which she recalls:

> Denis' stallion, Sahara, lived in a separate stable. He was an interesting animal. Denis bought him from a Mrs Betty at Cardiff Show, who in turn had bought him in Gibraltar out of a load of ponies from Morocco. He was so named because he had the Sahara Desert brand on him, a double necklace of black dots round his neck and girth. Apart from this, he was rather a mystery. He did not carry his tail at the height typical of an Arab so the authorities considered him a Barb and this made him ineligible to be shown in Arab classes.[2] As he was of perfect formation and temperament, they inaugurated a new class in 1921 for 'Eastern stallions' to enable him to be shown at the annual National Pony Show at Islington. He won first prize, the Prince of Wales taking most of the other prizes. Denis also bought Sahara's young stock and some mares to foal to him, and with other mares being sent to him Sahara eventually built up a reputation, and his progeny achieved considerable renown as polo ponies. Mrs Inge, known as the queen of the Atherstone country, sent all her grey Welsh mares to Sahara from Thorpe Constantine, and they would return to Sketchley to foal and to be mated again. One of these foals, a son of Sahara, was the famous Tan-y-Bwlch Berwyn, his prefix being the name of Mrs Inge's estate in Merionethshire. He was sire of many outstanding ponies in the show ring. Another of his sons was called Wiseman: he always carried Captain Allison, the starter of the Jockey Club, down to the start and if it sometimes happened Captain Allison was given a lift in someone's car back to the paddock, Wiseman would return on his own

---

2 Although some people have referred to Sahara as a Barb stallion, Peter Upton, world authority on the eastern horse breeds and Arabs in particular, has little doubt that Sahara was in fact Arab in origin. According to Upton, he showed all the characteristics of an Arab horse and not a Barb, which he had judged on many occasions in Africa.

to find his master. A brother of Wiseman was sold for 1,000 guineas to one of the Rothschilds to go to France as a polo pony. Sahara was so gentle the farm children could climb on his back and play round his legs and he would never hurt them.

Aldridge himself summed up his stallion thus:

When I got him home I rode him every day and he was the perfect hack ... When I reluctantly gave up breeding I leased Sahara to Mrs Bolam in the New Forest, where alas he ended his days. His life was like that of many artists whose work was not appreciated until they had passed away.

Aldridge mentions Mrs Inge, 'queen of the Atherstone country', who, by selecting Sahara for her Welsh mares, unwittingly established this grey Arab as a major influence on the developing Welsh Pony within the Welsh Stud Book. She and her family lived at Thorpe Hall, a grand house built by Richard Inge in 1651 at Thorpe Constantine near Tamworth. Records show that the Reverend George Inge and his family were living at the Hall in 1881 when there were 21 members of staff employed and it was his son, William, who inherited the estate and lived there with his wife Mary and three daughters. They were a great hunting family, William Inge having been Master of the Atherstone Hunt from 1891 to 1895 following Edward Oakeley, the Master for the preceding 20 years. Mary Inge was Oakeley's sister and, following the untimely death of her husband in 1903, she became Master of the North Atherstone between 1914 and 1920 (when the hunt was divided into North and South for a while). For four seasons she was assisted by her daughter Margaret, who died in 1919. Her younger daughter, Hilda, inherited Thorpe Hall, and took over as Master during the seasons 1945 to 1950.

By all accounts, Mrs Inge always rode side saddle mounted on beautiful horses, often grey, and was generally considered to be a formidable lady but one with a 'kind heart', according to Kathleen Aldridge. The respect she commanded within the hunt is borne out in an article by J N P Watson in 'Country Life' on 11 December 1975, entitled 'Sassoon's "Tip-Top" Country':

So we return to the vivid personality of Mrs Inge, resplendent in side-saddle habit and silk hat. In 1920, when she spoke of retirement, 1000 farmers requested that she carry on in their interest. But, because of ill-health she retired.'

It was around 1919 that Mrs Inge inherited the south side of the Oakeley family's estate of Tan-y-Bwlch in Merioneth which, over a 400-year period, had belonged to some of the wealthiest families in North Wales. The north part, along with the large house (The Plas), had been left to her brother Edward de Clifford Oakeley, but was purchased for £26,000 in 1915 by Mrs Inge's daughter, Margaret, when her uncle fell on hard times. On Margaret's death her sister, Hilda, managed this land for 30 years before the whole estate passed to her mother in 1953 following Hilda's own death.

Situated within the Snowdonia National Park, close to the small towns of Maentwrog and Ffestiniog, Tan-y-Bwlch was first named in the will of Robert Evans in 1602 and we know that a house named Plas was built there in 1634. It remained in the same family for almost 200 years, the ownership passing to Oakeley when a

Plas Tan-y-Bwlch, Mrs Inge's home in North Wales (postcard circa 1930 by F Frith & Co).

granddaughter of William Griffiths (grandson of Robert Evans) brought much-needed funds into the estate by marrying William Oakeley, a wealthy landowner from Staffordshire. He took over the estate and did much to improve the land during the period from 1789 to 1811. At the same time he maximised the value of the nearby mountains by developing an extensive slate quarry in the area and it was during his son's time (1879 to 1912) that the famous Ffestiniog Railway was constructed to transport slate from the quarries.

It is interesting that both the Inge and Oakeley families should be short of male heirs, particularly during the latter part of their history. On Hilda Inge's death in 1953, Thorpe Hall was left to a distant relative, Lt Commander George Inge-Innes-Lillingston, DSO, a former High Sheriff of Staffordshire, on condition that the name Inge remained as part of the surname. While the name Lillingston was well known in the Atherstone Hunt (Captain Luke Lillingston had been a former Joint Master), the Innes-Lillingston name is associated with the north-west coast

FOR SALE.

1. CAIRO, Grey Arab Stallion, 14 hands, 6 yrs, very quiet, good-looking, sound, and a sure foal-getter.
2. ARIAN, Dark Grey Mare, 12 hands 3in, 4 yrs.
3. SIABOD, Dark Grey Gelding, 12 hands 2in, 4 yrs.
4. RHOS, Grey Mare, 11 hands 3in, 4 yrs, winner of several prizes, ridden by small children.

All the above Ponies were bred by Mrs. Inge, and with the exception of No. 1 have been ridden by children, and are quiet.—Apply MRS. INGE, Thorpe, Tamworth.

Left: an advertisement in 'Horse and Hound', 3 November 1933, for the Arab stallion Cairo and three homebred Welsh ponies Tanybwlch Arion, Siabod and Rhos.

Below: Mrs Inge, aged 85, with her favourite, Bluebird.

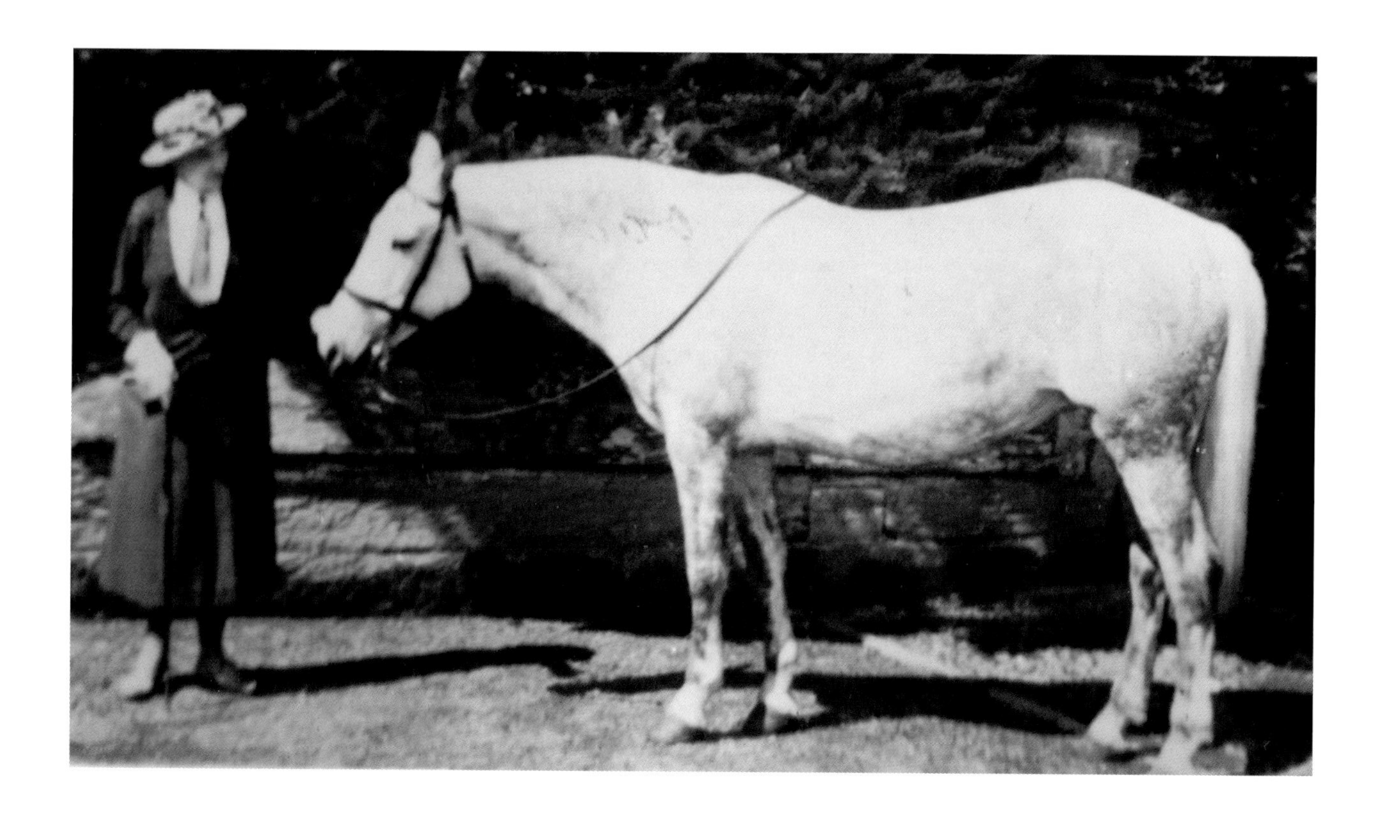

of Scotland, in particular Lochalsh. Sir Hugh Innes built Balmacara House on an estate there, which was later purchased by Sir Alexander Matheson, co-founder of the famous Hong Kong trading company Jardine-Matheson. (Incidentally, history tells us that this company itself has links with the Welsh Section B through the famous stallion Solway Master Bronze.) Meanwhile, Sir Hugh's niece, Katherine Lillingston, acquired land on the estate and built a large house at Lochalsh, later to become the home of Commander Hugh Innes-Lillingston, father of Thorpe Hall's new owner, George.

The Hall is currently owned by former rock star and company director Hugh Inge-Innes-Lillingston, a drummer in the 1970s punk band Rikki and the Last Days of Earth. Back in North Wales, on Mrs Inge's death the Tan-y-Bwlch Estate passed in 1961 to a member of the Oakeley family, John Russell, later known as Lord de Clifford. Crippling death duties left the estate untenable and, following a brief period in the ownership of the well-known Bibby family, it was purchased in 1969 by Merioneth County Council as part of the Snowdonia National Park and the huge house of Plas was made into a study centre in 1975.

Tanybwlch Rhew (second right) held by Peggy Pacey, whose sister Betty is also seen holding her ponies.

Mrs Inge fell in love with Wales and with the small ponies that were to be found in abundance on the hills and which were much in demand for work in the mines. So it was no surprise that she selected Welsh Mountain mares as mounts for her daughters. When they grew too big for the ponies, Mrs Inge was left with the quandary of what to do with them. Her ambitions for them were not aimed at furthering the Welsh breed at that time – far from it. Following a trend established by the polo pony breeders at the last part of the 19th century and continued by the National Pony Society during the early part of the 20th, Mrs Inge decided that by crossing her Welsh Mountain Pony mares with a larger stallion, she might breed a pony ideally suited for a child, the likes of which she had sought many years previously for her own children. In this case, she opted for the grey Barb, Sahara,

famed for his good temperament and owned by fellow huntsman, and good friend, Denis Aldridge.

Little did Mrs Inge realise at the time what a momentous decision this would be for the development of the Welsh breed. Not only would she help fashion the type of pony we now know as the Welsh Section B, but she would give the gene pool of the Welsh breed an injection of blood that would have an immediate and lasting effect on future generations. It was one of her first foals by Sahara, Tanybwlch Berwyn, which would later be hailed as the 'Abraham' of the Welsh Riding Pony.

This colt's dam was one of Mrs Inge's early purchases for her estate in North Wales. Her detailed records show that in 1921 she bought two mares, bred relatively locally, to which she gave her own prefix, namely Tanybwlch Gwynn (f. 1916) and Tanybwlch Arenig (f. 1917. She added a mare, breeder unknown, which she called Tanybwlch Ceridwen (f. 1915). In 1923, she turned to Buckinghamshire, where she purchased, at £20 each from their breeder, Walter Glynn, the full sisters Brynhir Black Star (f. 1917) and Brynhir White Star (f. 1918 Bleddfa Shooting Star x Brynhir Flight, by Grove Ballistite). The foundation of her emerging stud was completed with the addition of a Grove Ballistite daughter, Kilhendre Celtic Rumour, and one from Lady Wentworth, Grey Princess by Wentworth Windfall out of Lady Greylight (both by Greylight). Like Ceridwen, Gwynn and Arenig were registered without details of breeding. However, it would be reasonable to assume that they had riding qualities, since the other four mares were full of the Starlight breeding previously discussed.

It was no surprise that, as a hunting lady, Mrs Inge once more chose not to breed purebreds but to use them as a foundation for breeding the more lucrative children's riding ponies. Sahara was just the sort likely to cross well with her quality Welsh mares. In 1923 Sahara produced fillies to Celtic Rumour and Gwyn. The same year, at a transport cost of £9, she sent her little mares by rail from Tan-y-Bwlch to Sketchley Hall, where again they were

Tanybwlch Penllyn, FS1 mare by Tanybwlch Berwyn.

covered by Sahara. From the resultant crop of foals, two fillies and two colts were registered; Tanybwlch Rhaiadr Ddu (f. 1924) out of Celtic Rumour was kept entire and later she used him on Brynhir Black Star, the dam of the better of the two Sahara colts born in 1924. Bearing in mind the influence she would play in the future of the Welsh Section B through this son, it is noteworthy that Black Star was measured at 12.1½ hands as a four-year-old when registered in 1922 as a Welsh Section A Part II in Volume XXI of the Welsh Stud Book. Mrs Inge named this colt Tanybwlch Berwyn.

For the next two seasons Mrs Inge enhanced the 'Starlight' stain in her breeding when she leased Grove Orion, a son of Bleddfa Shooting Star out of a mare by Dyoll Starlight. With Sahara now in the New Forest (he went there in 1925 and died in 1927), she set about breeding her own 'Arab' by turning to yet another hunting acquaintance, Colonel Patrick Stewart, whose 'Eastern' sire Crosbie was establishing a reputation based on his showing record, temperament and, most importantly, progeny. In a change of direction, she used him on a grey Arab mare called Jordania, originally imported in 1922 as a gift to King George V from King Abdullah of Jordan. The records fail to tell us how Mrs Inge came by her in November 1922, but she put Jordania to Crosbie in 1926 to produce a grey son, Cairo, born in 1927. He added yet another infusion of Arab into her stud. Sadly, his use was short-lived as he was killed by lightning at a young age – but not before he had produced four high-class fillies – Tanybwlch Penwen (f. 1930) and Serenol (f. 1935), both out of Wentworth Grey Princess, and Snowdon and Bera out of Brynhir White Star and Brynhir Black Star respectively. Ineligible for entry into the main Stud Book, all four mares were registered in the Foundation Stock Appendix of the Welsh Stud Book.

By this time the prefix of Tanybwlch was finding favour in the show ring as Mrs Inge herself exhibited successfully during the early 1930s. She became a member of the National Pony Society in 1931 and showed ponies at Islington, sharing honours with the likes of Lady Wentworth. Her own successes were somewhat eclipsed by others, particularly those who had purchased ponies from her for the children's ridden classes. One of these was Mrs Mabel (May) Pacey, who firstly selected Llwyd (£25), which Mrs Inge had bought in as a two-year-old as

Tanybwlch Berwyn (f. 1924), aged 24.

well as the original Sahara progeny, Rhew (£25) and Gwynedd (30 guineas), and the colt Rhaiadr Ddu (£25), Seren (£18) by a Section A stallion Lidbury Touchstone and Prysor (£30) by Grove Orion. Mrs Pacey was well acquainted with the Aldridge family and Sahara, as she lived nearby at Burrage Hall in Hinkley. Her father, A E Hawley, was founder of the Sketchley Dye Works which was central to the hosiery industry for which that part of Britain was famous. As a breeder, exhibitor and judge of renown, May Pacey was a world authority on white West Highland terriers, about which she wrote extensively. Her daughter, Peggy, was an accomplished rider and won with the ponies her mother had purchased from Mrs Inge. Among many wins recorded, at the age of ten she rode Tanybwlch Rhew (f. 1923) to victory in the Under 13 hh class at the National Pony Society Show at Islington in 1928. Miss Pacey was a senior officer in the RAF during World War II, after which she took up the breeding of racehorses; she was also in demand as a judge and often took charge of major championships for both the National Pony Society and Ponies of Britain.

Miss M. Brodrick's Welsh Pony, *Tan-y-Bwlch Prancio*, winner of the Country Life Perpetual Challenge Trophy for the best mountain or moorland pony. It was ridden by Martin Peter Brookshaw.

Tanybwlch Prancio pictured in Riding Magazine in 1938, the same year he also won the class for riding ponies under 12.2 hands at Islington.

The next chapter will look at the path the developing Welsh Riding Pony would take within the Welsh Stud Book, but suffice to say here that Tanybwlch Berwyn was accepted into the Stud Book in 1927 at a time when new blood was needed. Mrs Inge used him extensively on her own mares with great success and these, too, were in great demand along with the geldings she bred by her Arab stallion, Cairo. Some of the top riding pony exhibitors of the day selected Berwyn offspring for the ridden classes – such as Mrs Hepburn from Sutton Coldfield and Mrs Inge's great friend, Miss Margaret (Daisy) Brodrick, a member of the NPS since 1934 and one of its Council members from 1939 to 1942.

Miss Brodrick had inherited in trust the extensive Coed Coch Estate near Abergele, North Wales (see Chapter 5) and started her Coed Coch Stud in 1924 with the help of her groom, John Jones. In the years to come, Miss Brodrick was a driving force behind the development of the Welsh Riding Pony within the WPCS. She was a member of Council from 1925 until her death in 1961 and Society President for the year 1936–37. Her original selection of ponies bore little fruit in the long term, but the purchase of the mare, Tanybwlch Prancio, followed by the gift in 1940 of her sire, Tanybwlch Berwyn, would set her firmly and successfully in the direction of the Welsh Section B. Miss Brodrick's success found universal acclaim in the equestrian community and was summed up by the former editor of 'Riding' magazine, Reggie Summerhays, who wrote in *A Lifetime with Horses*:

> As the name Crabbet Park is insolubly identified with Arab breeding, so is Coed Coch associated with the Welsh Mountain pony. There was a time when it was expected of judges of riding ponies that they should all be ridden and as a lightweight I was in much demand. Years before the war I was judging at the National Pony Show at Islington and rode a Welsh mountain pony, Tan-y-Bwlch Prancio, a lovely little grey mare, and gave her first prize. This was not only one of the first successes gained by this famous Stud but her descendants are considered by Miss Brodrick to be amongst the best of ponies ever bred by her at Coed Coch. I wonder how many ponies I have judged of this strain over the last thirty years or so! In front of me as I write, on a bracket all to himself, is my model of Coed Coch Madog, a gorgeous little fellow if ever there was one.

With the popularity of the Welsh Riding Pony as a child's pony on the rise, the scene was set for a major effort to establish criteria for its inclusion into a section of its own within the Welsh Stud Book. Miss Brodrick, among others, saw the legislation through to its current form but it was not without difficulty along the way. In the next chapter we follow the development of the Welsh Stud Book and the role this remarkable breeder would play in shaping the sections within, which are so familiar to us today.

Chapter IV

Stud Book Developments

Illustration from the 1922 NPS Stud Book showing four 'Welsh Ponies of Riding Type'.

HAY

The Welsh breeds have come a long way since the establishment of a stud book recording their ancestry. Fashioned by the demands (whims, some might say) of the time, all four purebred Sections of the Welsh breeds that we have come to know and love in many ways reflect the social and industrial history of Britain in the 20th century. This development has endured two World Wars and the Great Depression and the peaks and troughs of demand from home and abroad, and responded to the needs of a country for which leisure time has become an influential feature. Breeders, exhibitors, politicians and animal lovers have all exerted an influence on the Welsh breeds, but none more so than the Welsh Stud Book, which has had the ultimately task of reflecting the demands of all while staying true to the ponies and cobs originating from Wales.

The development of the Welsh Stud Book over the years has been anything but plain sailing, trouble free or without controversy. There have been individuals who have battled hard to steer it their own way, whether for the common good or for their own benefit. Mistakes were made and had to be rectified; the direction frequently changed – at times dramatically, at others with a mere tweak; and circumstances beyond the control of all concerned, such as the outbreak of war, meant that progress – or lack of it – was frustrating, to say the least. Despite all this, the Welsh breeds endured, much to the credit of WPCS Council, members and breeders alike. Their collective efforts have brought the Welsh breeds to the healthy state they currently enjoy as the leading native pony breeds in Britain, if not the world.

A summary of the developments of the Welsh Stud Book over its lifetime (see Appendix 1) reveals that it took almost 60 years before the 'Sections' now so familiar to us fully evolved and stabilised. Bursts of activity characterised the changes that took place – in particular the evolution of the Welsh Riding Pony as the Welsh Section B attracted more attention than any of the other Sections. Many important decisions were made along the way, such as the introduction of 'foreign' blood from the Arab and Thoroughbred, which were not universally popular but were instrumental in the progress of the breed in the long term. Certainly, the pattern of pony accommodated in Section B of the early Stud Books bears little resemblance, if any, to that found in the same Section today, and it is this development that concerns us in this chapter.

As outlined in Chapter I, the formal recording of Welsh pedigrees, be they pony or cob, was initiated by the Polo Pony Stud Book Society in 1894 with the

Opposite: Dyoll Starlight aged 19 at the National Pony Society Show, Islington, in 1913.

publication of the society's first Stud Book. In fact, some cobs had been registered in the Hackney Stud Book as early as 1884. Luckily for the breed, some of the principal supporters of the newly-formed Polo Pony Society, notably John Hill, also shared a keen interest in the Welsh breeds, so it was little wonder that in the space of only a few years, having established the need for written records, these enthusiasts would be responsible for the formation of a stud book dedicated to the Welsh breeds alone.

The Preface to Volume I of the Welsh Stud Book states:

> It has long been felt by those interested in the Welsh Mountain and other Ponies and Cobs for which the Principality has for so long been celebrated, that the establishment of a Breed Society similar to those which have done so much for other breeds was necessary.

Following a preliminary meeting of eleven Welsh enthusiasts at Llandrindod Wells in mid Wales on 25 April 1901, it was unanimously decided to hold a meeting in July of that year at Cardiff, the venue for the Royal Agricultural Society of England's Show. It was here that 25 attendees decided to create a society dedicated to the Welsh breeds. Lord Tredegar was elected President, and Gwynne Holford and Colonel Platt were elected Vice Presidents for South Wales and North Wales respectively. A Council was elected on a representational basis from the counties of Wales as well as Monmouthshire, the only county outside the Principality, and a set of rules was adopted at a Council meeting held in September 1901.

Thus the WPCS was born and along with it a determination to form a stud book. This Stud Book was published in 1902 and was divided into four sections determined initially by height under Rule 8:

> That the following Rules and Regulations apply to the entry of Ponies and Cobs.
>
> (a) The limit for the height of Foundation[1] Mares and Stallions in the Welsh Mountain Pony Section 'A' shall not exceed 12.2 hands, the Welsh Pony Section 'B' shall not exceed 13.2hh, in Cob Section 'C' shall be from 13.2 hands to 14.2 hands and in Cob Section 'D' shall be from 14.2 hands to 15.2 hands. All heights and markings be registered with entry. [Section D was abolished in 1905]
>
> (b) The Council shall appoint Judges in different districts who shall inspect, and no Foundation Stallions or Mares shall be entered in the Stud Book until a Certificate has been signed by an Inspection Judge, and in the case of a stallion, the Certificate must be accompanied by a Certificate stating that the Stallion is sound, and signed by a qualified veterinary surgeon. The produce by a registered sire and from a registered dam will be eligible for entry.

---

1 The term 'Foundation' was used in the first set of rules to describe the mares and stallions which would lay the foundation of the newly-formed Welsh Stud Book. This term was soon abandoned as the Stud Book developed; however, it was resurrected almost 30 years later in 1930 as a means to boost numbers by accommodating unregistered ponies and cobs of obvious Welsh ancestry – but more of that later.

(c) Judges are requested not to grant Certificates to any animal they consider to be affected with any hereditary disease, and in the case of Mares and Stallions not being full grown, to reject those which, in their opinion, are likely to grow over 12.2 hands and 15.2 hands. Animals two years old may be provisionally registered with complete registration at five years old. Should the Judges have any doubts as to the soundness of any mare, they may call in a Veterinary Surgeon.

(d) All Members of Council are Inspection Judges, two of whom have the power to appoint other judges in the districts when they may be required.

(e) The entrance fee for Mares shall be, Members 2s; Non-members 5s; each year's produce, 2s; and Stallions, Members 10s; Non-Members 20s each. All Fees to be sent with entry.

Due to the high number of animals seeking entry, the deadline for the first volume was extended to 31 January 1902. The entries in the different sections were as follows:

	Stallions	Mares
Section A	9	273
Section B	7	118
Section C	10	113
Section D	12	67
Provisional	29	187
Total	67	758

Appropriately Dyoll Starlight, belonging to Meuric Lloyd, was selected for one of the four illustrations in Volume I. Very few of the provisional entries were subsequently registered, but 22 re-entered in Volume II and 21 in Volume III, and it was in the third volume that provisional entries were abandoned.

At the first Annual General Meeting of the WPCS, there were 12 life members and 243 annual members. In addition, honorary membership was bestowed upon John Hill in appreciation for all his efforts in setting up the Society and the Stud Book. Subsequent meetings would witness debates, sometimes quite heated, where new rules were adopted and the Stud Book fashioned according to the needs of the day.

It cannot be overemphasised that the Sections of the original Stud Book bore little resemblance to those of today, but they did reflect the ponies and cobs that were recognisably Welsh in both type and origin. No one would argue that it was a very good starting point or that height was as good a criterion as any for the basis of the Stud Book during this embryonic phase, and the difference between pony and cob was also highly significant. These factors are pointed out in the Introduction to Volume I of the Welsh Stud Book which reads:

> Although the Welsh Ponies and Cobs have been of late terribly neglected and the original beauty and distinctive character in many districts spoilt by injudicious crosses for the purpose of gaining size, still there is (sic) evident signs of awakening appreciation of the pure old blood, which has at all the time been carefully guarded by men who value it as much as Eastern breeders do the high

caste Arab. There are several distinct types of these Ponies and Cobs, the small hardy original Mountain type, those somewhat larger, bred on the lower grounds, those more of Cob type, and lastly the larger Cob. Although quite distinct in appearance and height, still they have the same family likeness, true pony character, hair, and action, which latter is remarkable for its freedom and dash …

Each class of pony has a separate section; the Mountain Pony has one to itself, so that breeders of that variety may keep their herds distinct and true to type. The larger pony again, such as those which are bred on the lower lands, or enclosed ground, has its own section; and the Cob has its separate position in the Stud Book. It is not a jumble of all sorts and conditions of Cobs and Ponies into one Stud Book, but as it were, a distinct Stud Book for each comprised in one volume.

During the ensuing years, much discussion surrounded all sections of the Stud Book. However, for the purposes of this account the Welsh Cob, interesting as it may be, will be left to one side. Just as 'type' within all the native breeds of Britain enjoys much discussion today, it was also a central theme in Welsh Pony circles. Dr Wynne Davies, in his book *Sixty Years of Royal Welsh Champions,* sums up the importance of type within the Welsh Section B:

For many generations the Welsh Pony (Section B) was the means of transport for shepherds and hill farmers. With the increased popularity of Children's riding pones around 1930, the ponies were bred on slighter lines whilst retaining adequate bone and substance, hardiness and constitution, combined with the kind temperament which is such an outstanding characteristic of all Welsh breeds. With the placing of riding type up to 13 hands 2 in into Section B, the previous Welsh Pony of Cob-type became Section C and the Welsh Cob took its place in section D.

Many pony enthusiasts were at pains to describe the 'mountain' pony in the context of the Section A, particularly as it was considered by many as the 'fountain' of the breed. This was an idea put forward again by John Hill, who wrote prolifically on ponies at the time; in 'Notes on the Welsh Pony and Cob', which appeared in the Welsh Stud Book, he wrote:

I have always held that the Welsh Mountain Pony is the foundation of all that is characteristic in the Welsh breed of horse or pony.

In the same article he considers type in both the Section A and Section B – his account provides us with an objective comparison within and between the Sections then, and allows us an opportunity to compare them to the same Sections known to us now. It certainly was an early pointer of the direction in which the pony sections of the Welsh breeds would go in the future. On the subject of the Section A, he continues in his 'Notes':

It is a great mistake to fill the pages of the Stud Book

> with animals which, although they may rightly be called ponies, and be under the regulation height of 12.2 hands, yet are not 'Mountain' ponies in the strictest sense of the term ...

> My first pony which I can remember was a 'Welsh Mountain' of the old fashioned type – beautiful quality, small, clean-cut head, delicate ears, and deer-like prominent eyes ... what was the type 60 years ago ought to be the type of the present day. The Welsh mountains are the same, the climate is the same, and the conditions on [sic] which the ponies have to live in is [sic] the same. It therefore follows that all attempts to alter the breed by crossing with those of a more delicate constitution should be avoided.

> There are two types of Mountain Ponies, each of which have a right to the designation. One is the lighter made and bloodlike sort, which may be described as the riding type, and the other stronger built and better adapted for harness purposes. The former are acknowledged as being eminently suitable as a foundation stock for the production of Polo or Riding ponies, when mated two or three generations with Thoroughbred or Arab stallions, and the latter cannot be surpassed as the foundation of high-class harness ponies and cobs, if mated with hackney-bred sires.

And of the Section B he remarks:

> They are practically as true Welsh as the smaller ones (Section A), and breeders should keep them in mind, and breed true to type, upon which so much depends, if their value as Welsh ponies is to be retained. There are two sorts of these ponies, as there are among the smaller ones. The bloodlike riding type and the stepping harness pony.

Another major early contributor to the emerging WPCS and at one time its Chairman was Charles Coltman Rogers (grandfather of Mrs Teresa Smalley, President in 1985) from Stanage Park, Radnorshire, who wrote extensively on Welsh ponies and cobs, particularly the latter. On the subject of breeding type in the smaller ponies, and bearing in mind the Commons Act of 1908 and the strong export trade during the early part of the decade, in 1918 he sounded a cautionary note to breeders:

> If your ambition is to breed cheap pit pones and suchlike cart shaft drudgers, so well back to your non-selection and admit that the note of progress has been sounded in vain. If you want to breed ponies for export, for the boys and girls of wealthy parents, they must be narrow, shapely and easy actioned of the Starlight type.

Before this, however, the Mountain pony was gaining popularity within the wider equestrian community as the beautiful little grey ponies from Wales wowed the audiences of London and beyond. One of the country's leading equestrians, Reggie Summerhays, in his collection of reminiscences *A Lifetime with Horses*, describes the opening of the very first show, held at Olympia in 1907:

> At 9 am on the morning of the 7th June 1907 the great doors of the arena at Olympia swung open to admit the first competitors, a class for Hackney Stallions 2-year old. This is the story of the International Horse Show which was conceived on lines of such unimagined elegance, that Olympia became at once the centre of fashionable London at the very height of its season. Ten dukes, nine marquises, thirty-six earls, fourteen viscounts, and seventy-eight lords became its vice presidents.

The medal winner that day was the much-garlanded stallion, Dyoll Starlight (led by Claude Holmes, stud groom at Dyoll from 1903 to 1915), one of many Welsh ponies to make an impact during the early days of the Welsh Stud Book. It was the influence of the show ring that would make one of the first significant changes in the Stud Book when, in 1908, to accommodate the smaller mountain types being exhibited at major shows, Section A was spilt into two parts. Part (i) catered for ponies not exceeding 12 hands with neither manes hogged nor tails docked. In addition, ponies registered in Part (i) had to prove unquestionable descent on one side or the other from ponies that were foaled or ran wild, or usually lived on the mountains or moorlands of Wales, or were descendants of ponies already registered in Section A. Part (ii) catered for ponies not exceeding 12.2 hands, which could have docked tails and hogged manes as a voluntary option. At this time there had also been an active political campaign at government level to improve the quality of stallions kept for breeding on the hills and mountains. In 1908, a bill for clearing the hills of undesirable stallions was passed by the House of Lords. The height ruling was further endorsed in 1932 when 12 hands would essentially become the definitive height for the Section A.

Despite almost 20 years of discussion over the closing of the Stud Book, by 1925 plans were well in hand to close it to further inspections for base stock. However, this was postponed for another two years due to the falling number of registrations – a fall attributed to the difficult economic times brought about by World War I and the General Strike of 1926. According to Dr Davies in *One Hundred Glorious Years,* registrations in 1926 (Volume XXVI of the Stud Book) had dropped to 36 stallions and 64 mares. Following an initial burst of enthusiasm to build up the numbers, many breeders had failed to register their ponies, largely on economic grounds, although a population of unregistered ponies was known to exist without proof of parentage.

This was partly the reason (as well as the closing of the Stud Book to ponies of registered parentage) for the introduction in 1929 of an Appendix for Foundation Stock (FS) which would appear in Volume XXIX. There were many influences on the breeding of the emerging Welsh Riding Pony but arguably this was the greatest – it not only allowed entry to the Stud Book of sizeable numbers of female breeding stock but it also brought an infusion of new bloodlines to the Welsh breed. Mares were accepted for the new Appendix only after inspection and being considered of true Welsh type. Fillies by registered Welsh sires were eligible after inspection as FS, and their female progeny in turn by registered Welsh sires after inspection would become FS2. Meanwhile, colts from these matings would remain ineligible for registration. It was only when an FS2 mare was mated to a fully-registered Welsh stallion that all the progeny, male or female, would be eligible for full registration. Thus, after three generations, a new population of ponies was brought into the Stud Book, maintaining the integrity of the established bloodlines while at the same time introducing, albeit in a diluted form, something new. The Appendix, having done its job and made a considerably contribution to the Stud Book over almost thirty years, was closed for further Foundation Stock entries on 1 January 1960.

The early 1930s witnessed a major overhaul of the Welsh Stud Book which had remained more or less the same for some thirty years. Captain Thomas Howson, Secretary of both the Royal Welsh Agricultural Society and the WPCS following his appointment to the latter in 1927, wrote in the 1931 *Journal of the Royal Welsh Agricultural Society:*

> The outlook for the Welsh Mountain Pony has changed almost completely within the last ten years. There was a day when there were many outlets for surplus and inferior stock ... however it is now necessary for the breeders of Welsh Mountain ponies to ponder this position and to lay new plans.'

> According to Dr Davies:

> Captain Howson goes on to describe the increased interest in mountain ponies as children's riding pones either in the pure state or as foundations from which to breed a larger riding pony. This undoubtedly sowed the seeds for the discussion which followed in 1932 regarding the setting up of the new 'Section B'.

At this point in the history of the Society several breeders entered the discussion – in particular, Miss Brodrick of Coed Coch, whom the Chairman of the day, Lord Swansea, held in very high regard. In a letter to Lord Swansea on 12 January 1932 she wrote:

> About the 12.2–13.2 or 3 size of pony. If it were possible to produce a really good one of riding type, I should be all in favour, as at the present time I think they are what is most in demand, and hardest to find. Most of them are chance bred, or polo ponies that have not grown.

> As far as show purposes are concerned I think the Thoroughbred cross with the Welsh would be the best; but as regards temperament, I believe an Arab cross is ideal. If a committee were appointed they could probably view Mrs Inge's ponies and see for themselves the produce of a pure Barb stallion and Welsh mares, also progeny of a stallion of the above cross out of Welsh mares – i.e. ¾ Welsh ¼ Barb. Seeing them might help to clear their minds.

> I am afraid that I do not know of any pure bred Welsh-Thoroughbred that we can see. I have two here, but they are both out of non pedigree mares – one 11.2, the other 12.2.

Their exchange of letters prior to the Council meeting at Shrewsbury held in January 1932 makes interesting reading and Miss Brodrick may well have been disappointed that the resultant changes to the Stud Book didn't altogether go her way. Miss Brodrick had a very simple view of how ponies and cobs should be organised within the Stud Book – her proposal to Council via Lord Swansea simply stated:

> That animals eligible for entry in WPC Stud book be divided into two groups
>
> 1. Cobs to include what at present are referred to as Welsh Ponies
> 2. Smaller ponies all of *riding type*. That is to say, having low straight action
>
> Height limit of the WM Ponies to be raised to 12.2 and this to be strictly adhered to: no pony being 'stretched' while measured.
> Shoes for all ponies to be of light hunting type.
> A small committee to be formed to discuss the proposals of type.
> Panel of judges to be revised.

It was obvious that Miss Brodrick had a vision of a Welsh pony which very much reflected that of Captain Howson and which fulfilled her own vision of the breed based on the purpose to which she considered it was best fitted at that time. Note in particular that she uses the terms 'riding type' and 'low straight action', and there is also a reference to 'hunting' where she describes the weight of shoes to be accepted. She also made a point of banishing the heavier, stocky type of small pony to a general category of cob, allowing this heavier type no room in her definition of a Welsh Pony – it was only the last mentioned of her proposals that found favour with Council.

Miss Brodrick's proposals may have been somewhat simplistic but it was very clear where her priorities lay. Needless to say, she was influenced by others, notably members of the Polo Pony Society, whose interest in breeding children's riding ponies had gathered momentum as the demand for polo ponies under 14.2

Captain Howson, secretary of a number of societies including the Royal Welsh Agricultural Society and Welsh Pony and Cob Society.

hands waned and increasingly ponies were required specifically for children, rather than for the gentry or the show ring, or for hunting and leisure.

The decisions of the Council meeting reflected a different outcome to that proposed by the owner of Coed Coch and it is fairly clear that the 'Mountain' pony breeders and enthusiasts influenced the way the Stud Book would be laid out. Lord Swansea himself may have had a hand in this, as his own interests lay with the Mountain Pony – his Caerberis Stud at Builth Wells in Mid Wales was well known in the Principality and beyond. He was one of the exhibitors who regularly showed a team of ponies at the National Pony Society Show in London and his group of Mountain ponies, led by the home-bred stallion Caerberis King Cole, won there in 1929.

It was the definition of action of the small Welsh pony that would prove most controversial since there was strong lobby that insisted the Mountain pony should have knee action and not the 'low straight' action advanced by Miss Brodrick. Consequently, in 1932 it was decided by Council to divide the Stud Book into three sections, namely:

Section A	Ponies of mountain type, undocked and not exceeding 12 hands
Section B	Ponies of riding type not exceeding 13.2 hands
Section C	Cobs and ponies of cob type exceeding 12 hands

The year proved a significant milestone in the history of the WPCS since decisions made at this time clearly set a template for the Section A and Section B and shaped the Stud Book into a form recognisable to the present Section B breeders. At the time the Section A was commonly referred to as the Welsh Mountain Pony, just as it remains today. However, the term 'Welsh Riding Pony', which was applied to the Section B during those early days, has not been maintained and the term 'Section B' is currently used. Thus the Stud Book hosted three new sections in 1933, for A, B and C. There were only three stallions to be found in the first new Section B, in Volume XXX (1931–1934): Dyoll Star Turn, Tanybwlch Rhiniog (Tanybwlch Berwyn x Celtic Rumour) and William (Criban Chief x Criban Pearl), who was exported to the United States in 1933 and 'lost' until Miss Brodrick recognised him in 1948 on a trip there – he became the foundation sire of Mrs du Pont's Liseter Hall Stud.

The new Section C included cobs of all heights, from the very small to the large. Section D was resurrected in 1935 and remained a place for registering geldings from all sections until 1948. Although cobs had been grouped together, by 1949 the pony of cob type (not exceeding 13.2 hands) would once more have its own place within the Welsh Stud Book. It would be registered as a Section C and the new Section D was formed to accommodate cobs exceeding 13.2 hands.

As Chairman, Lord Swansea led the Society with great skill. He also had much to do with the breeding of Mountain ponies on the Gower Peninsula and the establishment of an Improvement Society there for local breeders. The Gower was close to his home of Glanrafon near Swansea, where he lived before moving to Caerberis in Mid Wales. However, it would be the restructuring of the Stud Book for which he would become best remembered.

TELEGRAMS: DOLWEN.
TELEPHONE: LLANDDULAS QUARRIES 28.

COED CÔCH,
ABERGELE.N.W.
Jan. 12.1932.

Dear Lord Swansea.

Herewith is what I propose to send to Howson to put on the agenda for the forthcoming meeting. Incidentally thank you very much for having it at Shrewsbury. It is a great help.

I do not think it would be wise to put more on the agenda, in fact I wonder whether I have not been too explicit already! What I really would like to work for, is to get a small Committee appointed to go into the matter. I think we should do more good that way.

About the 12.2 - 13.2 or 3 size of pony. If it were possible to produce a really good one of riding type, I should be all in favour, as at the present time I think they are what is most in demand, and hardest to find. Most of them are chance bred, or polo ponies that have not grown.

As far as show purposes are concerned I think a thoroughbred cross with the Welsh would be the best; but as regards temperament, I believe an Arab cross is the ideal. If a committee were appointed they could probably view Mrs Inge's ponies, and see for themselves the produce of a pure Barbe stallion and Welsh mares, also the progeny of a stallion of the above cross out of Welsh mares – i.e $\frac{3}{4}$ Welsh $\frac{1}{4}$ barbe. Seeing them might help to clear their minds.

I am afraid I do not know of any

Above and opposite: the correspondence between Miss Brodrick and Lord Swansea in January 1932 outlining her recommendations to WPCS Council on changes to the Stud Book.

The momentum Lord Swansea had started continued after his sudden death in 1934, when J J Borthwick took over as Chairman and served in that capacity for 30 years. Miss Brodrick became Vice Chairman and her great friend, Mrs Inge, President. All three had an interest in hunting, children's riding ponies and the breeding of a riding type of Welsh pony. It was just as well, as the Council had a bit of a dilemma on their hands when two stallions, previously entered in Section A – Tanybwlch Berwyn and Craven Cyrus – failed to fulfil the criteria for registration to allow transfer to the new Section B, due to their Eastern origins. Fortunately the Council saw the benefit of introducing this blood to gain both height and riding quality into the Section B and allowed the two stallions entry. Had the decision gone the other way, who knows what the destiny of the Section B would have been without their considerable influence?

Meanwhile, the show ring at local and national level had played an important part in the development of the

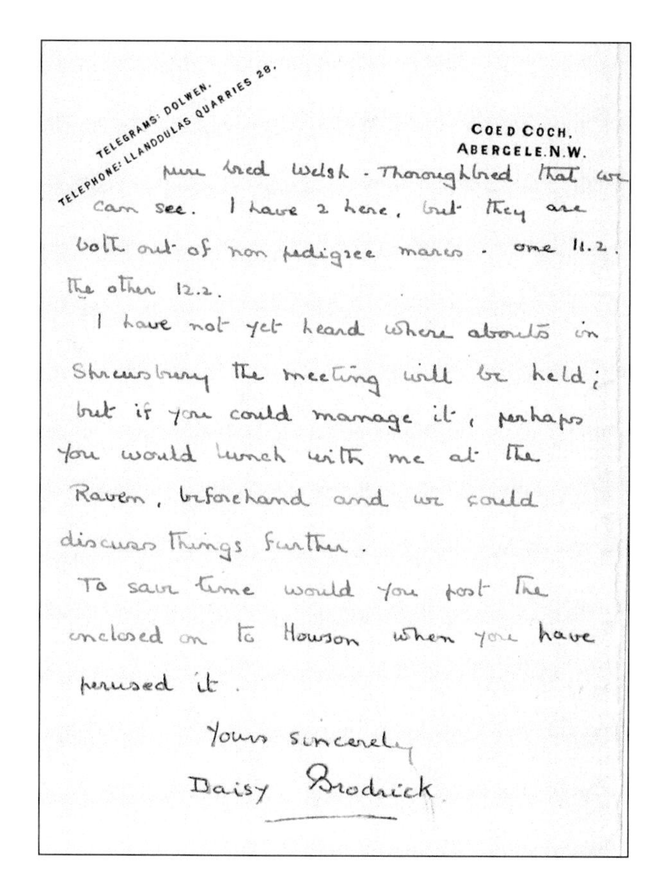

TELEGRAMS: DOLWEN.
TELEPHONE: LLANDDULAS QUARRIES 28.

COED COCH,
ABERGELE.N.W.

pure bred Welsh - Thoroughbred that we can see. I have 2 here, but they are both out of non pedigree mares - one 11.2, the other 12.2.

I have not yet heard whereabouts in Shrewsbury the meeting will be held; but if you could manage it, perhaps you would lunch with me at the Raven, beforehand and we could discuss things further.

To save time would you post the enclosed on to Howson when you have perused it.

Yours sincerely

Daisy Brodrick

Welsh Pony of riding type as people increasingly wanted to show their ponies, both in hand and under saddle. The Royal Welsh Show and the Welsh breeds have gone hand in glove throughout the breeds' history. However, while the Royal Welsh Agricultural Society had embraced the sections of the Welsh Stud Book from its earliest days, it did not immediately take up the new classification for the newly-formed Section B in 1933. The innovation occurred in 1934, when a single class for brood mares or fillies of 'riding type' was introduced and attracted six entries. First and second places were filled by Lord Howard de Walden, whose fillies' sire, Gray Cross, was a son of the Arab, Crosbie; third place was taken by Mrs Inge's filly by Tanybwlch Berwyn.

Not everyone was in favour of the new class, as Captain Howson reported in the 1934 *Journal of the Royal Welsh Agricultural Society*:

> Opinions differ, even amongst members of the WPCS themselves, as to the need for, or the wisdom of, this innovation and many of the older breeders are disposed to shake their heads and view it with disfavour. And when we contemplate the wonderful achievements of the small hill pony as foundation stock, today and in the years gone by, there does appear to be some cause to question whether well had better not be left alone.

By 1935, there were so few entries that the Welsh pony (Section C equivalent) stallion class had to be cancelled – only five mare entries and three for the class for mares and fillies of riding type were received. In the words of Howson (1935 *Journal*), this was 'a lamentable state of things indeed'. In the 1936 *Journal* he states, 'The position of the Welsh ponies of cob type is critical to say the very least of it'. He adds:

> The Breed Society has set itself the task of endeavouring to establish – largely on the basis of Welsh Mountain pony blood – a race of self-reproducing animals of riding shapes and ranging up to 13.2 hands in height, while still retaining all those points of excellence for which the 12 hand Mountain ponies are renowned. And it is with the object of assisting in the furtherance of

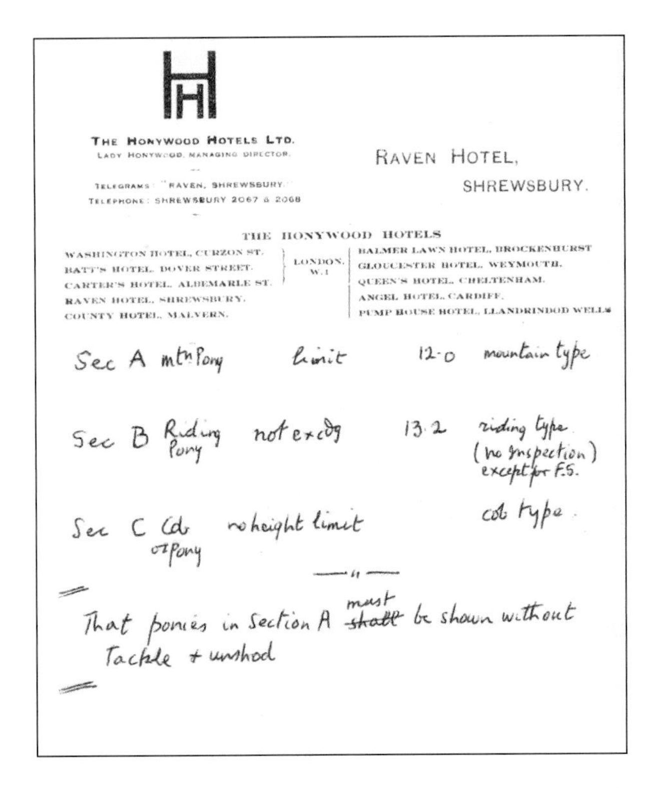

THE HONYWOOD HOTELS LTD.
LADY HONYWOOD, MANAGING DIRECTOR.

TELEGRAMS: "RAVEN, SHREWSBURY."
TELEPHONE: SHREWSBURY 2067 & 2068

RAVEN HOTEL,
SHREWSBURY.

THE HONYWOOD HOTELS

WASHINGTON HOTEL, CURZON ST. | BATT'S HOTEL, DOVER STREET. | CARTER'S HOTEL, ALBEMARLE ST. — LONDON, W.1
RAVEN HOTEL, SHREWSBURY.
COUNTY HOTEL, MALVERN.
BALMER LAWN HOTEL, BROCKENHURST
GLOUCESTER HOTEL, WEYMOUTH.
QUEEN'S HOTEL, CHELTENHAM.
ANGEL HOTEL, CARDIFF.
PUMP HOUSE HOTEL, LLANDRINDOD WELLS

Sec A mtn Pony limit 12·0 mountain type

Sec B Riding Pony not excdg 13·2 riding type (no inspection) except for F.S.

Sec C Cob or Pony no height limit cob type.

— " —

That ponies in Section A ~~shall~~ must be shown without Tackle + unshod

Miss Brodrick's suggestions on restructuring the Welsh Stud Book.

> this laudable design that classes for stock of this description have been instituted at our show.

So, although the Royal Welsh Agricultural Society was playing its part in supporting the Welsh pony of riding type, entries remained low. Those who did come forward, although entered as Foundation Stock in the Welsh Stud Book, caused disquiet due to their lack of Welsh character. The 1937 *Journal* tells us that the winner was Cuckoo – 'at least three parts Thoroughbred'; the second was of unknown breeding but 'if looks are to go by, blood played a large part in her composition'; the third, Brown Sugar by the 'Premium sire Royal Pom'; and fourth, Craven Bess of 'Arab, Thoroughbred, Welsh Mountain pony and Welsh Cob extraction'. One piece of encouraging news from Howson's reports is recorded in the 1938 *Journal* when he says of Tanybwlch Berwyn, 'sired by Tanybwlch Berwyn – who is proving a fine sire of stock of riding stamp'. This was praise indeed from someone who had voiced such criticism of the developing Section B.

In terms of development of the Welsh Section B, there was no change throughout the following decade – understandable, since the country was enduring the hardships of the Great Depression and World War II. During this period, registrations in the Stud Book remained extremely low, as did the level of membership. No Stud Books were published between 1938 and 1948; once publication resumed, Volume XXX took in the years 1931 to 1934, while Volume XXXI covered 1935 to 1938 and Volume XXXII the years 1939 to 1949.

The structure of the Stud Book then bore little resemblance to the current model, but by 1947 plans were in hand to create a structure more relevant to the modern types found within the Welsh breeds. Heights would feature, particularly for the cobs, and there was an obvious need to address the lack of suitable stallions available to fulfil the demand for the Riding Type of Welsh Ponies (Section B). To this end, a sub-committee of prominent breeders was set up and their recommendations were accepted at the 1948 Annual General Meeting. The major restructuring of the existing four sections introduced a fifth with wording as follows:

Section A:	remain as at present
Section B:	to be completely restructured
Section C:	for Welsh Ponies of Cob type exceeding 12 hands but not exceeding 13.2 hands
Section D:	for Cobs exceeding 13.2 hands
Section E:	for geldings from all sections

Section B unquestionably caused the most concern to breeders, as those in favour of breeding the up-to-height pony had made use of outside blood, mainly Thoroughbred, Arab and their crosses, to achieve this goal. Traditionalists were sceptical and anxious to maintain the purity of their ponies, especially the Mountain types which seemed to have established some form of uniformity over the years. In the 1948 *Royal Welsh Agricultural Journal* Captain Howson summed things up when he wrote:

Major E G E Griffith, Council member of the Royal Welsh Agricultural Society, responsible for promoting additional classes of Section Bs at the Royal Welsh Show.

> Some time ago the WPCS determined to embark upon their task of trying to establish a self-reproducing breed of Welsh ponies of purely riding type and ranging up to 13.2 hands in height. But it is now considered that the almost complete dearth of registered Welsh stallions of a type which would be likely to assist in this attainment of the end in view presents a serious obstacle to progress.
>
> It has, therefore, been decided to add another section to the Stud Book (beginning with Volume 33) and to accept for registration in it stallions of extraneous blood, provided they have one registered Welsh parent and have been inspected when not less than four years old by three judges appointed by our Society for that purpose and have been approved by them.
>
> Attempts have been made in the past to fashion breeds of hunters and of polo ponies and have failed. It now remains to be seen whether the efforts of the WPCS will prove successful. Personally, we have our doubts.

The restructuring of the Section B evidently caused a bit of an uproar and stallions in the new scheme would be restricted to use in this Section alone. The need to control the influence of the stallions in the new Section B was obvious and a glance at the rules helps explain why.

1. That Section B be reconstituted under new regulations to admit the registration therein of stallions not exceeding 14 hands high which have been inspected and approved as of true riding type and which are of registered Welsh parentage on both sides, or are by a registered Welsh stallion or from either registered Welsh mares or mares which are in the Appendix to the Book as FS1 and FS11 animals.
2. The stallions registered in Section 'B' unless also registered in some other section of the Book, and the progeny of such stallions, have the letter 'B' placed after their names in the Stud Book and all other official publications.
3. That 'B' stallions and their progeny be not eligible for registration in any other section of the Book.
4. That no stallions other than those registered [in] Section 'A', whether registered in another section of the Book or not, be accepted for registration in Section 'B' unless they have been inspected by three of the Society's judges appointed for the purpose and approved by them as being of true riding pony type.
5. That if their owners so desire, stallions may be inspected at three years of age, and if they do not exceed 13.2 hands high, may be provisionally approved for Section 'B' but shall not be accepted [for] registration in that Section until they have been inspected and approved on attaining the age of four years.
6. That the height limit of 13.2 hands for mares registered in Section 'B' to be retained.
7. That a panel of judges be appointed for the purpose of inspecting and approving or rejecting stallions submitted for registration in Section 'B' and that the panel be composed of – Miss Brodrick, Mr T Jones Evans, Major E G E Griffith, Captain T A Howson, Mr A R McNaught, Mrs Pennell, Mr H Ll Richards, Mrs Tabor and Mr Matthew Williams.
8. That the foregoing regulations, so far as they relate to Section 'B' of the Stud Book, be

Menai Queen Bee (f. 1937), transferred from Section C to Section B in Vol XLIII of WSB.

operative for a period not exceeding five years from their adoption and that they be reviewed on the expiry of that period.

9. That no mare will be eligible for inspection or entry in the Appendix to any section of the Stud Book until she has attained the age of two years.

Howson's pessimism proved well founded and it soon became clear that the new regulations did little to help the Section B, so in 1950 the regulations were rescinded and those in force prior to 1948 were reinstated. In a move to accommodate ponies which might otherwise have been found in the proposed Section B, a Welsh Riding Pony Stud Book (P) was formed, as well as a Welsh Part-bred Register (P1) restricted to a height of 13.2 hands, which required 50% of registered Welsh blood and had the facility to upgrade to full registration in three generations where purebred stallions were used. These latest proposals also posed difficulties, as the merits of one against the other were far from obvious to the breeders. In a letter to E S Davies (Ceulan Stud) from Kathleen Cuff in October 1955, she sought advice of registering her ponies:

> Can you advise me whether it is any advantage to transfer ponies from Section B (or the Appendix of Section B) into the Part-Bred Register? If a pony is eligible to go straight into Section B – for example a 3 yr old filly by a registered Section A stallion out of an FS1 mare, which is it best to do – put her into Section B or make her P1? If the object of the Part-Bred Register is to up-grade into Section B, there

A L Williams with Pam Hutchings in 1964 at Lampeter Stallion Show, of which he was Secretary and a driving force for many years.

is surely no point in the pony going through the P1 stage unnecessarily?

I know that I am absolutely dense about this new section, but so far I have not discovered its advantage. It seems to me that it is so tied-up that an animal in it is likely to be more nearly true Welsh than any in the Appendix to Section B. If this is so, do you think it will be to the ultimate advantage of the Stud Book if Riding Type ponies are transferred into it? From my own point of view, when once out of Section A, I don't mind where I put my ponies – as once you get to 13.2 and show ponies even 12.2 I never find people asking me if they are registered in any Stud Book.

Welsh Ponies (Section B).

Blandy, Mrs. Raleigh, Dolaubran, Llandovery, Carms.
Borthwick, J. J., The Foxholes, Bishop's Castle, Salop.
Brodrick, Miss M., Plas Llewelyn, Abergele, Denbighshire.
Cuff, Mrs., Downland Pony Stud, Norton Manor, Presteigne, Rads.
Darby, Mrs. C., Kemerton Court, Tewkesbury, Glos.
de Beaumont, Miss M., Shalbourne Manor, Marlborough, Wilts.
Deed, Mrs. S. C., Lavington, Putney Heath Lane, London, S.W.15.
Griffith, Mrs. E., Plasnewydd, Trefnant, Denbighs.
Griffith, E. G. E., Plasnewydd, Trefnant, Denbighs.
Griffith, Moses, M.Sc., Ponc Taldrwst, Amlwch, Anglesey.
Havard, J. M., Crud-yr-Awel, Gorseinon, Glams.
Hepburn, Mrs. A. R., Delaware, Birmingham Road, Wylde Green, Birmingham.
Jones, John, Coed Coch Pony Stud, Abergele, Denbighs.
Morley, Miss E., The Old House, Rearsby, Leics.
Pennell, Mrs. N., Tweenhills Farm, Hartpury, Glos.
Richards, H. Ll., Allt, Bwlch, Brecon.
Richards, R. J., Carr Grange, Dinnington, Newcastle-on-Tyne.
Thomas, W. J., Cawdor Cottage, Llandilo, Carms.
Williams, Dr. Arwyn, B.Sc., M.R.C.S., Y Fron, Tregaron, Cards.
Yeomans, Mrs. I. M., Planners Farm Stud, Bracknell, Berks.

Some famous names appear on the 1954 official list of Welsh Section B judges.

There was so little uptake in the Welsh Riding Pony Stud Book that it soon folded. However, the model was later adopted for the new Foundation Stock Register. The Part-bred register had minimal success but in 1960 it was reconstituted to include all heights with a minimum 25% of registered Welsh breeding in its pedigree; this would change to 12.5% in 1997. Such was the success of this change that by 2009 the registration of Part-breds (2,570) was greater than that of the combined registrations of Section B (1,166) and Section C (1,171).

It was also during 1960 that the rules governing the acceptance of stallions from FS and FS1 mares was temporarily rescinded in Welsh Section B to allow much-needed blood into that section. These stallions had to have a registered Welsh sire and were only registered after inspection. Notable among those accepted were Coed Coch Pawl and Reeves Golden Lustre (previously also registered in the Part-bred Register), who in due course would sire the Royal Welsh champions Cusop Hoity Toity and Springbourne Golden Flute respectively. Little did the Section B breeders of the day know that the Foundation Stock Register would soon yield its greatest harvest when, in 1959, four colts were born which would make their mark on the breed. All four, by registered stallions, were the product of FS2 mares and therefore eligible for full registration within the main Stud Book. We will hear of them later, but suffice to note here that they were the great stock-getters Solway Master Bronze, Downland Dauphin, Brockwell Cobweb and Chirk Crogan, whose influence can still be felt some fifty years on.

Welsh ponies enjoyed considerable support from other societies, particularly the National Pony Society, which initially held a record of breeding in its Stud Book. They indirectly received an additional boost during the early 1950s when breeding of Thoroughbreds and Arabs to native breeds was actively encouraged. As the number of registrations for polo ponies decreased and crossbreds increased, registrations in the 1953 National Pony Stud Book included for the first time a section headed 'Riding and Children's Pony Register'. With a pool of excellent Thoroughbred and Arab stallions available to pony breeders as well as an abundance of top-class native pony mares, the value placed on crossing either of them with the native pony was such that the resultant progeny enjoyed special demarcation – they were referred to as 'First-Cross'. Volume XXVII of the National Pony Society Stud Book makes the following reference:

> Animals which are First-Cross will be specially indicated by the addition of 'F-X' after their names … One parent must be a registered Mountain and Moorland pony, the other to be registered in a Stud Book recognised by the National Pony Society.

The National Pony Society staged classes for the First Cross pony, and the Ponies of Britain Club, under the guidance of its charismatic leader Mrs Glenda Spooner, gave Foundation Stock classes of their own at its Summer Championship Show.

The closing of the FS Register in 1960 coincided with a decision that would tighten up the pattern of breeding within the Welsh Section A, B and C. In Section A all new entries would be restricted to ponies entered in this section alone and in Section B no more ponies would be accepted by or out of a Section C or D. This meant that Section A could only be achieved by crossing Section A on A while Section B could be achieved by crossing Section B on B, or Section A with B. Although there have been intermittent calls over the years to restrict the Section B register to ponies of Section B breeding only, this has yet to be agreed. The other change concerned the Section C, when the lower height limit of 12.2 hands was removed and Ponies of Cob type were given an appropriate place for registration.

The Stud Book reveals a shuffling of ponies between volumes at this time in order to comply with the new regulations. One breeder who did just this was Willie Jones, whose family farmed at Pantydefaid near Llandysul in Ceredigion, West Wales. His family had bred ponies and cobs and registered them in the early Welsh Stud Books under the prefix Tyssul but Jones himself became a member of the WPCS in 1958, whereupon he registered them with the Menai prefix – one that is well known in cob circles today through his son, Peter, and grandsons, Richard and Tomos.

It is in Volume XLI (1958) of the Stud Book that we notice that Jones had registered in Section C a 13.2 hands grey mare born in 1937 by the name of Menai Queen Bee, whose grandsire was the Section A Bleddfa Shooting Star. Prior to breeding, she had been used for shepherding but late in life she was put to the cob stallion, Caradog Llwyd, to produce, in 1953, a Section C filly, Menai Ceridwen and in 1955 the Section D colt, Menai Ceredig. Ceridwen would have an enormous influence over the Section C worldwide through her sons Menai Fury (by the Section A Gredington Oswallt) and Llanarth Cerdin (bred at Menai although registered with the Llanarth prefix). Meanwhile at the Menai Stud itself, Ceridwen's influence appears in both Sections C and D through a wonderful female line whose names begin with 'C'. Menai Ceredig would have a huge influence on the Section D following his sale to the Llanarth Stud where he sired a dynasty of top class stallions.

Welsh Part-Bred Register

STALLION

REEVES GOLDEN LUSTRE (No. 1 P.1.) (W.B.R.) [2755], chestnut roan, blaze, snip, bottom lip white, near fore sock white, hind stockings white. Foaled 1945. Height 12.1. Measured May 1948.

Owner, Mrs. Gordon Gilbert, Reeves Stables and Stud, Penn, High Wycombe, Bucks.
Breeder, E. S. Davies, Central Stores, Talybont, Cardiganshire.
Sire, Reeves Ceulan Revoke 1720 (W.S.B.).
Dam, Ceulan Silver Lustre No. 191 F.S. (W.S.B.), by Incoronax (A.H.S.B.)

Reeves Golden Lustre (f. 1945) was registered as a Welsh Part-bred in the NPS Stud Book prior to his acceptance into the Welsh Stud Book in 1960. His picture appeared in 'Horse and Hound' on 8 May 1954 (right).

By 1960, as the rules on breeding had changed, Menai Queen Bee was transferred to Section B in Volume XLIII (1960) of the Stud Book, whereupon she bred two foals. In a remarkable demonstration of versatility in her breeding capacity, Queen Bee bred her last foal in 1962, a Section B colt by Llanarth Night Star (Section A) named Menai Shooting Star. A prolific prize-winner at major shows, his blood continues to flow through the veins of the Cloigen Section B ponies belonging to Janet Morgan, who took over the breeding of Section Bs on the death of her father, Willie Jones. Recently, her brother Peter has reintroduced the breeding of Section Bs to the Menai Stud itself. Little did the family realise that their shepherding pony, Menai Queen Bee, would have such an impact on three sections of the Welsh Stud Book.

All this time, the show ring was having an enormous impact on the Welsh Stud Book, as rivalry between breeders was played out through competition at local and national level. In the early days, Mountain ponies and Cobs held centre stage but, as fashions changed, slowly but surely classes emerged to meet the demand. We have already mentioned the influence of the Polo Pony Society and its successor, the National Pony Society, on the showing of Welsh ponies through its acclaimed Spring Show at Islington. This show, along with the Olympia and Royal International Horse Shows, made London a Mecca for the equestrian élite. The Polo Pony Society also provided silver medals for competition at provincial shows, of which the Royal Agricultural Society was one, as it moved its site annually around England and Wales. PPS and NPS medals were hotly contested and are still highly sought-after by today's collectors.

The medal scheme was also adopted by the WPCS and still exists, if slightly devalued by its lack of rarity. In 1990, the scheme was restructured to reflect competition at the more important shows – silver medals are awarded at the more prestigious shows, while bronze medals are commonplace at shows with Welsh sections across the length and breadth of the country. On the other hand, the introduction in 1990 at the Royal Welsh Show of 'Gold' medals for male, female and youngstock champions in each section of the Stud Book places the event above all others in the eyes of Welsh pony and cob enthusiasts across the world.

The importance of the Royal Welsh Show, one of the most popular agricultural shows in the British calendar, means that the development of the classes available for competition has reflected the demands of the time. Section B and Section C have seen the greatest development, but it must be remembered that they were initially classified as Section B, although the entrants were of today's Section C type. From modest beginnings, with only a few entries in Section B in 1934, the competition for Section Bs enlarged to three classes to include youngstock, mares and stallions – in 1951 there were four classes, attracting 23 exhibits. It remained this way for almost two decades until Council members of the WPCS put pressure on the authorities to enlarge the classification.

One of those members exerting pressure was Harry Chambers, whose Ardgrange Stud of Section Bs in West Wales had become a significant addition to the Stud Book. The minutes of the WPCS AGM held at Malvern on 14 March 1970 reveal his efforts to achieve this:

> Mr H Chambers referred to the Royal Welsh classification for Welsh ponies, Section B, the number of classes being four and although the Show was the 'Shop Window' for Welsh breeds, even the Aberdeen Angus Cattle Society were allowed five classes. Mr Chambers suggested that it was time that Council Members who were also members of the RWAS Livestock Committee used their influence and made sure that the Welsh breeds had a full complement of classification.
>
> Mr H Chambers added that the Section 'B' classification could also be increased by the inclusion of separate classes for Yearlings, Two-year-old fillies and Three-year-old Fillies.
>
> Mr E G E Griffith asked for full details of the classification required and promised to approach the Show Society concerned to try to influence them to include and increase their classification for the Welsh Pony and Cob Breed.

True to his word, Griffith took the proposal to the Livestock Committee of the Royal Welsh and by 1972 there were yearling classes for colts and fillies. A year later, the two- and three-year-old classes had also been split to accommodate colts and fillies. In 1995, youngstock competed in classes split by age and gender. Further changes were made to cater for barren mares and geldings in 1983, while brood mares would be split into novice and open in 1989. The format changed again in 1996, when junior and senior mare classes appeared on the schedule as well as separate classes for colt and filly foals; stallion classification has remained unaltered. In 2012, Section B geldings were recognised for the first time in their own section – this proved popular and was a crucial development in light of the expanding performance market.

At times, the market put pressure on the Council of the WPCS. It sometimes responded positively and at other times resisted the temptation to react to the prevailing fashions. An example of this within the Welsh Section B has been the question of height, consistently limited to 13.2 hands since the establishment of the Stud Book in 1902. As we have noted, the other Sections have also been subject to height changes over the years but in this respect the Section B, whether in Cob type or riding type, has been constant.

It would be fair to say that the debate about Section B has largely been driven by financial interests rather than an intrinsic interest in the development of the breed. In all cases, the argument has been centred on raising the height of the Section B, to anything from 14 hands to 14.2 hands, and in all cases it has been based on overseas demand and thus an overseas market. The American market, in particular, which had been the greatest importer of Welsh ponies and cobs during the early part of the 20th century, looked for the bigger pony in the absence of suitable breeds in their homeland. Combined with the fact that they saw the sections as interchangeable, with type remaining constant and height the variable, there was a demand for the larger pony of riding type and refinement. As far back as 1958, there is evidence that there was political pressure to increase the heights to suit the North American market, but this was resisted by a strong 'home guard' faction in Council.

The minutes of the 1967 WPCS AGM held at the Royal Mews in London outline a request from a Swedish enthusiast to raise the height of Section B to 14 hands. This was referred on the grounds of procedure to the following AGM where a member from Bristol, W E Higgins, proposed a motion:

> That as from 1st January, 1969, the height limit of Welsh Ponies, Section B, and Welsh Ponies (Cob type), Section C, be extended to 14.2 hands high.

Lord Kenyon, the highly-respected Chairman of the time and a prominent breeder of both Section A and Section B ponies, spoke against the motion and stated that 'breeders must be more settled in their ideas regarding type and conformation of Welsh Ponies (Section B)'. Another prominent Section B breeder, Mrs Gilbert of Reeves Stud, stated that 'Welsh Ponies (Cob Type) at 13.2hh were useful as mounts for the family and if the Motion was accepted and the animals allowed to be 14.2hh the Welsh Pony (Cob Type) would lose its purpose'. Mrs Gilbert was also of the opinion that 'height bedevilled the Section B Pony'.

There was no appetite for a change in the heights of the Section B at the meeting and the motion was rejected. Thus, Council had once more fended off the challenge and the decision on height in Section B made then, just as it had been 1902, has stood the test of time and remains at 13.2 hands (137.9 cm).[2] In practice, we find the Section B has a very important role to play in the full range of its heights – from the under 122 cm pony suitable for the youngest riders to the full height of 138 cm for adult riders in performance classes. The decisions made by Council over the years have placed the Welsh Section B in a strong competitive position within the modern equestrian world. In the following chapters we will see how the breeders have interpreted their role in bringing the breed to its present-day state.

2 Modern ridden classes are described in metric terms.

Chapter V

Coed Coch (Miss Brodrick)

From left to right: Shem Jones, 'Big' John Jones and John Jones proudly stand beside the trophies won by ponies from Coed Coch in 1964.

Miss Brodrick with, Shem Jones (centre), John Jones (right) and three Coed Coch ponies. From left to right are Siaradus, Madog and Pwyll at the 1955 National Pony Society Show at Roehampton.

In the opening chapter on the Coed Coch Stud in *One Hundred Glorious Years,* Dr Wynne Davies confidently states:

> The Stud which has had the greatest influence on the Welsh section A and B ponies during the 20th century, if only in terms of numbers, is undeniably the Coed Coch Stud started by Miss Margaret (Daisy) Brodrick in 1924.

It could be argued that he should have omitted the phrase 'if only in terms of numbers', for there is much evidence to suggest that the impact of the bloodlines that Miss Brodrick developed during the four decades of the stud under her guidance reached far and wide and the likes of which knows no equal.

Miss Brodrick had inherited the extensive Coed Coch Estate near Abergele, North Wales in 1929 in trust from her mother whose husband, Major General Edward Lloyd Wynne, died shortly after the birth of their son, Edward Henry John Wynne, in November 1893. Like his father before him, Edward joined the 3rd Grenadier Guards, with which his father had a distinguished career, and he would have inherited the Coed Coch Estate but for his tragic death at the Battle of the Somme in 1916. This meant that the estate was left in trust to his mother, who continued to live there and subsequently married the Hon Lawrence Brodrick, second son of the 8th Viscount Middleton. In 1897, she gave birth to a daughter named Margaret, affectionately known to all her friends as Daisy. Like her mother, whose sense of public duty led her to political life as a parliamentary candidate for West Denbigh, in time Margaret would also immerse herself in public duty during two World Wars. During World War I she served in the French Red Cross, for which she was awarded the Croix de Guerre and the French Red Cross Medal, and in World War II she worked for the Transport Branch of the Royal Army Service Corps.

However, it was for her devotion to the Coed Coch Estate and its workers that she was affectionately best remembered. It was no surprise, then, that her legendary groom and great friend John Jones, with his daughter, Lil accompanied her to Buckingham Palace in 1961 when she received from the Queen Mother the MBE for her services to the WPCS and to Wales in general. Another great friend, Major 'Eddie' Griffith, wrote of Miss Brodrick in an appreciation in the WPCS *Journal* of 1963:

> Though she travelled widely, Daisy's true and abiding love was the Welsh countryside and its people. She gave her heart to Coed Coch and all

that it stood for, the fields and woods and hills familiar from her childhood, the friends and neighbours in all walks of life. Thus it all came about that, on the death of her mother in 1929, Daisy succeeded to Coed Coch. She inherited no great wealth, but she inherited a great tradition of service, which she upheld faithfully throughout her life.

Of her ability as a breeder he continued:

From modest beginnings she built up what is undoubtedly the finest collection of native ponies that the world has yet seen.

Concentration on the best strains and a measure of line breeding produced a fixed type of great beauty and outstanding quality; each year saw a new string of victories in the show ring; Coed Coch ponies became world famous. And not in Section A alone, for Tanybwlch Berwyn was indeed the 'founder father' of the Section B.

Daisy Brodrick will be remembered as one of the great livestock improvers; she will be remembered as an enlightened and generous landowner, who made a real contribution towards the welfare of rural Wales. But those who knew her best will always think of her as a kind and cheerful friend with a mission in life to help other people.

When Miss Brodrick duly took over the estate of nearly 3,000 acres, she lived at Plas Llewelyn, a 19th-century farmhouse situated approximately four miles from Colwyn Bay and surrounded by the Coed Coch Estate, of which a considerable amount is hill land. A feature of the house was the uninterrupted view from her bedroom of the ponies grazing in the adjoining fields. Coed Coch boasted a large Grade II listed mansion house sitting in park land, which for approximately thirty years up to the 1970s was used as a preparatory school for boys. It has since been renovated and brought back into family use by the present occupants of the estate, Mr and Mrs Harry Fetherstonhaugh.

The story of how the Coed Coch Stud began was not one of a love affair with Welsh ponies, which has often befallen many a Welsh enthusiast, but one based on the practicality of farming an estate such as Coed Coch on a tight budget with a large amount of land and insufficient stock to graze it. Later it would be based on sound economics during hard financial times, but at first it was the suggestion of groom John Jones while hacking back from a Saturday's fox hunting in 1924. Observing that there was much too much grass in their fields, the discussion lent towards the purchase of a few Welsh Mountain ponies to help resolve the problem. Filled with enthusiasm, the next day five mares were purchased locally and the foundations of the Coed Coch Stud firmly established from somewhat sketchy beginnings. Miss Brodrick threw herself into her new-found interest, becoming a member of the WPCS in 1924 and a Council member from 1925 until her death in 1962; she became President in 1936, the same year that she was elected to Council of the National Pony Society.

From being Miss Brodrick's first groom, John Jones was associated with Coed Coch for over 50 years, as he and his famous family worked at the stud, and it would be Jones who did much to fashion the worldwide success of the Coed Coch Stud. In many ways the stud's success belonged as much to the Jones family as it did to Miss Brodrick herself – and she would have been the first to acknowledge it.

John Jones' father was a cowman on the estate while it was owned by Miss Brodrick's grandfather, Sir John Lloyd Wynne. The youngest of ten children (Jones would have the same number of children), his family lived in a small house in the Coed Coch farmyard, which later would become the estate office. John Jones first became a stonemason at Coed Coch in 1888, before moving away to take up an apprenticeship in the trade in Colwyn Bay. He had always had an interest in and natural flair for horses and he eventually took up a job in the area as groom to Lady Ward, who was a prominent breeder and exhibitor of Hackneys. After a brief spell in England when his employer moved back to Chigwell in Essex, Jones returned to his young family and home in Wales, eventually taking up a job once more at Coed Coch in 1915, as stud groom. For seven years he travelled stallions (Thoroughbreds) throughout Britain, including Scotland, for which he had a particular affection.

During his life at Coed Coch, John Jones worked for five owners within the same family – Sir Lloyd Wynne, his son, Maj Gen Wynne, his grandson, Edward Wynne, Edward's mother, Mrs Brodrick and her daughter, Margaret. The connection did not end there, as various members of the Jones family worked at Coed Coch over later years until its closure in 1978. John's son, Gordon, later to become a famous stud manager of the Gredington Stud, started his career as groom under his father at Coed Coch but moved on, whereas John's grandson, Shem, became stud manager in his own right. The small wiry figure of Shem Jones would become a very familiar figure in show rings throughout Britain, and it was Shem who handled each and every pony in the sale ring on the day the stud was finally dispersed in 1978. Shem's son, Wyn, who had been immersed in the ponies at Coed Coch from the time he was a boy, also worked there and later started his own successful stud of Welsh Mountain ponies carrying the Nerwyn prefix.

It was John Jones who spotted the potential of a young lad from the village of Dolwen who was quite small and slight of build, a great runner and fiercely competitive. He was just what he needed for the show ring so he persuaded him to come to work at Coed Coch, where he remained for the rest of his working life. The lad, John Jones, a distant relative, surprisingly grew in height and was affectionately known as 'Big John'. In the tradition of all the Coed Coch 'boys' Big John was a great showman – so much so that he and the Gredington Stud showman, Dil Powell, were paid by the well-known Hackney impresario, Cynthia Haydon, to travel to the Hackney Show at East Grinstead, where they ran the Hackney ponies and horses from her famous Hurstwood Stud. Following the Coed Coch dispersal sale, Big John started up his own highly successful Nantdywyll Stud. The Jones 'Boys' were all great enthusiasts for the Welsh breed, great showmen and totally dedicated to Miss Brodrick and her successor, Lt Col Edward Watkin Williams-Wynn.

Given that the Hackney was the most prestigious of the breeds of horses and ponies shown during the late 19th and early 20th centuries, it was little wonder that the expertise that John Jones had built up while working for Lady Ward would stand him in good stead when he moved to Coed Coch. In his autobiography, first published in 1961 in Volume 30 of the Royal Welsh Agricultural Society's *Journal*, John Jones remarks:

I showed hackneys at the first Welsh National in Aberystwyth in 1904 and also in the last horse show held in the old pavilion in Caernarvon, and later at the big show at Conway Marsh, when Tom Jones Evans was the judge. From Conway on to Holyhead and to the great Irish Show at Ballsbridge, Dublin!

Jones' show ring preparation and training skills were second to none and typical of the 'Hackney boys' of the time, the likes of whom have since disappeared. It was a tradition maintained during the lifetime of the Jones family during their association with Coed Coch and one which took the emerging stud to the very top in the show rings throughout Britain. If Miss Brodrick had the eye for a good pony and the skill of breeding them, then it would be her grooms who would show them to their advantage during the life of the Coed Coch Stud. As Jones noted later in his story, 'Gradually as numbers increased we went more and more into the showing business – and very successfully, largely because my boy, Gordon, took a tremendous interest in the ponies.'

Miss Brodrick's first show ring successes came with the dark brown Welsh mountain mare, Coed Coch Seren (f. 1925), an extensive winner in the early 1930s when shown in hand, winning no fewer than 20 first prises at leading shows during the period from 1931 to 1937, including the Royal of England, the Royal Welsh and the National Pony Society Show. She was a daughter of Coed Coch Eirlys (f. 1919), one of three mares Miss Brodrick purchased in 1924 from a horse dealer, James Berrow, having spotted them in a railway siding at Shrewsbury Station. Seren proved to be an exceptional breeder, and was responsible for the highly successful 'S' line within the stud. However, even she had a price, as was the case for many of the Coed Coch ponies and she was offered for sale with 33 others at auction in August 1937. Apart from three, all the ponies were purchased by a Mr Walton from Leeds, who had made no arrangements to graze them, which led to the ponies remaining at Coed Coch until a new sale could be arranged on 11 September. Along with Coed Coch Ebrill (f. 1933), Seren was bought back by Miss Brodrick, fortuitously along with a young colt named Coed Coch Glyndwr (f. 1935), for which she paid 30 guineas. Little did Walton know that he had been, albeit for just a fortnight, owner of some of the best ponies ever registered within the Welsh Stud Book. And little did Miss Brodrick realise that her young colt would become one of the most famous in the world of Welsh Mountain pony breeding and a name which has a special place among Section Bs, since he is sire of the very influential stallion Solway Master Bronze, and grandsire of Criban Victor.

As we have seen in the previous chapter, in her capacity as Council member of the WPCS, Miss Brodrick had a hand in the development of the Welsh sections within the Stud Book. By 1931 the heights and type for Mountain ponies had been defined in Section A up to 12 hands (neither hogged nor docked), while the riding type of pony was now established up to 13.2 hands with the small ponies of Cob type banished to Section C. At the same time a newly-constituted Foundation Stock Register had been introduced to boost numbers of mares in a numerically-depleted Stud Book.This

The famous Welsh Mountain Pony stallion Coed Coch Glyndwr.

provided a golden opportunity for breeders with vision and foresight to bring into the Stud Book mares which had the potential to deliver the type and height of Welsh pony of riding type that were increasingly in demand. Coincidentally, Tanybwlch Berwyn and Craven Cyrus (both by Arab stallions out of registered Welsh mares) had been allowed entry into the Stud Book, thus allowing a much-needed infusion of height while retaining Welsh breed characteristics.

By this time, the breeding of Welsh ponies, mainly the riding type of mountain pony, was firmly established at Coed Coch and numbers steadily grew in Section A. Due to her success and popularity, many breeders now flocked to Miss Brodrick for her ponies – which were fast becoming, like the Grove and Dyoll ponies before them, foundation stock for a great many developing studs for the next 40 years. Her ponies were also in great demand for an overseas market which had flourished during the early years of her stud. The United States took large numbers of ponies and other markets included Canada, Australia, South Africa and New Zealand as well as many of the countries of continental Europe. Miss Brodrick was at the forefront of this trade. In his appreciation written in the 1963 WPCS *Journal*, Eddie Griffith wrote:

> Study the pedigrees of the best ponies at home and abroad, and you will find that most of them are full of Coed Coch blood. Study the story of the expansion of the breed to so many countries, and you will find that it is largely due to Daisy Brodrick's initiative and enterprise. She went out to get the export business and she got it. She was our great ambassador.

Coed Coch Blaen Lleuad as a yearling in 1955 at Coed Coch.

The type of pony she preferred to breed coincided with a swell of interest by like-minded breeders who were keen to promote the Welsh pony as a suitable mount for children. Until the turn of the 20th century, riding for pleasure was not generally a pastime in which children engaged, other than those from rich families and possibly country children who hunted with their parents. Of these, it would be predominantly boys who rode in the show ring and consequently took part in equestrian sports such as polo or jumping. Times were changing, however, and girls increasingly took part. All of a sudden the need for good-quality ponies initially required to breed polo ponies (as set up by the Polo Pony Society) would translate into a need for the same quality ponies for children to ride.

Coed Coch stud card 1973, showing Coed Coch Berwynfa at a Personality Parade at the Royal Welsh in 1972.

Enjoying the financial fruits of her labour, there was a marriage of her own breeding preference to that of marketplace demand, and Miss Brodrick's interests soon extended to the ridden discipline, so she exhibited in both the Welsh breed and child's ridden pony sections. Her first main ridden success came with Tanybwlch Prancio (f. 1932) by Tanybwlch Berwyn, both bred by her great friend from Merioneth, Mrs Inge. Interestingly, although just over 12 hands, Prancio was registered as a Section A like her dam (bred by Lady Wentworth), Grey Princess, a mare brimming with Dyoll Starlight blood on both sides of her pedigree. Prancio won the Country Life Cup for the best mountain and moorland pony in the breed classes at the National Pony Society Show at Islington in 1938 before going on to take the 12.2hh ridden class, and again in 1939 when she was reserve. Her success was met with widespread approval – in his book on native ponies, Fell pony breeder and acclaimed pony expert, Roy Charlton wrote:

> I think the world would say that the present day Welsh mountain pony, as seen at shows like the London Pony Show, is the most perfect of all. Think of that delightful little mare, Tanybwlch Prancio, owned by Miss Brodrick, of Coed Coch, Abergele, North Wales. Where is there another pony breed that can show such a charming quality, and such uniformity of riding type.

At the Royal Show in 1938 Prancio, with a foal at foot, won both the ridden and in hand classes. She was ridden with great success by Peter Brookshaw, whose father Stanley produced Prancio from his farm in Shropshire. This little grey mare and the Welsh breed in general had many admirers and inspired others to buy Welsh ponies to show under saddle. One such exhibitor was Mrs A R Hepburn from Birmingham, whose enthusiasm for the breed would span 30 years. She took third prize in the ridden Welsh class at Islington in 1938 with Tanybwlch Rhos (f. 1929), a grey mare by Tanybwlch Berwyn, and also exhibited Craven Nell (f. 1933), a five-year-old bred by Tom Jones Evans from Craven Arms in Shropshire. Mrs Hepburn's time would come after World War II, when her gelding, Coed Coch Powys, a son of Tanybwlch Prancio, would sweep the boards just as his dam had done before him. Among more than 300 prizes, Powys took the supreme of the children's ridden classes at the Royal International Horse Show in 1949. Following remarriage, as Mrs Price, she judged the Royal Welsh Show in 1958.

Mrs Inge had a hand in the success of the Coed Coch ponies through the influence of her Tanybwlch ponies. 'She was a lovely lady, and more like a mother to Miss Brodrick than a friend,' according to 'Big John' Jones; she kept an eye on things at Coed Coch during the World War II while Miss Brodrick was away on army duties. In fact it was Mrs Inge who arranged the sale of Coed Coch Glyndwr for £400 to Lady Wentworth in 1943, while Miss Brodrick was abroad. Lady Wentworth also bought Tanybwlch Prancio, but sold her in 1946 to Violet Nichols from New Zealand along with a colt foal by Glyndwr named Wentworth Southern Cross – they were the first Section As sold to New Zealand after the war. The Tanybwlch ponies were already well established in the show ring and Miss Brodrick's success in breeding a bigger Welsh pony with riding quality obviously resonated with her wishes. The services of Tanybwlch Berwyn were selected for some of her mares before he finally came to stay at Coed Coch. In her personal stud record, Mrs Inge records that he (Berwyn) was measured

Tanybwlch Prancio shown in hand at Windsor Horse Show in 1939.

at '13.1½ hands' and was 'given to Daisy 1940'. Of Mrs Inge's mares, other than Prancio, one of the first mares registered in the Foundation Stock Register – Tanybwlch Penwen (f. 1930) by the Arab, Cairo, and out of Grey Princess (the dam of Prancio) – was secured for the Coed Coch herd. Now, through Sahara's son, Berwyn, and through Cairo's daughter, Penwen, two infusions of Eastern blood started to run freely through the veins of the Coed Coch ponies, most interestingly in both Sections A and B.

In an interview with the then British Prime Minister, Margaret Thatcher, the television personality, Terry Wogan, asked the question, 'Do you think that you have been a lucky Prime Minister, Mrs Thatcher?' She replied calmly, 'Mr Wogan, people who work hard always appear to be lucky'.

With this in mind, one wonders whether or not luck was on Miss Brodrick's side when she was given the mare Berwyn Beauty (f. 1942) – or was it a case of good judgement that she accepted her on the retirement of her

The many-times champion, Coed Coch Pryderi, ridden by Jennie Bullen.

breeder? She knew the mare and her family as they had often visited Coed Coch to be put to the stallions, and indeed it was there that she was conceived. Beauty was bred by Tudor Jones who lived near Ruthin in North Wales, not far from the beautiful Berwyn Mountain Range which is bounded by Llangollen in the north-east, Bala in the south-west and Oswestry in the south-east. Just as he had named his ponies with the Berwyn prefix, perhaps these mountains were also the inspiration for Mrs Inge, who must have passed them many times on her way to The Plas at Tan-y-Bwlch. Berwyn Beauty was by Tanybwlch Berwyn and out of Dinarth Wonderlight (f. 1936), which Jones had bought at the Dinarth sale in 1937 for 8 guineas. Her dam Irfon Marvel (f. 1916 by Dyoll Starlight) was also dam of Dinarth Henol (f. 1927), who in turn was the dam of the famous Mountain pony stallion, Coed Coch Glyndwr.

Beauty was shown from Coed Coch with success – in 1949 she came second at the Royal Welsh to Mrs Cuff's Criban Heather Bell, a mare which would etch out her

COED COCH PONY STUD

CATALOGUE

OF HIGHLY IMPORTANT

Sale of 50 Ponies

THE PROPERTY OF MISS M. BRODRICK

TO BE SOLD AT

Coed Coch Home Farm, Abergele

(under usual Conditions of Sale)

On Friday, September 26th, 1952

PONIES ON VIEW DAY OF SALE.

Auctioneers :

FRANK LLOYD & SONS

53, KING STREET, WREXHAM (Tel. 2041).

The 1952 Coed Coch Sale catalogue when Coed Coch Berwynfa was bought back into the stud for 45 guineas.

own piece of history within the Section B. That year Beauty had a colt foal at foot by Tanybwlch Berwyn named Berwynedd and, unbeknown to her owner, was carrying another that would have a major impact on the Welsh Section B, not only in Britain but across the world. In a shrewd act of line breeding, Miss Brodrick put Berwyn, then aged 26, back on his daughter, Berwyn Beauty, to produce the famous colt Coed Coch Berwynfa (f. 1951), the last foal registered at the stud by Tanybwlch Berwyn, who died in 1953. As successful as his sire had been in the preceding years, Berwynfa would prove to be a worthy successor, if not an equal. However, all that may have never come to pass if plans to sell him along with Berwynedd in 1952 had gone ahead. In September that year Miss Brodrick held a sale of 50 ponies at the home farm of Coed Coch, the site of two very important sales in the years to come. The foreword to the sale read:

> This is a genuine reduction sale – owing to the fact that export trade has been temporarily killed through the severe outbreak of foot and mouth disease. No country overseas will at the moment risk importing any animal from Great Britain,

in case they should carry infection with them, even though horses themselves cannot contract the disease.

Berwynfa, Lot 3, was described as dark grey with height 12.3 hands. The brief description read, 'A promising colt who will make either a very good Pony of Riding Type stallion, or a Show Riding Pony.'

How right Miss Brodrick was. From the time of entry, she obviously had a change of heart, and arranged for Emrys Griffiths of the Revel Stud to bid on her behalf, so Berwynfa was bought back for Coed Coch at the price of 45 guineas. This was among the best prices on the day, but top price by a long way was the FS1 mare, Coed Coch Pluen (f. 1947), a five-year-old by Berwyn out of Tanybwlch Penwen. 13.1 hands in height, she had a good show record with wins at the Royal Show and the National Pony Society Show to her credit. A selling feature was her 'sensational low riding action', as described in the catalogue – perhaps this is what inspired Mr Fitzroy to buy her at 135 guineas. Fortunately, she returned to her home some years later, where she produced two fillies and a colt, including the previously mentioned Coed Coch Pawl, born in 1958. Pluen reappeared on the 1959 sale, when she was purchased by Mrs Crisp.

Meanwhile, hot on the heels of Berwynfa was his half-brother out of Berwyn Beauty, Coed Coch Blaen Lleuad (f. 1953), a roan like his sire, Criban Victor, which had been selected to put a bit more substance into the offspring. Victor stood at the Gredington Stud belonging to Lord Kenyon, Miss Brodrick's great friend and rival whose bloodlines favoured those of Coed Coch. Beauty would have another two colts by Victor, named Ballog (f. 1956) and Barwn (f. 1954), the latter sold to the Wornanas Stud in Sweden. Unfortunately, Miss Brodrick had limited luck with fillies, which was perhaps the reason why she decided to cover her in 1956 with the chestnut Arab stallion, Jordan, bred in Arabia by Sheikh Abu Sarar of the influential Majali tribe. He stood at stud at Coed Coch and, although Miss Brodrick registered another mare, Coed Coch Desert Gold (f. 1952), by him in the Foundation Stock Register, Jordan was largely used to breed commercial riding ponies at Coed Coch and was popular with farmers in the district, with his stock regularly appearing in sale catalogues. The decision proved to be fruitful and a filly, Bronwen, was foaled the next year and duly entered in the Foundation Stock Register. Fired with enthusiasm to breed another filly, Beauty was put to the homebred Section A stallion Coed Coch Sandde (f. 1955) and luck was on Miss Brodrick's side when a filly, Berin, was foaled in 1958. Reverting back to the Arab, Midnight Moon was used on Beauty. However, this time yet another colt arrived – Coed Coch Blue Moon, sold like his half-sister, Berin, and his mother, Beauty, at the 1959 sale, was destined for the ridden show classes.

Although competition wasn't at the same level as we know today, Tanybwlch Berwyn's progeny had an amazing record in the show ring soon after World War II, with the Royal Welsh Show providing a good yardstick of success. In 1945 a colt was born out of the Section A mare, Coed Coch Sirius (f. 1937, a daughter of the show mare, Coed Coch Seren). He, like his half-sister, the illustrious Coed Coch Siaradus, was bred by Cyril Lloyd Lewis and, like her, sold to Miss Brodrick as a foal. She subsequently registered them using her own prefix (as was allowable at the time) and the colt,

1955 Royal Welsh Section B Champion Coed Coch Silian (f. 1947) by Tanybwlch Berwyn out of Coed Coch Seirian.

named Coed Coch Siabod (f. 1945), would record four consecutive male championships, two of them overall, at the Royal Welsh Show before his export to Australia in 1953. This was a remarkable achievement by any standards. However, the fact that Siaradus also took the championships there during the same years for Miss Brodrick speaks volumes for Lewis, who was the breeder of both – a feat never repeated in the history of the Show.

The first of these was staged locally at Abergele in 1950, when Matthew Williams of the Vardra Stud selected Siabod along with a Tanybwlch Berwyn daughter, Gem, a two-year-old FS1 filly bred and shown by Mrs Eddie Griffith from Trefnant in Denbighshire. There was no overall championship at the Show, but the following year a championship was initiated which favoured the filly Revel Nance (f. 1949), of Mountain Pony breeding, over Siabod. It is noteworthy that all the progeny of Revel Nance were subsequently registered in Section A, including Revel Newsreel (f. 1954), which had such an influence at the Cusop Stud. She was one of two fillies bought by Emrys Griffiths of the Revel Stud following the death of their breeder, Matthew Williams, in 1951; the other was Revel Choice (f. 1949). Both fillies were given the Revel prefix: Revel Nance was out of Vardra Nance (f. 1929) which, like Revel Choice, was out of the 1947 Royal Welsh winner, Vardra Charm, who carried several crosses of Dyoll Starlight on both sides of her pedigree. Choice was not only a champion at the Royal Welsh herself (1961), but the success of her offspring would also ensure her a special place in the Mountain Pony hall of fame in years to come.

Siabod had his day in 1952, when he took the championship over a mare related to him on the dam side called Coed Coch Silian (another by Tanybwlch Berwyn). She proved to be a marvellous show mare and breeder. Having been sold to Mrs Griffith, two of the foals she bred by Criban Victor came to the fore at this time. Her filly, Verity (f. 1952), stood champion over Siabod at Cardiff in 1953 and again over her sire, Criban Victor, the following year. In 1955 it was her brother, Valiant (f. 1953), who impressed to take the championship from his mother, Coed Coch Silian (f. 1947). He was later sold to South Africa.

1956 proved to be not the most successful year for Miss Brodrick and Tanybwlch Berwyn, but both stormed back the following year when Coed Coch Blaen Lleuad, Berwyn's grandson through his dam, Beauty, took the male award. In 1960 Downland Lavender took the female championship, the last major award at the Royal Welsh for Miss Brodrick before her death in 1962. Lord Kenyon's ponies dominated the Royal Welsh takings for the next few years, mainly through his five-time champion, Criban Victor, and Victor's daughters Gredington Daliad (f. 1949) and Milfyd (f. 1956), full sisters out of the FS mare Silver. The young usurper, Solway Master Bronze (f. 1959) by Coed Coch Glyndwr, stole the show for three years.

Under the new ownership of Lt Col Williams-Wynn, the stud's name rose to the top once more in 1964 with Coed Coch Penllwyd (f. 1953), a daughter of Pendefiges (f. 1946) by Tanybwlch Berwyn, and again two years later with this mare's Berwynfa daughter, Coed Coch Priciau (f. 1960), who bettered her dam's result by taking the overall title. The foal Penllwyd produced that year (1966), Coed Coch Dawn, a full sister to Priciau, would ensure that Williams-Wynn's name was engraved once more on the Royal Welsh trophy, when his mare stood champion in 1974. By this time the show ring was witnessing the major effect of Coed Coch Berwynfa, whose influence, particularly on the dam lines within Section B, became a major force with which to reckon. The last offspring of Berwynfa to triumph at the Royal Welsh was Gredington Blodyn (f. 1964), who was Emrys Bowen's choice as female champion in 1971.

Coed Coch Blaen Lleuad also proved his worth as his stock came through the ranks, especially via his son, Chirk Crogan (f. 1959). One of two full brothers bred by Lady Margaret Myddelton, Crogan made his mark on the Weston Stud, which had burst into the headlines in the early 1970s with a stable full of lovely Section Bs based

The commemorative slate at Coed Coch dedicated to Miss Brodrick's favourite hunting pony First Flight as well as Tanybwlch Berwyn, Coed Coch Berwynfa, Coed Coch Madog and Coed Coch Siaradus. The most poignant of inscriptions reads: "The air of heaven blew between their ears".

on Mountain pony bloodlines but elevated to Section B by the crossing of the Criban Victor son, Gorsty Firefly (f. 1965), as well as Firefly's son, Weston Chilo (f. 1970) and Chirk Crogan. The Royal Welsh champion fillies Weston Rosebud (f. 1975), Picture (f. 1978), Mary Ann (f. 1974) and Glimpse (f. 1971) carried combinations of these bloodlines, demonstrating another genius of breeding of which more will be told later. Like his brother, Blaen Lleuad would keep the Tanybwlch Berwyn blood flowing through the Section B through their grandchildren and great-grandchildren.

As important as it is to record the influence Tanybwlch Berwyn had on the emerging Welsh Section B, it is important also to record the amazing effect he had on the Welsh Section A. The Stud Book reveals that Miss Brodrick registered six Section A stallions by him (Coed Coch Erlewyn, Coed Coch Sadyrnin, Coed Coch Samswn, Coed Coch Moelwyn, Coed Coch Seren Du and Coed Coch Peris), as well as five Section A mares (Coed Coch Morfa, Coed Coch Sensigl, Coed Coch Silian, Coed Coch Sydyn and Coed Coch Trysor). Through them and his daughter Tanybwlch Prancio, he would feature in the pedigrees of some of the most successful Mountain pones ever to have graced the show rings – Coed Coch Planed, Coed Coch Pryd, Pelydrog and Coed Symwl to name but four. Through them and many others, his name features in the pedigrees of Mountain ponies across the world.

Meanwhile, in order to follow the development of the Welsh Section B, it is to the other end of Wales that we need to travel to find the breeders who would play such a pivotal role in taking the breed forward. The product of their labours, in themselves quite diverse but nonetheless related, would prove to complement ideally the bloodlines being established in North Wales by Miss Brodrick and others following her lead in the development of the Welsh Pony.

Chapter VI

Criban

Lady Reiss' famous champion under saddle, Criban Biddy Bronze, dam of Solway Master Bronze.

Criban Heather (f. 1954), a Foundation Stock mare by the celebrated riding pony stallion, Bwlch Valentino, out of Criban Red Heather by Criban Loyalist.

The efforts of Miss Brodrick to establish a Welsh pony with riding qualities can be described as nothing less than remarkable, the more so since she and her family had absolutely no tradition with the Welsh breeds of ponies and cobs prior to her purchase of the Mountain mares in 1924. Indeed, perhaps it was this lack of connection with the Welsh breeds that enabled her to attend to the task of breeding her own type of Welsh Riding Pony. This was mainly based on sound principles of conformation, action and temperament required in the ridden animal of which she did have great experience through the hunting field. (It is also the case that the rigours of the marketplace made demands on the breeding programme at Coed Coch.) Most importantly, her partner and mentor, John Jones, was like-minded and together they forged a pony which was Welsh in character but which also conformed to the modern demands of a child's riding pony with its sloping shoulder and long low action.

Mention has already been made of a stallion which would prove an ideal outcross for the Tanybwlch Berwyn blood which dominated the Coed Coch mares. This was the bay roan stallion, Criban Victor, the sire of the successful Coed Coch brothers, Blaen Lleuad and Barwn out of Berwyn Beauty. By stark contrast to the amalgam of bloodlines infused to create the ponies at Coed Coch, Victor came from a family of Welsh ponies steeped in the history of the breed at the other end of the Principality on the beautiful Brecon Beacons, which had been home to hundreds of hardy ponies running feral or semi-feral for centuries. It was the nursery for many of the ponies destined for the mines in the nearby coal-rich valleys of South Wales. Criban Victor was bred by the Richards family, whose ancestors can be traced back over three centuries to landowners in the Taf Fechan Valley who farmed at the adjoining properties of Coed Hir, Ystrad and Abercriban. They are direct descendants of Howell ap Richard, who was born in 1697 at Coed Hir and whose descendant Howell William Richards, born at Ystrad in 1865, was the founder of the famous Criban Stud (although prior to 1910 he registered the ponies with the Ystrad prefix).

The Criban stream flows into the River Taf, situated in a steep-sided valley which was flooded to create the Taf Fechan or Pontsticill Reservoir, opened in 1927, situated just north of the mining town of Merthyr Tydfil which, along with Newport, it supplies with water. Following compulsory purchase of the lands around the valley, the Richards Family had to move, relocating some eight miles away to Brynhyfryd in Talybont-on-Usk. They maintained their grazing rights on the Beacons, so the sight of Criban ponies roaming there remained, consistent with their 200-year history. In the early days of the stud, the ponies were mainly of the hard, solid colours of bay and brown dominating, along with a few black and dun; bay roan and chestnut roan were not uncommon. It is said that there were no greys among the old Ystrad strain of ponies until 1850, when the blood of the Crawshay Bailey Arabian crept in due to his availability on the south side of the Brecon Beacons; one of the old mares was bred nearby by Mr Thomas of the Pentre and is thought to be related to Moonlight, the dam of Dyoll Starlight.

In the early part of the 20th century, the bloodlines brought together produced a number of top-class and highly influential stallions which were extensively used at Criban. One such stallion was Klondyke (f. 1894), William Miller's champion stallion who stood nearby at Miller's Forest Stud at Brecon. A small, quality bay standing around 12.2 hands, he was typical of the Hackney cross which was favoured at the time and was very influential within the Welsh Stud Book – he appears in many of the pedigrees of the Criban ponies directly or through his son, Ystrad Klondyke. At much the same time, just prior to World War I, a name that appeared in the pedigrees of the best of the riding pony types was Invincible Taffy, whose dam, Chocolate Lass (f. 1905), came from Ystrad Jewel, registered in Volume I of the Welsh Stud Book. It is believed that a Thoroughbred stallion comes somewhere into the distant history of this female line, which accounts for the ability shown by Taffy – who was ridden and hunted by Dick Richards as a boy. In time, Invincible Taffy would appear in the pedigrees of the best of the Criban riding ponies to be found within the Stud Book, such as Criban Victor, Whippy, Bumble, Heather Bell, Nylon and Biddy Bronze.

The early Stud Book entries made by H W Richards and Sons show that they consistently bred within Section A. At the time Secton A allowed entry of ponies up to 12.2 hands. Interestingly, there are Criban ponies registered in Section A Part II (up to 12.2 hh) which accounted for the larger ponies including Criban Bay Boy, Bay Lad, Gold Dust, Orion, Satan, Shot and Wild Wonder among the stallions, as well as 25 mares including Criban Busy Bee, Betsy and Raspberry and the famous Criban Socks. There were a few registered in Section B (the modern Section C), which gave an early indication of height and substance in the Criban ponies, but these were few and far between. Criban Victor (f. 1944) was one of them and, like his dam, Criban Whalebone (f. 1936), he was registered in Section B. Whalebone was by the Welsh Cob Mathrafal Broadcast, a great performer under saddle standing 14.3 hands, which was hunted by Dick Richards during his stay at Criban prior to his sale to the United States. On Whalebone's dam side were Raspberry (f. 1920) and Orion (f. 1912), mentioned above, as well as Klondyke (f. 894) and Invincible Taffy (f. 1910), who were known to be a bit larger in height. Victor's sire, Criban Winston (f. 1940) was by the famous Coed Coch Glyndwr, one of the very few infusions of Coed Coch breeding to be found in the Criban Stud's breeding programme over the years. In 1939 Winston temporarily left his home for Abercriban, where he was leased for a season in a swap for Mathrafal Tuppence. Winston's dam, however, was full of old Criban breeding, bringing more evidence of height through Criban Betsy (f. 1920), Busy Bee (f. 1920) and Orion (all Section A Part II) and, once again, Invincible Taffy and Klondyke. With roan colour peppering throughout both sides of his pedigree, it was

Criban Heather Bell is the dam of Criban Red Heather, who is seen here ridden by Rosemary Cuff at the National Pony Society Show at Roehampton.

no surprise that Criban Whalebone should produce a big strong bay roan colt foal in 1944, which was registered in Volume XXXII of the Welsh Stud Book. By 1949, a colt of this breeding and type would have to be registered in what is today Section C.

Criban Victor was sold as a two-year-old to Mrs Cuff, who used him very briefly before selling him as a three-year-old to Lord Kenyon, with whom he remained for the rest of his life. He was a great champion in the show ring, and started as he meant to finish by winning

From left to right: Criban Sweetly, Criban Nylon and Criban Heather Bell ridden by the Cuff children.

at the Royal Welsh in 1947 – he would go on to win the male championship there in 1954 and the overall championship in 1956, 1958, 1959, 1960 and 1964. Even at the ripe old age of 25, he recorded a championship win at the North Wales Association Show at Caernarvon in 1969. He had a remarkable temperament, which was part of his appeal as a sire – and indeed he sired many winners, both in hand and under saddle. He would become a figurehead among the emerging Welsh Pony, the Section B, and was one of many ponies bred by the Richards family that made the Criban name famous both at home and abroad.

Howell Richards formed a farming partnership with his three sons, Llewellyn, the oldest, Richard (Dick) and William (Bill), the youngest. Over time, Llewellyn took over the Criban prefix and Bill, with his wife, Betty, registered their ponies with the prefix Cui. Llewellyn and Bill continued to farm locally, while Dick chose to join the Ministry of Agriculture and Fisheries in 1934

Criban Viola, FS mare by Bwlch Valentino, a big winner under saddle for the Bullen family before becoming a foundation mare at Mrs Egerton's Treharne Stud.

as a livestock officer, which involved a move north to Northumberland, where he was held in very high esteem. Bill's untimely death in 1954 led to a major dispersal of the Cui ponies; a few were retained by Betty and some by her daughter, Libbie (Cantref Stud), while 20 mares were purchased by Dick and his wife Sheila (long-time Welfare Officer for the Welsh Pony and Cob Society) to form the Criban (R) Stud.

The Richards family farmed sheep, cattle and ponies, with sheep outnumbering the other two by a long way. Shepherding was done on horseback and it was with this purpose in mind that the type of pony bred at Criban was developed over the years. All the work was done with a pony and even in old age Howell famously preferred to ride rather than drive a pony, continuing to ride well into his 90th year. All the Richards men rode and, shunning the heavier ponies bred for the mines, their preference was for the active type of pony suited to riding. At that time, heavier men rode the stocky type of pony which today we would call the Welsh Section C. Given that during their early history the Section B as we know it had yet to be registered, they added a 'dash' of Thoroughbred to their hill ponies in order to add a bit of size and speed. In an article on the Criban ponies in the September 1974

Mr and Mrs E H Richards and family: from left to right Dick, Marjorie, Llewellyn and Willie.

issue of 'Riding' magazine, Elwyn Hartley Edwards, one of the greatest writers of his generation, described the type of ponies they required:

> These ponies carried a full-grown man for a whole day and could gallop if need be up and down the Brecon hills without putting a foot wrong whilst doing their job of herding the wild ponies, the sheep and cattle and driving them across the rough mountain ground, and along unfenced roads and tracks. This sort of pony also provided relaxation for the hill farmers and shepherds who hunted them enthusiastically, crossing seemingly unrideable ground at a gallop in pursuit of a good fox setting his mask for some rocky mountain earth. In addition the ponies were raced at local meetings which were then a feature of the country life. Unsoundness, apart from that caused by accidents, was virtually unknown and not even thought about.

They were a great hunting family, like so many of the breeders who promoted the development of the Welsh Riding Pony. Dick Richards was Master of the Talybont Hounds between 1929 and 1934, while his brother Bill took over for the season 1935–1936 and Llewellyn acted as whipper-in. Llewellyn also enjoyed riding and hunting and possessed an instinctive eye for a pony. His interest stood him in good stead as a cavalryman during World War I, completing his service as an officer with the 12th

Llewellyn Richards at his home near Brecon.

Lancers and developing his riding skills under the best of regimental riding masters. The knowledge acquired at this time would prove very beneficial later in life, when he took up duties as a judge for many societies, all of which held him in the highest regard. On returning home after the war, he continued farming with his father, breeding ponies better suited to riding. Hartley Edwards continued:

> Many South Wales breeders at the end of the last century and, indeed, well into the present one, bred ponies specifically for use in the industrial areas of Merthyr, Dowlais and Cefn Coed and particularly for the pits. Often such breeders kept a 'pitter' stallion for the purpose. This, however, was not the case of the Richards who always inclined towards the active riding pony and held somewhat aloof from the industrial markets. In this, in part, lies the secret of the phenomenal success of the Criban ponies. The Richards were riding men through and through and the fact is declared in the well-worn volume of breeding records which Llewellyn Richards keeps in [the] study of his neat house near Bwlch, some eight miles from Brecon. On the first page of that volume, virtually a stud book in its own right, are written lines from Psalm 20, 'Some put their trust in chariots, and some in horses…'

With ponies still in demand for shepherding on hill farms, the family was already breeding ponies which were a bit larger than the Welsh mountain pony but lighter than the small cobby type which would eventually evolve as the Welsh Section C, the pony of cob type. Using a dash of Thoroughbred such as the chestnut stallion Lally, bred by their neighbour and great friend, Mrs Nell Pennell, they produced a deep-bodied pony with a good length of rein standing on strong legs. They could carry adults all day long and gallop up the Brecon Hills if required when rounding up sheep or ponies. They were also ideal ponies for leisure, in demand as hunting ponies and for racing at local country meets, which were popular at the time. They were sound and hardy, and over time would become the backbone of the pony we know today as the Welsh Section B.

In a sale of Criban ponies held at the cattle market in Brecon on 7 May 1937, prices demonstrated the wisdom of breeding these larger ponies. Within the 'Remarks' section of the sale catalogue, the following statement was made on behalf of the vendors:

> The riding ponies are bred from Welsh foundation by suitable sires; bred for:-
> CONSTITUTION - To live at high altitudes, stand hard riding and breed sound stock.
> CONFORMATION of true pony character with good shoulders, quality and balance.
> TEMPERAMENT – Great care has been taken in breeding from stock of good temperament suitable for children.
> The bigger ponies are bred by first and second crosses from Welsh mares mated to Thoroughbred and Polo Pony sires.

Riding ponies sold that day, many by Lally, reached a top price of £40 and the riding ponies averaged £20, compared with £11 for the Welsh mountain ponies, some of which had the best of breeding and had been winners in the show ring. The demand for a bigger pony was evident, thus providing an impetus to join the few breeders who saw a future for the larger riding type of Welsh pony, and it was no time at all before the Criban prefix joined that of Coed Coch and others in the show ring.

Llewellyn Richards moved to Oundle in Northamptonshire following his marriage to Irene, the daughter of a well-known owner of a company of bargemen operating on the Monmouthshire and Brecon Canal, along which they transported lime from the quarries in the hills of the Brecon Beacons to the network of canals around Britain; at Oundle he farmed and continued to breed and sell ponies. In order to breed the more 'modern' type of pony which was steadily growing in demand, he took advantage of the abolition of height regulations for polo ponies and purchased several of the smaller type, which were no longer popular with the polo pony breeders. He evidently struck up a friendship at that time with Herbert Bright who, among others, was coming to the end of his time as a breeder of polo ponies. Herbert Bright was the last surviving grandson of the extensive family textile business, John Bright & Brothers, synonymous with the cotton industry in Rochdale. By 1920, Herbert Bright had all but retired from the business, having decided to move from the industrial heart of Lancashire to the small village of Silverdale near Carnforth, overlooking Morecambe Bay. Very soon Silverdale would become synonymous with the very best of polo pony breeding in Britain and its address, The Cove, would appear in the London show catalogues.

The main feature of the Silverdale Stud was the type of polo pony that had been established – they were Thoroughbred in appearance with that all-important mixture of native pony blood which their breeder maintained was essential for their quickness, manoeuvrability and intelligence. Here was a breeder who was carrying out the ideals of the breeding policies of the National Pony Society – it was no wonder that Herbert Bright was highly respected within the Society and held prominent office within it, including membership of the Editing, Finance and Show committees as well as serving as President in 1937, the year of King George VI's Coronation. Without question he would be known to Miss Brodrick and Mrs Inge, as well as all the well-known native pony breeders of the day including Llewellyn Richards.

Hence, prior to the outbreak of war in 1939 when Bright retired and sold his famous Polo ponies, Richards purchased the outstanding stallions of the time including Silverdale Loyalty (f. 1923) and Silverdale Bowtint (f. 1926), while he leased Silverdale Tarragon (f. 1930); later the family gained the services of Silverdale Aquila (f. 1946) and his dam, Silverdale Aquitania. The first three of these stallions were the leading show horses of the day, every one a champion at the National Pony Society Show at Islington and winning Society Medals for a decade spanning 1928 to 1938 – they were household names among pony breeders across the land. Taking advantage of a remarkable opportunity to influence the ponies bred at Criban, it was through Richards' foresight that he brought into his breeding programme stallions of pony height with substance, good Thoroughbred limbs, movement and exceptional conformation. Combining their attributes with those of the hardy Welsh ponies from Criban, he would provide the riding pony world with a type of pony

Silverdale Loyalty, the champion Polo Pony stallion used to cross with Welsh ponies at Criban.

just right for the market of the time and, unwittingly, with an exceptional gene pool for future generations of riding ponies registered in the Welsh Stud Book.

Eventually Llewellyn Richards returned to Wales: he settled at The Allt, a property near Bwlch, only a few miles from his father's home at Brynhyfryd, to continue breeding children's riding ponies and the developing Welsh riding pony, the Welsh Section B, which had been established by this time. The Foundation Stock Register, established by the Welsh Society in 1930, allowed into the Stud Book the valuable Thoroughbred bloodlines, largely through the polo ponies selected by Richards for the Criban ponies.

Llewellyn Richards also seized upon another opportunity to introduce Thoroughbred blood into his ponies when he was offered on lease for a season, the grey, Bwlch Valentino (f. 1950), later to become a legend in the history of the British Riding Pony. He was bred by Mrs Pennell, the daughter of James Gwynne Holford, the Breconshire Member of Parliament from 1870 to 1880 and the first Vice-President of the WPCS. The family lived near the village of Bwlch at Buckland, which looked towards Brecon on the road from Herefordshire into Wales; for many years Criban ponies grazed the parkland on the estate. It was a place that provided the young Nell Gwynne Holford with fond childhood memories and the opportunity to enjoy riding with her near neighbours, the Richards brothers – together they had many an adventure with her small pack of Basset Hounds. Following the death of her father, the family moved to her mother's family home at Hartpury House (now Hartpury College) in Gloucestershire where Mrs Pennell was encouraged by her mother to breed some Welsh ponies. A move to nearby Tweenhills Farm followed shortly after World War II, along with the opportunity to continue breeding Welsh ponies and riding ponies carrying her Bwlch prefix, previously founded in 1910. With a keen eye for a good horse or pony and regarded as one of the most knowledgeable breeders of her day and since, it is most likely that it was Mrs Pennell's childhood that played a huge part in her future success. A keen horsewoman who loved hunting, she struck up a lasting friendship with the Richards family, one of whom, the oldest child, Llewellyn, particularly shared her love and interest of the riding pony.

Mrs Pennell had fortuitously spotted Valentino's dam, Goldflake, grazing in a field near Oxford as she drove past with her husband in 1931 and purchased the four-year-old for £35 from a dealer, Jack Castle, who had bought her along with a load of cattle at a sale near Llandeilo. Goldflake (born 1927) was by the Thoroughbred Meteoric out of a racing pony, Cigarette, believed to be a Thoroughbred/Welsh cross bred by Mr Davies of Carmarthenshire and, according to Mrs Pennell, 'was a legend as a racing pony in Wales after the First World War'. Valentino's sire was the good moving grey stallion, Valentine (f. 1933) by the Argentinian Polo Pony, Malice, an Anglo Arab introduced from South America as an outcross for the polo ponies being bred at the time. Malice upset some of the established breeders by standing reserve to Silverdale Loyalty for the Gold Medal for Best Stallion at Islington in 1928.

Without mares for him to cover at Tweenhills, Mrs Pennell was at a bit of a loss to know what to do with Valentino, and was persuaded to lease him to Llewellyn Richards as a three-year-old in 1953. While at Criban he covered a few mares, the offspring of which would be in the Foundation Stock Register in the Welsh Stud Book. Criban Lily (f. 1954), Criban Heather (f. 1954), Criban Viola (f. 1956) and Criban Ninon (f. 1956) were the mares entered into the Register after inspection. The list of Foundation mares registered at Criban makes interesting reading (see Appendix 3) as it more or less charts the development of the Welsh Pony (Section B) as seen through the eyes of quite different breeders to those previously mentioned earlier. The lack of Arab breeding is evident and the preference for the Thoroughbred quite emphatic.

To start with, the back breeding of the Criban ponies lacks the heavy influence of the Dyoll Starlight strain which had been so fashionable and plentiful elsewhere during the early part of the 20th century. It must surely have been a deliberate choice by the Richards family not to infuse their ponies with Eastern blood, just as they had little use for the Cob with its preponderance towards the Hackney. Apart from the use of Klondyke, the small pony of Hackney breeding belonging to Mr Miller, the Criban Stud had almost become an island sitting within a sea awash with Arab and Hackney blood.

The same could not be said for the Thoroughbred, which the family held in high esteem. It obviously suited the model of pony that they aspired to breed and which was functional in nature – one which stood around 13 hands with the best of limbs that would stand up to work and stay sound; it had to be quick on its feet and, above all, hardy in order to live on the Criban hills and carry shepherds across them. This can be seen directly by studying the mares carrying the Criban prefix in

Taf Fechan Reservoir Dam. The creation of the reservoir forced the relocation of the Criban Stud.

Section B of the Foundation Register. Of the 24 FS mares registered from 1939, the year when the first of the Silverdale stallions was used, to 1958, the year of the last Foundation Stock entry, we see that four were by the polo pony Thoroughbreds from Silverdale and one by Golden Cross, another popular small Thoroughbred of the day . As noted, Bwlch Valentino – with his high preponderance of Thoroughbred blood – accounted for another six, while a son of Silverdale Loyalty, Criban Loyalist, sired eight of the mares. This meant that almost one fifth of the FS mares carried 50% Thoroughbred blood, one quarter carried approximately 40% and one third carried 25%. This was carried by the FS mares through to FS1, where over half carried between 25% and 12.5%, and in FS2, almost two-thirds carried 12.5% or more.

By any standards this was a great commitment to the use of Thoroughbred blood in the riding type of Welsh pony coming into the Stud Book from Criban. Some people liked it, while others preferred the prettier types emerging from the Arab bloodlines. One prominent breeder who made use of it to meet her own standards was Kathleen Cuff, whose Downland ponies would carry a great deal of Criban blood through their veins.

Chapter VII

Downland

Kathleen Cuff (left) with Mary Bowen (Cennen Stud) at the Will Jones' President at Home Day in 1999.

Craven Bright Sprite ridden by Rosemary Cuff to win at Aldershot in 1950.

The circumstances through which the Downland Stud of Welsh Ponies came about could not have been more different from those surrounding either Coed Coch or Criban, although all three were driven by very knowledgeable people with a sound understanding of horsemanship. In addition they possessed a love for the classic conformation and quality of the Thoroughbred, while all the time looking for the characteristics of a Welsh pony. At Criban there was an age-old tradition based on the virtues of the Welsh pony, while at both Coed Coch and Downland these had to be acquired as neither founder had grown up with them. But all three studs had one characteristic in common – the vision of breeding a child's pony. In the case of Downland, Mrs Kathleen Cuff, having based her foundations on mares bought for her children to ride, set about a breeding strategy that remained unaltered to the very end, as any visitor would very quickly become aware. It was the model of a quality 13.2 hands child's pony with its long front, neat head, sloping shoulder and free action which inspired her and the type of pony for which she became famous worldwide.

The Stud's name originated from the rolling Downs north and south of Horsham in West Sussex, where Kathleen Cuff was the youngest of six children. Her Canadian father worked for the Royal Engineers, while her Kensington-born mother, Mrs Jessie Monro Higgs, was a well-known personality in the area; helped by four of her older daughters, she ran the private Causeway School from 1917 to 1957, where riding became the most important part of the curriculum. Standards were high and the school competed successfully at shows such as Windsor and Richmond, with rosettes won displayed in the dining room. According to local researcher Elizabeth Vaughan, 'Where most schools had a sports day, Causeway had a gymkhana with show-jumping'.

With this background, it is little wonder that Kathleen would have a taste for showing with her own children and through her equestrian interest met her husband, Sydney Cuff. They farmed in his home county of Suffolk for some time before moving from Bury St Edmunds to a rented farm on the Badminton Estate, home of the Duke of Beaufort. Referring to her father's love of Red Poll cattle and Suffolk Punch horses, which his family had bred and shown with success for many years, Rosemary Rees, the younger of the Cuff daughters, said of her father, 'He bred the most unpopular but the loveliest of breeds'. At their new home, as well as sixteen Suffolk Punch mares and three stallions, the Cuffs kept the best of horses to hunt with the Beaufort, one of the most prestigious packs of foxhounds in the world.

At Badminton the Cuffs raised a family of four who were all keen on riding – Anthony, Gillian, Rosemary and Simon all enjoyed enormous post-war success with their ponies, many of which were homebred following the foundation in 1946 of their mother's Downland Stud of Welsh ponies on the Badminton Estate. However, the children were initially mounted on ponies of unknown breeding, possibly most famous of which was the bay 12.2 hh gelding The Nut, champion for Gillian Cuff at the Royal International Horse Show in 1949. Along with Anthony riding Pip, Rosemary riding Craven Mona and Gillian riding Downland Pheasant, the children and their ponies dominated the 12.2 hh class at all the major shows across the country, including the London shows at Richmond and Roehampton.

Mrs Cuff had a very good eye for a horse and chose well from the Welsh studs which were within easy access from the Welsh border. One of her best early purchases was Criban Sweetly (f. 1934), which had been previously purchased by Moses Griffith for his Egryn Stud at Cwmystwyth, where she remained pretty wild, refusing to be caught and, as a result, avoided a sale overseas. The story goes that while on a buying trip in Mid Wales, Mrs Cuff spotted her through her binoculars from a seat in Griffith's Land Rover; she duly purchased Sweetly for the ridden classes in which she enjoyed great success during the period 1947 to 1949. Sweetly was a daughter of Criban Socks (f. 1926), the mare described by many at the time as the most perfect example of a Welsh Mountain pony. Socks, a dark chestnut by Criban Shot, was out of Forest Lass, a mare belonging to William Miller. Lass spent most of her days grazing with the Criban mares, and was eventually purchased by Howell Richards along with her attractive filly foal, Criban Socks. Socks first produced a dun filly called Criban Rally to Criban Bumble Bee in 1932 and then Criban Sweetly, dark like her mother, to Criban Marksman (f. 1926) in 1934. Socks passed through a few hands during her lifetime but went from Criban to Dinarth Hall in 1936 following a win in the brood mare class at the Bath & West Show held that year in Cardiff, but only on condition that her colt foal, Criban Cockade (f. 1936), return to Criban at weaning. By Ness Commander, this dun was used extensively at Criban and is considered by many as the stallion which kept the pony character in the ponies while they were bred up through the Foundation Stock Register to Section B. He appeared to have a particular influence on the dam side of the pedigrees, including Criban Brenda (f. 1940); when owned by Miss Frank, she was put to her well-known Thoroughbred stallion, Potato, producing in 1955 the FS mare, Miss Crimpy Peek-a-Boo, the champion riding pony which became a foundation mare at the Weston Stud. Through Brenda's son, D-Day (f. 1944), came the line that produced Criban Biddy Bronze (f. 1950), the dam of Solway Master Bronze.

In 1935 Criban Socks was a big winner for the Dinarth Hall Stud.On the death of the stud's owner, T J Jones, she was sold at the dispersal sale in 1937 to Mrs Sivewright, who had her for a short time before selling her to the West Country, where a great pony enthusiast and Dartmoor breeder, Miss Sylvia Calmady-Hamlyn, provided Socks with her last home. Apparently Miss Calmady-Hamlyn was overwhelmed by a sense of patriotism prior to the outbreak of World War II when, along with a number of her ponies, she had Socks destroyed, depriving the Welsh breed of one of its most acclaimed ponies of pre-war years. For Miss Hamlyn she bred to Coed Coch Seronydd (f. 1931) a bay colt, Peter Pan, a leading 12.2 hands riding pony for his tiny five-year-old rider, Scarlet Rimell, from 1950 to 1951, during which time he famously galloped off with the championship at the Royal International Horse Show.

Initially the Downland Stud consisted of Mountain ponies – the stallion, Coed Coch Sidi (f. 1941), a full brother to Coed Coch Madog's sire Seryddwr, was purchased from Miss Brodrick, with whom Mrs Cuff was friends. By Coed Coch Glyndwr, Sidi was used for a couple of seasons before coming to an untimely end due to a broken leg sustained in a kick from one of Sydney Cuff's hunters. The children's winning riding pony mares, Craven Mona (f. 1943) and Craven Sprightly Twilight (f. 1943), as well as Craven Good Friday (f. 1936), all came from the Craven Stud owned by the aforementioned Tom Jones Evans. Dr Wynne Davies would argue that he was one of the most influential men in the WPCS throughout the period from 1910 to 1950, not only for his work on Council but also in his role as advisor to many important studs, including Grove and Ness. These mares carried the best of Mrs Greene's breeding, full of Starlight blood and of riding type; Mrs Cuff was one of the greatest advocates of the riding type of Welsh Mountain pony, and demonstrated its success extensively through the show ring.

As Welsh ponies were able to be shown both led and under saddle at many of the shows, catalogues soon revealed Mrs Cuff's appetite for in-hand showing. One of her notable early successes came at the Bath & West Show in 1948 in Cardiff, when she won the brood mare class with one of the children's ridden ponies, Craven Bright Sprite (f. 1937); Matthew Williams's Vardra Nance stood second, with Miss Brodrick's Coed Coch Prydferth third. Craven Mona was unplaced in the barren mare and gelding class won by Vardra Charm, the famous dam of Revel Choice, foundation mare at the Revel Stud of one of the most famous Mountain Pony families in the Stud Book. Coed Coch Sidi was unplaced that day in the stallion class won by Tregoyd Starlight. Incidentally, Sydney Cuff and the Duke of Beaufort were also exhibitors that day, in the Suffolk Punch and Percheron horse classes respectively.

As the Cuff children grew in age and size, bigger ponies had to be found; the Criban pony Bowman and his sister Heather Bell (f. 1943 by the Welsh Mountain stallion Criban Cockade) were among the most famous for their successes under saddle. Bowman was another pony spotted in the rough on the mountain when still a young stallion – he was brought home, gelded and broken before embarking on a very successful career under saddle. A chestnut standing just over 13 hands, his breeding brought together a high level of polo pony breeding as his sire was Criban Loyalist by Silverdale Aquilla and his dam, Criban Bowbell (f. 1939), a daughter of Silverdale Bowtint. Mrs Cuff purchased Bowbell in 1952 at a sale in Talybont-on-Usk for 47 guineas. The following year, Criban Heather Bell recorded a win under saddle at the National Pony Society Show for Mrs Cuff in 1953, one of many victories for her family over the years.

Mrs Cuff had demonstrated shrewdness in her choice of ponies, as she had not only selected a group of ponies for her children to ride but also a group of mares which, in time, would have a major impact, through their grandchildren and great-grandchildren, on the Welsh Section B, the Welsh Pony of riding type. The Cuff children would later pursue their own interests –

Criban Heather Bell, Champion at the Royal Welsh in 1949.

Anthony in the agricultural industry, Simon on the stage in London, Gillian as a horse dealer and highly-acclaimed trainer of showjumpers (her son Andrew Davies was an international showjumping competitor) and Rosemary, whose son Timothy is a successful event rider, following in her mother's footsteps by breeding Welsh and British Riding Ponies at her Small Land Stud in West Wales.

This allowed Mrs Cuff to develop her own interest in breeding, which by 1950 was rapidly moving away from Section A to the newly-constituted Welsh Section B for Welsh ponies of riding type up to 13.2 hands, for which she saw a good market. This obviously suited the type and size of pony that she was aiming to breed and her thoughts on the Part-bred Register (see her 1955 letter to E S Davies in Chapter 4) showed no advantage over the Foundation Scheme. Appendix 4 demonstrates how she shrewdly used the Mountain pony stallions that she had bred to great affect by crossing them with her Criban mares in order to secure the Welsh type and a sound mare base. It was a stroke either of luck or of genius that she purchased an old-fashioned Section B colt, Star Supreme, which she crossed with some of her Section A mares, giving her another string to her bow for crossing both on the top and bottom lines of the pedigrees in her stud. Both strategies would prove their worth and their combination would provide their breeder with unparalleled success within the breed at a critical time when others were sticking to more traditional lines. The emergence of the next generation of Downland stallions registered in Section B of the Stud Book proved very popular with the new cohort of Welsh Pony breeders, who were waiting in readiness for a new direction in which to move the established bloodlines.

Mrs Cuff had wisely selected as her foundation mares bred at Criban – a nursery for some of the greatest matrons witnessed by the next few generations of riding ponies – and Criban Bowbell, a chestnut like her sire (the champion polo pony Silverdale Bowtint), arguably held centre stage. Criban Bowbell had already shown her worth as a saddle pony for Mrs Cuff, as had her son by Criban Loyalist, Criban Bowman, and her daughter by Criban Cockade, Criban Heather Bell. Prior to the move of Bowbell and Heather Bell to Downland, both mares had been used for breeding at Criban, with Bowbell producing another daughter in 1945 called Criban Belle, sired by the Welsh stallion Bolgoed Squire (f. 1938). Belle would later influence the British Riding Pony through her

WELSH BORDER RIDING HOLIDAY at the DOWNLAND PONY STUD . . .

Norton Manor, with its own grounds of 150 acres of hill and woods on the borders of Radnor Forest, provides a perfect holiday centre. Reliable horses, fishing in own trout pools (river by arrangement), tennis, boating, games room and billiards. Centre Hawkstone Otter Hound country. Children's holidays with own family a speciality. Adults 10 gns., children 8 gns. per week, inclusive (reduction for non-riders).

For further details apply:

MR. & MRS. CUFF, NORTON MANOR, PRESTEIGNE RADNORSHIRE (Presteigne 358)

Norton Manor and Downland Pony Stud advert, 'Riding' magazine, 1955.

famous daughters Criban Activity (f. 1949) and Criban Biddy Bronze. Meanwhile, however, obviously happy with the progeny of his cross-bred pony stallion, Criban Loyalist, Richards put him to Criban Heather Bell in 1947 to produce Criban Red Heather the following year.

Red Heather's offspring would play a major role in the developing riding pony, particularly when crossed with a new stallion which arrived at Criban on lease – Bwlch Valentino. Bowbell was shown as a broodmare and had three Welsh foals including Downland Coral shown in hand and sold to Mrs Crisp; her best filly was Downland Shepherdess by Downland Dominie which, in turn, had 13 riding pony foals, all quality show ponies.

When Heather Bell's days under saddle were over, Mrs Cuff turned her attention to the paddock, where she used her homebred Welsh Mountain stallions on both Heather Bell (FS1) and her dam, Criban Bowbell (FS). Downland Dicon (f. 1947 Revel Brightlight x Longmynd Donna) is the first stallion registered with the Downland prefix; when put to the former saddle pony Criban Sweetly, they produced a colt which would make his mark on the stud. This was Downland Serchog (f. 1951), sire of two of Heather Bell's daughters, Dragonfly and White Heather, born in 1955 and 1956 respectively. Prior to this, Heather Bell had produced a filly in 1953, Downland Red Heather, by another homebred Mountain Pony stallion, Downland Imp (f. 1950), a son of Coed Coch Sidi and Craven Mona.

The Royal Welsh Show in 1956 was a memorable one for the Cuff family. Mrs Cuff won in hand with both Downland Lilac (f. 1949) and Red Heather, at a time when three-year-olds could also be shown under saddle. In the children's ridden pony section, the Cuff children had a clean sweep, with Red Heather taking the 12.2 hands class, Downland Bowtint the 13.2 hands class and the small Thoroughbred Lady Airy the 14.2 hands class and championship. Lady Airy was used to breed Part-breds at Downland to service the bigger height market, one that would particularly appeal to the overseas buyers whose focus was on the breeding of performance ponies. Lady Airy was purchased when in foal to Blue Domino; the result was a beautiful bay colt, Downland Dominie, which was used at home for a limited period and ridden successfully by Gillian Cuff, both show jumping and in under 15 hands point-to-point races. As it happens, the first of the Section B stallions to be born was the product of Serchog and Red Heather, when Downland Roundelay arrived in 1958; Roundelay would in due course sire two colts, Downland Drummer Boy (f. 1961), which went on to the Kirby Cane Stud, and Downland Romance (f. 1961), a mainstay of the stud and a major player in its success. At an early stage we witness the crossing of the former riding ponies to produce the type of pony their breeder was seeking.

Over the years Mrs Cuff tried to bring outside stallions into her stud but, by her own admission, it never seemed to be much of a success no matter how hard she looked and tried. The homebred Downland stallions always seemed to work well together. Perhaps her early experiences were somewhat off-putting. On 5 November 1946, she bought for 40 guineas a two-year-old colt at the Criban sale which she thought she might use on a few mares and later castrate for use under saddle – the bay roan was called Criban Victor. He was never the type to fill the eye of his new owner, so it was no surprise that he was little used and quickly sold on for good money to Lord Kenyon. (Mrs Cuff always had a nose for a good deal.) Given the impact Criban Victor had on the Section B later in life, we will never know how he would have bred to the Downland mares had Mrs Cuff persevered – suffice to say that she, personally, had no regrets.

The same could almost be said for her purchase of a grey colt in 1953 from 'A L' Williams of Llanwrda, Carmarthenshire. Standing 13.1 hands, Star Supreme (f. 1949) was registered in Section B of the Stud Book and was typical of the old-fashioned cob-type pony. He was by the chestnut Welsh Cob-type pony Welsh Echo (f. 1943), Captain Howson's champion at the Royal Welsh Show in 1949, which stood 13.2 hands, the same height as Supreme's Royal Welsh-winning dam, Lady Cyrus (f. 1941), a grey by Craven Cyrus (by the Arab King Cyrus). Again, Star Supreme failed to fill Mrs Cuff with confidence, so she covered a few mares with him in 1953, and then gelded him before he embarked on a successful ridden career prior to sale. Rosemary rode him at this time and recalls that he was every inch a performance pony, more of a hunter type with a great shoulder, and a good mover. The gods were obviously on Mrs Cuff's side when she decided to cover two of her Section A mares with Star Supreme – firstly to Downland Grasshopper (f. 1949 out of Craven Good Friday) she produced the colt Downland Gay Star (f. 1954), whose daughter out of Criban Heather Bell, Downland Cameo (f. 1958), provided Mrs Cuff with a plethora of offspring whose names appear in the best of Downland breeding, including, three generations later, the stallions Krugerrand (f. 1979) and Arcady (f. 1981).

By this time the Cuff Family had left the fertile pastures of Badminton to live at Norton Manor, a large country mansion near Presteigne, Radnorshire, which had once belonged to Sir Richard Greene Price, but which had fallen into disrepair following its requisitioning by the Army during the war. It was at this time that the stud expanded, but Mrs Cuff also broke and schooled ponies for clients and offered riding holidays for children, which proved popular and with which her girls were well equipped to assist. Miss Elspeth Ferguson (later famous for her Rosevean Stud of British riding ponies) was brought in to help occasionally, but the mainstay of support was Jackie Knott (née Ross), who worked at Norton Manor for five years. She has since become a well-known show pony judge. In typical style, Mrs Cuff used the resources available to her to full effect, and the show ponies were called upon for duties with the holiday children as well as the show ring.

Star Supreme covered the Mountain mare Craven Sprightly Twilight, registered by Tom Jones Evans but bred by Arthur Pugh, one of the earliest members of the WPCS and its President in 1944, who had moved from Wales to farm in Leicestershire near the Aldridges (of Sahara fame) at Hinkley, where he continued to breed Welsh Mountain ponies. Twilight had several owners before her sale in 1952 to Mrs Cuff, who bred only a few foals before passing her on to the Ankerwycke Stud belonging to Campbell Moodie. The filly by Star Supreme would be sufficient to send Twilight's name into the Hall of Fame, as her daughter not only became a major prizewinner under saddle but went on to breed some of the most influential ponies known to Downland and to the Welsh Stud Book.

This grey filly, born in 1954, had the wonderfully romantic name of Downland Love-in-the-Mist and, by some quirk of hybrid vigour that could only be attributed to her Welsh Cob and Arab antecedents, grew tall and competed for Rosemary in the 14.2 hands classes. She was many times a champion under saddle including at the Royal Welsh in 1958 when, judged by Eddie Griffith, she qualified for the Horse of the Year Show at Harringay on her first appearance she came fourth. She returned to HOYS

Above: Downland Dauphin in front of Plas Llangoedmor.

Below: Downland Romance.

in 1960 and was actually tested in foal on the way to the show by Emrys Bowen. However, on this second occasion, having bolted in the collecting ring, she frightened her young rider so much that she refused to ride her in the main ring. After a great panic, another rider was found at short notice; she went beautifully but was unplaced. Kitted in long riding boots befitting the showjumping classes in which she was competing, it was the future (1961) Junior European Show Jumping Champion Sheila Barnes (later Baigent) who saved the day.

Love-in-the-Mist had a highly successful career when shown in hand but it will be as a matriarch that she will be best remembered. Put to Roundelay, she produced Downland Romance in 1961, a grey like his mother with height, substance and riding qualities – it was no wonder that he was chosen as a stallion for the stud. In 1964, on the same day that Dauphin won the Section B stallion class, Romance won both the Section B and Riding Pony colt classes at Ascot. He had the temperament desired for breeding good children's ponies and, indeed, was ridden at home and in the show ring – in 1968 Romance was the best ridden of all Welsh sections at Glanusk Stallion Show and in 1970 was champion both in hand and under saddle at Glanusk. The next four years witnessed the arrival of four very significant offspring – Downland Chevalier (f. 1962), Misty Morning (f. 1963), Water Gypsy (f. 1964) and Madrigal (f. 1965). They were all by a recent product of the Cuff pony combinations, the bay, Downland Dauphin, born in 1959 – a memorable year for Welsh Pony breeding as it heralded the arrival of Brockwell Cobweb, Chirk Crogan and Solway Master Bronze, three stallions which, like Dauphin, would have a lasting affect on the Section B.

Dauphin was only lightly shown, twice winning at the Ponies of Britain Stallion Show at Ascot (1961 and 1964); in 1965 he won the ridden Welsh at Glanusk and the stallion class at the Royal Welsh. His success as a sire was immediate, but tragically cut short when he died of grass sickness later that year after four seasons. (Some sixty years on, this equine disease – which affects the central nervous system – continues to have scientists baffled, despite extensive research.) Dauphin's dam was Downland Dragonfly (f. 1955), a daughter of Criban Bowbell like Red Heather but, unlike her, by Downland Serchog, a son of Criban Sweetly. It is interesting to note that Dr Wynne Davies had purchased Dragonfly for £120 for export to the United States but, when it was discovered that she was Section B and not Section A, the sale fell through. Had this not been the case, Dauphin would have been lost to the Welsh Stud Book in the United Kingdom. Dragonfly had a major influence on the stud, breeding colts and fillies of outstanding merit to form the famous 'D' line at the Stud. Dragonfly's progeny included her sons, Dauphin, Drummer Boy, Dandini, Dragoon, Dalesman, Dualist and Dubonnet and her daughters, Demoiselle, Debutante, Dryad and Delphine.

While Dragonfly's offspring showed little of her Section A heritage and much of the Thoroughbred breeding that came down from Criban Bowbell, the pony characteristics remained true. Most were the result of crossing her with both Downland Romance and Chevalier, but it was an unlikely mating that produced the influential Dauphin. His sire, Criban Pebble (f. 1942), a small Mountain pony, had come to Mrs Cuff following a successful career breeding in South Wales, where he

Downland Chevalier.

Downland Mohawk, Royal Welsh Champion 1972.

One of Mrs Cuff's most famous mares, Downland Camelia.

Downland Love-in-the-Mist with Chevalier at foot.

was very popular with Section A breeders and where he sired, among other good ponies, Vardra Sunstar, the sire of Revel Choice. It was following his time as a Premium stallion on the Black Mountains that he went to Downland, which had now relocated to a small hill farm near Llanddeusant, near Llandovery.

Dauphin's cross with Love-in-the-Mist gave the Welsh Section B one of its undoubted most famous sons, the chestnut Downland Chevalier (f. 1962). Chevalier's cross with Romance mares at home became legendary and his use as an outcross was regarded as outstanding for the Coed Coch bloodlines being created in North Wales by Miss Brodrick – as portrayed by the beautiful Downland Jamila out of the Coed Coch Berwynfa mare, Lydstep Jasmine. Another stallion with impeccable temperament, he enjoyed a limited but successful show career when he was Supreme Champion Youngstock at the Ponies of Britain Stallion Show at Ascot in 1963 (he then outgrew the classes, so was not shown in subsequent years). There was no need to publicise his name through the show ring, as his sons and daughters soon did this for him. Like Tanybwlch Berwyn, his offspring took the show ring by storm with the Royal Welsh championships going to a variety of champions including the mares Rotherwood Honeysuckle (in 1970 and 1978), Rotherwood Lilac Time (1981), Lydstep Lady's Slipper (1977) and the stallions, Baledon Squire (1978), Glansevin Melick (1980), Downland Gold Leaf (1981 and 1985) and Paddock Camargue (1991).While many people preferred Chevalier to his half-brother Romance, the combination of the two produced a duet hard to beat in the annals of the WPCS.

Chevalier's two full sisters would also step up to the plate and produce future stallions for the stud. His year younger sister, Downland Misty Morning, was an outstanding individual and among the prettiest of the mares to be found at the stud following a brilliant career under saddle. She was all that Mrs Cuff had hoped to breed, and held her own at county shows across Britain when produced by Harry Dawson, an extraordinary character of the show world with a keen interest in ponies. Mrs Cuff herself also showed her with great success in riding pony breeding classes as a brood mare. Her breeder desperately tried to find outcrosses for Misty Morning during her life in the stud with a view to breeding a future stallion, but without success. Her best son came from a mating with Romance when the exquisite dark brown/ black colt Manchino was foaled in 1974. He was used for several years at Downland and produced the famous mare Downland Ripple (f. 1977) out of Camelia from the Criban Red Heather line. Put to Chevalier, Ripple produced a future stallion in Chivalry (f. 1984), whom many considered the most like his sire in type and looks. Manchino then went to the Sunbridge Stud, where he was a big success, but it was Misty Morning's daughter by Krugerrand, Minuet, which produced one of the latest stallions to be kept at Downland – this was another dark brown colt, Downland Night Rider (f. 1993).

The mare lines became so strong within the stud largely due to the birth of exceptional stallions which suited one another and which seemed to bring out the best in the Welsh Mountain/Thoroughbred cross from which they originated. Love-in-the-Mist would once more provide another male line, this time through her daughter, Water Gypsy (f. 1964 by Downland Dauphin). A big mare like her dam, she bred well to Romance and in 1969 produced a colt which would take the Downland

Downland Jamila, one of Mrs Cuff most beautiful mares.

Downland Misty Morning – a champion under saddle.

Downland Titlark shown for Clive Morse by Len Bigley to take the Section B championship at The Royal Show in 1986.

ponies into another era. This was Mohawk, a quality dark brown pony with height and the most beautiful of heads – he was Emrys Bowen's choice of champion at the Royal Welsh Show in 1971. He proved to be the ideal cross for the Chevalier mares and even doubled up successfully on mares by Romance. Like Chevalier, his progeny would enjoy a high level of success at the Royal Welsh – Rosedale Mohican (f. 1973) was male champion in 1977 and overall in 1982, while Downland Edelweiss (f. 1979), daughter of the prolific Downland Eglantine, was overall champion in 1988.

Good temperament has always a been a trademark of the Downland ponies – which suited Mrs Cuff and her long-standing friend, Phoebe Sandford, who joined her at Cardigan in 1969 to help with the ponies when Rosemary married. Phoebe, as she was respectfully and affectionately known by everyone – as opposed to her close friend, who was always referred to as Mrs Cuff by everyone other than her very closest friends – had a marvellous way with the stallions, which she absolutely adored, Mohawk in particular. The foals seemed to be attracted to her like a magnet. The daughter of a vicar,

The young stallion Downland Wild Fowler, part the entire crop of Downland foals purchased in 2000 by Mats Olsson from Sweden.

she was well-spoken like Mrs Cuff and, also like her, mad-keen on riding ponies, which she also bred in a small way – Phoebe Sandford bred Eclipse, the dam of the Pony of the Year Snailwell Charles, bred by her great friend Mrs Waddilove from Newmarket and ridden by Nigel Hollings. Mrs Cuff and Phoebe were two of a pair – devoted to the breeding and rearing of the Downland ponies, which they both enjoyed into their later years. Mrs Cuff moved from Carmarthenshire to a large Georgian house, The Plas, situated at Llangoedmor near Cardigan; she then moved nearby to her final home, Wellelwyd, the place where many modern breeders will remember visiting them.

There are too many factors to consider when evaluating the success of the Downland Stud, and it was considerable by any standards. Most would agree that Mrs Cuff's initial choice of mares, both Mountain ponies and the Thoroughbred crosses from Criban registered within the Foundation Stock Register, was one significant factor. Then there was Downland Dauphin, who crossed so well on Mrs Cuff's prominent mare lines that the future stallions were able to have such an impact and such a solid mare base. This brings us to the stallions themselves – all homebred, which had such a complementary effect on one another through the years. It was John Davies (Rhoson Stud) who summed up the essence of the magic conjured up by the Downland stallions in an appreciation of fifty years of the Downland Stud published in the WPCS 1999 *Journal*:

> It would be remiss of me not to mention perhaps the most famous stallion of all, namely Downland Dauphin, who was struck down at an early age, but set in train the bloodline which was to influence the renaissance of the Welsh Stud Book Section 'B'. Roundelay has also left his mark via Romance, but it was the golden cross of Chevalier on Romance mares and the reverse, Romance on Chevalier daughters, which laid the foundation of so many studs today. This, coupled with the infusion of Mohawk, proved a heady mixture, and produced generations of ponies which shone in the show ring and topped the sales.

Probably most important of all was the model of pony which Mrs Cuff aspired to breed – it was driven by purpose rather than type, although their breeder always admitted to liking a nice Welsh head. While most of the early breeders would have agreed with her principles, not all of them agreed with the product that she bred and some were sceptical about the path along which it would lead the Section B. This view may have been considered in some ways protectionist – nothing new within the politics of breeding circles – but it was short-lived as more and more of the studs around Britain and the rest of the world gravitated towards Downland for stock on which to base their own breeding programmes.

Time would prove, in fact, that the Downland ponies would blend in beautifully with bloodlines being developed quite differently with others. To the very end, they remained in great demand and close examination of today's winning pedigrees will reveal a Downland pony somewhere in the breeding, albeit perhaps several generations back.

Chapter VIII

Plasnewydd (Griffith), Trefesgob, Gredington, Rhyd-y-Felin

Gredington Milfyd (f. 1956) at the 1960 Royal Show at Cambridge.

Mrs Borthwick's Cusop Sheriff produced by Jennie Bullen.

North Wales remained a fertile location for the developing Welsh Pony 'of Riding Type', with several notable breeders following a pathway etched out by Miss Brodrick at Coed Coch but, like her, with no roots in the Welsh breeds. One such person was Mrs Mary Griffith, whose husband, Major Eddie Griffith, born in 1900, had inherited the substantial Plasnewydd estate near Trefnant in Denbighshire following the death of his parents. His father, Lt Col E W Griffith, formerly of the Royal Dragoons, was Deputy Lieutenant of the county, as his son would be later in life.

Major Griffith's mother was the sister of Lord Daresbury, both children of Sir Gilbert Greenall of Walton Hall, Cheshire, whose fortunes had been amassed through the Warrington brewing family for which the Greenalls are still well known. They were also a great fox-hunting and racing family, with three generations holding Mastership of Sir Watkin Williams-Wynn's Wynnstay Hunt. Sir Gilbert's grandson, the current (4th) Baron Daresbury, has a huge reputation within the racing industry and is Chairman of Aintree Racecourse, home to the Grand National; all four of his sons have been champion amateur jockeys, like himself, and one of them, Tom Greenall, rode Trust Fund to victory in the Foxhunters Chase at Aintree in 2009.

Mrs Griffith was a great supporter of fox-hunting, having hunted with the Ledbury before her marriage. Their nephew, Michael Griffith, President of the Royal Welsh Agricultural Society in 1994, who took over the Plasnewydd Estate, also enjoyed hunting with the Wynnstay and Denbighshire Hunts and was an amateur jockey like his Greenall relatives. It seemed to be a family tradition to become immersed in public service and he, like his uncle, was very popular and held posts within the local community. Currently his son, Anthony Griffith, maintains the family tradition and runs Plasnewydd.

Major Griffith was no stranger to breeding livestock as he owned one of the leading herds of Friesian cattle in Britain, dispersed in 1964. He had become a member of the WPCS in 1929 and became a prominent and well-respected Council member, serving on the influential Editing Committee for many years and as Vice Chairman in 1935. Griffith registered the Plasnewydd prefix the following year, but Plasnewydd Madcap, born in 1950 by Tregoyd Starlight out of Coed Coch Pendefiges, is the only pony that appears to carry it – the couple preferred to use single names for their ponies. Mrs Griffith would add to their reputation by breeding and showing some of the very best Welsh Ponies when they took up their interest in them just before World War II following their marriage in 1937. In a tribute to Mrs Griffith following her death in 1962, J J Borthwick wrote:

> The development of the Section B was the immediate objective, and notwithstanding the limitation in sires at the time, it resulted in produce which was seldom out of the ribbons and in exports to the [United] States and South Africa.

Major Griffith, who became a member of the National Pony Society in 1934, served on its Council at the same time as Miss Brodrick – he was in great demand as a judge and was regarded as one of the very best. Both he and his wife judged at the Society's Spring Show at Islington in 1938, when he gave two firsts to the famous Tanybwlch Prancio. Soon he and his wife would fall for the charm of the Welsh pony and purchased from their friend, Miss Brodrick, two fillies by Tanybwlch Berwyn, Coed Coch Pendefiges (out of Tanybwlch Penwen) and Coed Coch Silian (out of Coed Coch Seirian).

It was at this time that their unregistered brown mare, Kinkie (f. 1939), which was very much Welsh in type, was put to Tanybwlch Berwyn and the outcome was the FS mare Gem, born in 1948. She provided the couple with their first sense of victory at the Royal Welsh, when she was Matthew Williams' (Vardra Stud) female champion in 1950 when the show was held at Abergele. Coed Coch Siabod was winning the first of a run of four male championships at the show and in 1952 he took the overall award from Major and Mrs Griffith's Tanybwlch Berwyn mare, Coed Coch Silian, whose foal at foot was the filly Verity, by Criban Victor. This filly would take the overall award at the Royal Welsh for the next two years, beating Siabod in 1953 and her sire in 1954. Mrs Griffith's run of success seemed unstoppable when Coed Coch Silian took the championship in 1955 at Haverfordwest, where she topped her son, Valiant (f. 1953), like the three others in the family by Criban Victor. The last of the full siblings, Vesta (f. 1955), was eventually sold to Lt Col Williams-Wynn at a time when he decided to build up the Section B stud again at Coed Coch after it was dispersed by Miss Brodrick in 1959. Verity and Vanity were both exported to Mrs Robert Chambers of New York in September 1955.

A great lover of the Thoroughbred, Major Griffith judged the Hunter Improvement stallions at Newmarket several times and was also a former Council member of the National Pony Society. In 1949 Major and Mrs Griffith decided to send Kinkie to the very successful small Thoroughbred stallion, Gay Presto, which stood at stud in Hampshire with Miss Jelley; using Thoroughbred or Arab on the native pony was a proven and popular recipe for breeding top-class children's riding ponies. The result was IX (f. 1950), which was successfully passed following inspection for the Foundation Stock Register. The Griffiths chose for Twinkle the Mountain pony stallions Coed Coch Samswn and Coed Coch Planed, resulting in the fillies Spangle (f. 1955), which was registered with a Kirby Cane prefix, and Pleiad (f. 1956), which was sold to Mrs Borthwick for her Trefesgob Stud on the Welsh Marches. That same year (again to Gay Presto) Gem produced Promise, shown by Miss Ferguson to stand Supreme at the Royal Show 1969 as a brood mare and the dam of Mirth, arguably one of the best riding ponies born in the 20th century. In 1959 Gem produced another daughter to Gay Presto, Prudence, which was bequeathed to Mrs Ailsa Pease to become a

PEDIGREE WELSH PONY

CRIBAN VICTOR 1775

Foaled 1944 13.0 h.h. Roan

Ministry of Agriculture Pedigree Licence No. 8634

Sire: **CRIBAN WINSTON 1705**
by **Coed Coch Glyndwr 1617**

Dam: **9138 CRIBAN WHALEBONE**
by **Mathrafal Broadcast 1502**

Bred by H. Llewellyn Richards, Esq., Brecon

Criban Victor, who is now twenty-five years of age, has been a most successful sire of children's Riding Ponies of 13·2 h.h.

A regular winner in the Show Ring, this stallion was Champion at the R.A.S.E. in 1953, and R.W.A.S. in 1956, 1958, 1959 and 1964. In 1961 he secured the Mountain and Moorland Champion at the 'Ponies of Britain' Stallion Show, standing reserve to Grey Start for the Supreme Championship. In 1962 he secured Championships at Leicester, Denbigh and Flint, and Altrincham; and in 1963 again at Denbigh and Flint. The following year he was awarded the Male Championship at The Royal Show, and was Champion Welsh Pony at the Royal Welsh. In addition, he took Championships at Liverpool, Peterborough and at the Denbigh and Flint.

1965 brought him First prizes in each of six appearances, with Championships at 'Ponies of Britain' Stallion Show, Ascot, where he secured the Merlin Cup over all other breeds and was adjudged Reserve Supreme Champion Stallion; R.A.S.E.; Denbigh and Flint; and Altrincham, and Reserve Champion to his daughter Milfyd at Liverpool.

In 1966, at twenty-two years of age, he secured the Welsh Pony Stallion Championship at Ponies of Britain Stallion Show, Ascot.

In 1968 he attended the Northern Counties Pony Association Show at Haydock Park, and at the age of 24, stood First in the Welsh Pony Stallion Class.

Since 1950 his progeny, which includes Gredington Daliad, Gredington Milfyd, Coed Coch Penpali, Trefesgob Lagus, Henbury Serena and Henbury Lucinda, have been consistent winners in the Show Ring, and many have changed hands at high prices. In South Africa, his son VALIANT out of Coed Coch Silian is the outstanding stallion in that country.

Available to a limited number of mares.

Fee: 25 Gns. **Groom's Fee: 1 Sov.**

PEDIGREE WELSH PONY

CRIBAN VICTOR 1775

Gredington Stud card 1969 featuring Criban Victor.

foundation mare at her Lemington Stud after the death of Mrs Griffith in 1962.

There had been a definite pattern emerging among the Welsh Pony breeders, who saw the financial rewards available from breeding riding ponies using their beautiful Welsh breeds as a foundation, a principle set out by the Polo Pony Society at the turn of the 20th century. Both Welsh Sections A and B fitted the bill perfectly and North Wales and the Welsh border counties became nurseries for a mixture of both pure and part-breds. The 1959 reduction sale at Coed Coch shows that Miss Brodrick followed this trend by using a variety of crossing stallions such as Jordan, Count Dorsaz, Midnight Moon, Naseel, the popular pony stallion Bubbly and the polo pony stallion Silverdale Madhatter. Her great friend Lord Kenyonwent one stage further by making available at his Gredington Stud stallions suited to all purposes, such as the riding pony Lemington Buckaroo, the part Arab Count Romeo, and always a well-bred Thoroughbred for the horse breeders.

Top-class Welsh stallions of Sections A and B were the mainstay of Lord Kenyon's stud, and it will be for Welsh Mountain ponies that the name Gredington will be indelibly linked. Lord Kenyon bought, bred and exhibited the best with the help of his stud groom, Gordon Jones, and in later years the interest of his daughter-in-law, the Hon Sally Tyrell-Kenyon, the present Lady Kenyon, maintained the momentum of the stud during Lord Kenyon's failing health. His successes at all the major shows are well recorded in the history books, particularly those of two Section A stallions bearing the Gredington prefix which led the prize lists over a 20-year period at the Royal Welsh. Simwnt, a son of the legendary mare Coed Coch Symwl (top price at the 1959 Coed Coch Sale), was overall champion there in 1973 and 1974; similarly in 1993 and 1994, Gredington Canol Lan by Revel Janus was judged champion by David Blair, who later became Editor of the WPCS 'Green' *Journal*.

'Grdington' is thought to derive from the name of an Anglo-Saxon settlement but is first recorded at the very end of the 17th century in the parish of Hanmer, near Whitchurch. Its Shropshire address belies the fact that the Gredington Estate lies within the boundary of Clwyd (formerly Flintshire), thus giving the owners a foot in both the English and Welsh camps, a feature which the 5th Baron, Lloyd Tyrell-Kenyon, embraced throughout his lifetime (1917–1993). The estate, including a house in parkland, was purchased by his family in the1670s, although a principal residence of red brick of classical style was built there by the 2nd Baron during the early part of the 19th century. This was eventually demolished to give way to a modern, two-storey brick house, built in plain Georgian style in the early 1980s, which sits at the south end of Gredington Park to the south-west of the village of Hanmer.

Many visitors to Gredington will best remember the period red-brick stables which formed a three-sided courtyard set in cobbles and the adjacent paddock with its corrugated tin shelter. Gordon Jones, son of John Jones from Coed Coch, was stud manager at Gredington for 40 years and for many represented the public face of Gredington through his showing successes round the major shows. He was a great showman – even at home, where he found great pleasure in showing off the movement of the stud's famous Mountain pony stallion, Coed Coch Planed, by rattling the side of the tin shed with his stick in order to produce a spectacular display. Jones was ably assisted by fellow employee, Dil Powell, an immaculate small wiry 'Hackney' man who could run with the best in the ring.

Lloyd Tyrell-Kenyon, the aforementioned 5th Baron, was educated at Eton and Magdalene College, Cambridge and succeeded to the title in 1927 when only ten years old. During a life which by any standards must be judged extremely busy, he held many public offices, which included a long association with the arts and museums. He was a member of the Standing Commission on Museums and Galleries and, later, the Royal Commission on Historical Manuscripts. As well as Chairman of the Friends of the National Libraries from 1962 to 1985, he was a Trustee of the National Portrait Gallery from 1953 and its Chairman from 1966 to 1988, during which time he was credited with developing it into one of the great national galleries of Britain. He was also a director of Lloyds Bank.

Locally, he was Deputy Lieutenant for his county and a Justice of the Peace. Further afield, he was President of the National Museum of Wales from 1952 to 1957 and President of the University College of North Wales at Bangor from 1947 to 1982. His public work extended into the WPCS when he became its Chairman in 1962, a position he held for almost 30 years – only ill health forced his retirement in 1991. It is generally accepted that he was the most influential Chairman of the society in the post-war years; having gained incomparable experience elsewhere, he showed clarity of mind and procedure while maintaining a firm hand on the tiller, which gained him great respect within Council and the membership.

In some ways it will be for his work in the WPCS Council that Lord Kenyon will be best remembered, but his prowess as a breeder also showed early signs of success with the astute purchase of mares from Miss Brodrick when the Gredington Stud was established in 1945. Coed Coch Seirian (f. 1937) was a crucial purchase for Section A and Coed Coch Brenhines Sheba (f. 1945) for Section B. Sheba was typical of Miss Brodrick's early breeding at Coed Coch with its Arab origins as she was by Tanybwlch Berwyn out of a mare by Craven Cyrus.

Intriguingly his mother, the Dowager Lady Kenyon, registered two mares in the WPCS Foundation Stock Register. These were the full sisters, Brenhines (f. 1950) and Cariad (f. 1952), both by the Royal Welsh champion Coed Coch Siabod out of Princess, a cream Section D mare of unknown breeding registered as Foundation Stock. Records show that they had virtually no impact on the stud. The same could not be said for the grey, Silver (f. 1938), of unknown breeding but suspiciously of Arab descent due to her grey colouring and large size. Little is known of her origin, but her name is etched on the pedigrees of some of the best ponies to emerge from Gredington, initially through the Foundation Stock Register, starting with Gredington Bronwen (f. 1946), a daughter by Tanybwlch Berwyn. Put to Criban Victor, Brownwen produced Gredington Dywenydd (f. 1949), which in turn produced the influential small-height stallion, Gredington Mynedydd (f. 1956), very much to the pattern of his Section A sire, Coed Coch Planed. This was the start of a long line of females which would provide Lord Kenyon with a valuable foundation for the Section Bs in his stud.

Four times Royal Welsh Male Champion (1950-1953) Coed Coch Siabod shown for Lord Kenyon by Gordon Jones.

It was at this early stage of the stud's development that a stallion arrived at Gredington which would forge the destiny of the Welsh Section B, not only in Wales but throughout Britain and indeed the rest of the world. The purchase of Criban Victor as a three-year-old from Mrs Cuff after his win at the Royal Welsh in 1947 must be regarded as one of Lord Kenyon's most astute purchases, as he became popular with breeders and crossed perfectly with the foundation mares at home. His remarkable show ring success and breeding was covered in Chapter 6; the fact that he remained at Gredington until his death in 1973 speaks volumes for the regard in which he was held in the stud. Added to this, in an unprecedented move in the WPCS, Lord Kenyon presented to the society Victor's head, which had been stuffed and mounted, for the wall of the new headquarters at Chalybeate Street in Aberystwyth. Furthermore, Victor was selected to represent the Welsh breeds on an official commemorative stamp issued by the Post Office in 1978 to celebrate British equine breeds.

In Criban Victor's obituary in the 1974 WPCS *Journal*, Gordon Jones observed, 'It was a pity that he wasn't used more in his younger days as I am sure that if he had been there would be much more true to type about today.'

To some extent this reflected what many considered to be an 'old-fashioned' viewpoint, which originates in the shepherding pony and does not reflect the modern trend in the breeding of the Welsh Pony of Riding Type. However I am sure that Jones would have strongly argued that Victor was of riding type, and proved this by winning under saddle. He had excellent conformation, with plenty of bone and substance standing on strong limbs – he had a riding shoulder and the best of quarters. Admittedly his head was a bit plain and strong, so he lacked prettiness and to some, such as Mrs Cuff, refinement; in addition, his dominant dark bay roan colour, while considered to be an old 'hard' colour of the Welsh breed, was fairly unpopular then, as it remains to this day. He very much appealed to those with a hunting background, as they could see his value as a sire of good, sound stock fit for work; in today's world he would be considered an ideal sire of working hunter ponies.

The first foals registered by Lord Kenyon at Gredington appeared in 1949, with the start of Victor's successful mating with Silver in the shape of the dark bay filly Gredington Daliad, a champion at the Royal Welsh in 1956 and then in 1957 when she won the overall title. The Griffith-bred youngsters, Valiant and Verity, both by Victor, took championship awards in 1953, 1954 and 1955, which gave the Criban-bred stallion a remarkable run of success during his introduction as a sire. This reputation increased when Gredington Milfyd, a full sister to Daliad by Victor out of Silver, took the championship as a two-year-old in 1958, followed by further championships at the Royal Welsh in 1962 and 1963, by now a mare. It was particularly good luck that her breeder refused 475 guineas for her in 1961 when she remained unsold at the Gredington sale.

Milfyd was a remarkable producer for her owner and will go down in the history of the breed as exceptional within the bigger picture of its development. She was first line covered by her sire to produce the filly Saffrwn (f. 1961), who in turn to Coed Coch Berwynfa would produce Gredington Blodyn (f. 1964), female champion at the Royal Welsh in 1971 in the ownership of Ann Bale-Williams, for whom she became a successful foundation

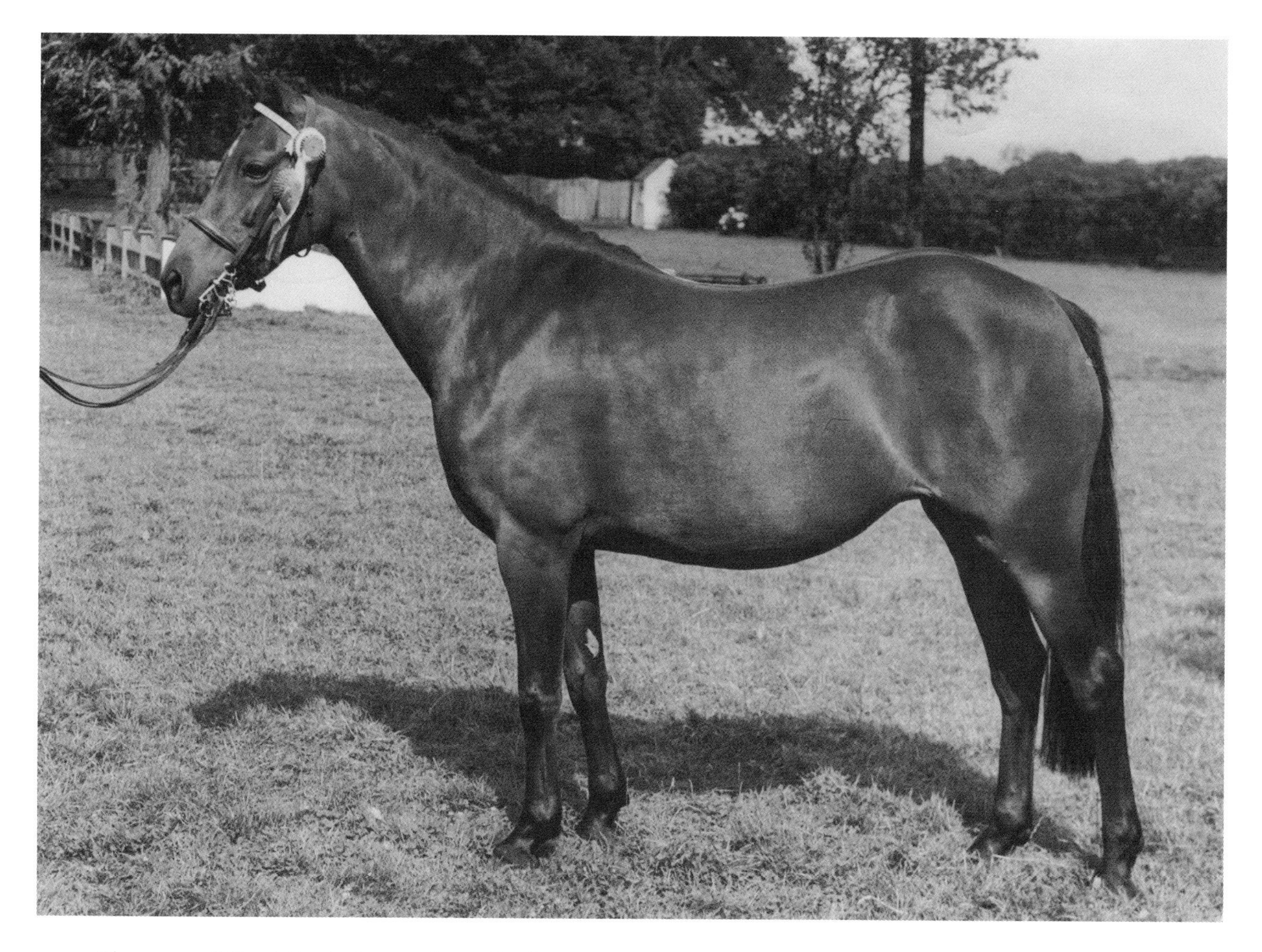

The 1957 Royal Welsh Champion Section B, Gredington Daliad.

mare. The cross of Coed Coch Berwynfa on Gredington Milfyd (f. 1956) proved to be a clever move, as Lord Kenyon doubled up the Tanybwlch Berwyn blood, both top and bottom sides of the pedigree. This was repeated several times over with great effect – Gredington Tiwlip (f. 1962) would become a major success story for the Rotherwood Stud (of which more later), while Gredington Beryn (f. 1964) became the dam of one of the stud's best show mares, Gredington Lily (f. 1972), a winner under saddle as well as in hand. Carolyn Bachman secured Lily for her Carolinas Stud in Hampshire when she bought her for the top price for Section Bs of 1,200 guineas at the Fayre Oaks Sale in 1985, when the Gredington ponies were finally dispersed.

Above: Gredington Milfyd in the stable yard at Gredington with Gordon Jones.

Opposite: the only known photograph of Silver, the foundation mare at Gredington, with her foal Gredington Bronwen in 1946.

Lily was by Bowdell Quiver (f. 1966), which Lord Kenyon purchased at the Fayre Oaks Sale in 1970 for 750 guineas. Quiver introduced a strong flavour of Downland to Gredington, as he was one of the few sons of Downland Dauphin out of Kirby Cane Goosefeather (f. 1961), whose own dam, Kirby Cane Grig (f. 1957), was a daughter of Coed Coch Berwynfa out of Mrs Cuff's old mare, Downland Grasshopper. Chapter VII showed how Grasshopper to Star Supreme produced Downland Gay Star, the sire of Mrs Cuff's prolific winner and breeder Downland Love-in-the-Mist. By contrast, Lily's dam was based on the traditional North Wales lines, as she was by Trefesgob Benedict (f. 1957 Criban Victor x Berenice f. 1948), a daughter of Tanybwlch Berwyn out of Gatley Stardust (f. 1923), bred like her dam by Mrs Monica Darby, a well-respected breeder and judge in the WPCS who came from Gatley Park, Herefordshire.

Gatley Stardust was by Stanage Planet out of Stanage Meteor, both full of Dyoll Starlight breeding and bred by Charles Coltman Rogers, one of the founder members of the WPCS who lived at Knighton in Radnorshire; Rogers was first Vice Chairman then Chairman of the society over its first 26 years. Stardust was a good winner under saddle for Lord Digby's daughter, Jaquetta, who competed against the famous Tanybwlch Prancio at Islington. Later Stardust went to Coed Coch, where she bred beautiful ponies including Coed Coch Sydyn (f. 1942), Sensigl (f. 1943) and Syndod (f. 1947), all by Tanybwlch Berwyn. It is worthy of note that Sensigl stood 11.3 hh while her full sister, Berenice, was much bigger at 13.1½ hh.

Berenice bred a host of good colts for Mrs Mari Borthwick, President of the WPCS in 1968, whose husband, J J Borthwick, was Chairman of the WPCS prior to Lord Kenyon from 1932 to 1962. They lived at The Foxholes, a small property on the Welsh border near Bishop's Castle in Shropshire. Since 2007 it has become a camping site which enjoys the beautiful countryside of the surrounding area called the Welsh Marches. The Foxholes is 20 miles east of the Llandinam Estate, where J J Borthwick was agent for the Davies family, whose fortune was amassed from very modest beginnings by David Davies, a railway builder and owner of the Ocean Coal Company based at the Cwmparc Mine at Treorchy in the Rhondda Valley on land that had been leased from the Crawshays. Knighted for his efforts, David Davies built Broneiron on the estate which lies on the A470 from Newtown in Powys to the pretty market town of Rhayader. The house provided a temporary home to Gordonstoun, the celebrated Scottish independent school, when it was evacuated there during World War II, and has since become the training centre of Girlguiding Cymru, the Girl Guide movement of Wales.

It was while Borthwick was agent for the estate that he must have come in contact with Major William Marshall

Rhyd-y-Felin Selwyn (f. 1956), bred by Cyril Lewis.

Dugdale as his daughter, Ruth Eldrydd, had married the second Lord Davies, who was killed in 1944, leaving the estate and title to his three-year-old son. Dugdale, whose family was among the biggest breeders of Welsh ponies and cobs which carried the Llwyn prefix, was President of the WPCS from 1924 to 1925 and Vice Chairman of Council for ten years. It was the Dugdale connection that influenced both Borthwick to take up office in the Society and his wife to take up the breeding and showing of Welsh ponies.

Mrs Borthwick was popular in Welsh pony circles and enjoyed showing her ponies just as much as breeding them – she was a regular exhibitor on the show scene but never travelled too far afield. She specialised in the Welsh Pony of Riding Type section, which was gradually enjoying popularity in show schedules. Probably her most successful wins came at the Royal Welsh with Norwood Starlet (f. 1955) – in 1959 Mrs

Mrs Borthwick showing her homebred colt Trefesgob Lagus, later to become a big winner under saddle for Gillian Blakeway.

Valiant, Mrs Borthwick's 1955 Male Champion at the Royal Welsh.

Pennell selected her to stand reserve champion to Criban Victor and in 1961 E S Davies made her reserve to the emerging star, Solway Master Bronze. Starlet was bred in Northumberland by Mrs Lizanne Bates and, along with another of the foundation mares, Wyrhale Fairy Tale, was part of the Trefesgob dispersal sale at Fayre Oaks in 1965, following the death of J J Borthwick in 1965, and like many of the consignment he was sold to a Mr Raymaker from the Netherlands.

Mrs Borthwick made shrewd choices in her breeding policy as, like Mrs Griffith, she kept only a few mares. Initially she bred Welsh Mountain ponies, selecting mares such as Craven Fairylight and Craven Sprightly Twilight, which she had purchased with a foal at foot by Coed Coch Sidi from Mrs Darby in 1948. Fortunately for Mrs Cuff (and arguably for the Welsh Section B) Twilight later became her property, not only winning under saddle but also breeding her Love-in-the-Mist. It was obvious that Mrs Borthwick favoured the more riding type of Mountain pony, but the grazing was too good for them at The Foxholes and her breeding policy soon turned to Section B.

The choice of mares could not have been better and she turned once more to Mrs Darby, who sold her the homebred Tanybwlch Berwyn mare Berenice, born in 1948. Unfortunately for her new owner, she proved to be a colt breeder, although the colts she bred were of the highest calibre – by Criban Victor, they included Trefesgob Lagus (f. 1955), Reuben (f. 1956), Benedict (f. 1957) and Hector (f. 1961). Hector was used in the stud before sale, Benedict went to Mrs Crisp's Kirby Cane Stud and Lagus became one of the most successful ridden ponies in Britain for the Blakeway family. Ridden by Gillian Blakeway, he was part of the British Show Pony Society Team that successfully competed against the Americans at the British Timken Show in 1961, and was consequently sold to the United States. Mrs Borthwick found an outcross full of the breeding she liked named Cusop Sheriff (f. 1959), whose dam was Coed Coch Brenhines Sheba by Tanybwlch Berwyn, while his sire was a grandson of Criban Victor, Cusop Call Boy (f. 1956) who, once gelded, made a big name for himself as a 12.2 hands child's riding pony. Sheriff was a big winner in hand before his sale to the GlanNant Stud in the United States, but before this Pleiad produced yet another colt by him, Trefesgob Septimus (f. 1962), which was exported to Australia. Mrs Binnie secured Berenice for her Brockwell Stud in 1965 at the Fayre Oaks.

It was at the dispersal that Trefesgob Plush (f. 1960) was bought – at 700 guineas, the top price for a Section B that day – by Sir Nigel Colman (of the mustard dynasty) for his wife to drive; their Nork Stud in Surrey was already world-renowned for Hackneys. Plush's dam, Pleiad, bred by Mrs Griffiths at Plasnewydd, was one of the foundation mares for Mrs Borthwick and did the same for Mrs Hutt at her Lydstep Stud in West Wales. Another foundation mare was Coed Coch Pws (f. 1950), purchased by Emrys Griffiths (Revel Stud) on Mrs Borthwick's behalf for 40 guineas in the 1952 Coed Coch Sale. A daughter of Gwyndy Limelight and Coed Coch Pilia (f. 1945 Tanybwlch Berwyn x Tanybwlch Penwen), she and Rhyd-y-Felin Seren Wyb (f. 1949 by Tanybwlch Berwyn) made valuable contributions to the stud. Next to Miss Brodrick herself, Mrs Borthwick was the principal breeder who maintained the link with Sahara through his son Tanybwlch Berwyn in Section B.

Rhyd-y-Felin Seren Wyb was also offered for sale at Fayre Oaks, but avoided export when knocked down to Mrs Alison Mountain for her Twyford Stud in Sussex. The 16-year-old mare came from one of the most famous families registered in the Welsh Stud Book, although better known for their successes in Section A through her dam, Coed Coch Sirius. Incredibly this bay filly, Sirius, while only a few months old, passed through three sales within a few weeks before settling with Cyril Lewis, who lived in Gwynedd. She was firstly offered on the 1937 Coed Coch Sale, where she was one of those purchased by Mr Walton, who, due to a lack of grazing, was forced to re-offer her for sale two weeks later, when she was purchased by Thomas John Jones of Dinarth Hall fame. His sudden death two weeks later forced her sale yet again and this is when she was bought by Edward Lloyd Lewis for his son, Cyril; she remained with his family for the rest of her life, establishing a dynasty of ponies carrying their Rhyd-y-Felin prefix.

Miss Brodrick bought two of her foals by Coed Coch Glyndwr and registered them as Coed Coch Serog (f. 1941) and Siaradus (f. 1942); both mares became an enormous success for their new owner, although Serog was exported to Dr Elizabeth Iliff of Maryland, United States, in 1948. Meanwhile, Sirius was also bred to the Section B stallions at Coed Coch, which complemented her scopy conformation and Starlight breeding. To Tanybwlch Berwyn she produced first the filly, Rhyd-y-Felin Seren (f. 1947), a successful breeder at home and then for Lady Myddelton.

The next foal by Berwyn was Seren-y-Boreu, a colt born in 1948, while the following year Mrs Borthwick's Seren Wyb arrived as the second of her Berwyn fillies.

Coed Coch Blaen Lleuad was the obvious choice for Seren Wyb, and Rhyd-y-Felin Selwyn was foaled in 1956. He was elected by Miss Brodrick for the Coed Coch mares but sold, as many others were, in her 1959 reduction sale when he went to Sweden for 320 guineas; much later he was brought back to Britain by Mrs Mountain, who used him before finding him a retirement home on the Isle of Wight.

The Trefesgob dispersal at Fayre Oaks provided breeders an opportunity to add valuable bloodlines to their existing stock, something that has been a tradition within the Welsh breeds and still exists to this day. Looking back at the early sales, we can see that some studs were more or less established on the back of sale purchases, and none more importantly than those ponies, like Rhyd-y-Felin Selwyn, secured in 1959 when Miss Brodrick decided to significantly reduce her Welsh riding ponies. The product of her efforts would soon be spread even more widely around Britain and, fortunately, other studs such as Downland were able to provide outcrosses for the Coed Coch Berwynfa and Criban Victor bloodlines which now dominated.

Downland Chevalier (far right) with his mares at Llangoedmor

Chapter IX

Brockwell, Nefydd, Kirby Cane, Chirk (Lady Myddelton), Hever

Lady Astor's champion stallion Hever Quiver.

Wynne Davies

Catalogue 1/6

CATALOGUE

OF

A Portion of the Coed Coch Stud

Bought 58
209
221.

bought later
maroon lot 212
dam of Minnet
(Downland Boy)

213
Dyrin Jennifer
bought as foal
Calypso 210

Champion Coed Coch Madog 1981.

TO BE SOLD BY AUCTION ON

Saturday, September 26, 1959

Printed by J. H. Williams, Visitor Office, Abergele. Tel. 2230.

1959 has been mentioned more the once in this book as a year which will go down in the annals of the Welsh as a 'turning point in the fortunes of the Welsh Pony Section B', as Dr Wynne Davies put it. Of course he was referring to the birth of four highly-significant stallions registered within Section B (Solway Master Bronze, Downland Dauphin, Brockwell Cobweb and Chirk Crogan), all upgraded from Foundation Stock. Few would argue with his statement; however, accompanying the birth of these stallions was a sale held in 1959 which would guide the development of the Section B not only to new pastures but to new heights in its development.

It was held on Saturday 26 September at the Coed Coch Home Farm at Abergele and the catalogue, priced at 1s 6d, had as its title, 'A Portion of the Coed Coch Stud'. The Foreword read:

> This is a genuine reduction sale owing to the fact that Miss Brodrick has been advised, for reasons of health, to curtail her activities for a time. She has, very reluctantly, decided to sell her Section 'B' ponies, riding type (height limit 13.2 [hands]) herd, with the exception of a stallion (Coed Coch Berwynfa) and three mares, and to concentrate in future on Section 'A' Welsh Mountain Ponies (height limit 12 hands).
>
> Do not miss this rare opportunity of acquiring ponies of a class that are not to be found on the open market.

Opposite: Dr Davies' catalogue of the 1959 Coed Coch Sale , at which several ponies were purchased to enhance Section B breeding at other prominent studs.

As expected, the sale attracted a great deal of interest from both home and abroad and many of the leading breeders were present to stake their claim on some of the best bloodlines to be found in the Welsh Stud Book at that time. While focusing on the Welsh Riding ponies Section B, it is worth noting that there was keen interest in the Mountain ponies, which found new homes at major studs after selling for big prices for the time. Lord Kenyon famously bought his stud groom's favourite, Coed Coch Symwl, for 1150 guineas, while Mrs Mountain added to her Twyford herd when she paid 900 guineas for Coed Coch Sws; another top price was 925 guineas, paid for Coed Coch Pwysi, which sold to Canada. It is also worth noting that Mrs Hope purchased for 220 guineas Coed Coch Gold Mair, a chestnut Section A filly foal by Coed Coch Madog out of Cefn Graceful by Ceulan Revelry.

When subsequently in the ownership of Mrs Cuff, she bred the well-known stallion, Downland Mandarin.

There were four ladies present at the sale that day who knew exactly which ponies they wanted for their Section B studs and cleverly arranged not to bid against each other in order to keep the prices at an affordable level. History shows that they chose wisely, and each of their purchases would feature in pedigrees of the champions they would breed in the years to come.

Unlike many of the early breeders whose experience was firmly based in the hunting field, Barbara Binnie came from a completely non-horsey family – in fact her father had a dislike for horses and refused to allow her to have one. With her bedroom walls plastered with posters and photographs of ponies and a bookshelf groaning with all sorts of pony books, it was fairly obvious that his daughter's interest was serious and not to be easily altered. It was only after saving every last penny that Barbara managed to accumulate enough money to buy a pony, much to her father's disapproval – although over time he grew to like the little Welsh pony that had joined the family. Their farm at Duddington was not far from Oundle in Northamptonshire, where Llewellyn Richards had taken a neighbouring farm and, much to Barbara's delight, brought with him some of his Criban ponies.

Married in 1937, her husband Anthony was a naval officer who played polo for the Navy, raced and point-to-pointed; their only son Christopher had an interest in ponies. Following a posting to Bath, where Binnie worked for the Admiralty, the family moved to West Kington, where Mrs Binnie and her son went everywhere in a pony and governess cart. In 1945 they bought the small dairy unit of Brockwell Farm near Minehead in Somerset; however, Anthony sadly died in 1947 only two months after retiring, leaving his wife to make a living from farming. Mrs Binnie chose breeding of ponies as a special interest, one she maintained until her own death in 1998.

Luckily for the breed, Criban Victor was purchased by Lord Kenyon from Mrs Cuff, who outbid Mrs Binnie at the Criban Sale, thus avoiding castration by both ladies, who otherwise would have gelded Victor for riding. Mrs Binnie had a strong preference for the Criban ponies, which she felt made perfect ponies for children, and managed to purchase for £40 Criban Ester as a mount for her son, who rode the bay mare along with a Part-bred, Criban Brecon. Typical of the ponies bred by Richards at the time, Ester was sired by a small Thoroughbred, Golden Cross, but later passed for the Foundation Register; she became successful at Brockwell before her sale at Fayre Oaks in 1961 for 445 guineas to Major and Mrs Hedley, who used her as an outcross for their renowned Arab stallions. The foal she carried by Brockwell Cobweb was Briery Starlet, later to become a foundation mare at the Millcroft Stud in Devon. The first pony to carry Mrs Binnie's Brockwell prefix, in 1955, was the Foundation Stock mare, Brockwell Butterfly, by an unregistered stallion out of Criban Briar, a daughter of Criban Loyalist (the Welsh Part-bred by Silverdale Loyalty) and Criban Bramble by Mathrafal Broadcast.

The Fayre Oaks Sale at Hereford became a familiar stamping ground for Mrs Binnie, firstly to buy mares for her stud and later, like many breeders of her era, as an outlet for the ponies she bred. Her first foray to the inaugural sale held there in 1957 saw her purchase the only Section B pony offered for sale, which she secured

Mrs Binnie's champion stallion Brockwell Cobweb.

for 40 guineas; this was the yearling filly, Fayre Ladybird, which, although registered within the Foundation Stock Register, bred foals eligible for full registration as she was FS2. Fate was on Mrs Binnie's side when Ladybird produced, as a three-year-old, a colt named Brockwell Cobweb in 1959. Without question one of the leading sires born that year, Cobweb proved to be a major asset to the breed in years to come – his siblings Brockwell Spider and Moonlight went to America and the Netherlands respectively.

Brockwell Cobweb was a great show stallion and a great stock-getter, although his unruly behaviour as a yearling almost led to castration had it not been for the advice of Mrs Pennell, who saw him at a show prior to his operation – which had been booked for the following

Harford Starlight, bred by Lady Wentworth, was later purchased in Wales for £25 by Mrs Binnie.

week. He won championships throughout his life; among his many wins at all the major shows, his male championship at the Royal Welsh in 1966 under Lady Margaret Myddelton was rated by his owner as one of the most memorable. Although Mrs Binnie did not win the overall title that day, she was compensated by the fact that she had taken it the year before with a yearling filly, Belvoir Tosca (f. 1964 by Belvoir Talisman), purchased in utero when Mrs Binnie secured her dam, Belvoir Tangerine, in 1963. Tangerine and her daughters went on to produce the 'T' line at Brockwell.

Two of Tangerine's granddaughters, Trinket (f. 1968) and Twiggy (f. 1969), both by Brockwell Cobweb, were purchased as foals by Christine Jones as future mares for her Bunbury Stud in Cheshire. She and Mrs Binnie became great friends, and other fillies followed such as Brockwell Puffin (f. 1969) and Muslin (f. 1971); Brockwell Briar (f. 1971) and Jemima (f. 1975) were given to Mrs Jones by her friend when she retired from breeding ponies. All the mares proved to be good breeders at Bunbury, with the Brockwell Cobweb daughter, Muslin, put to Downland Beechwood, producing the colt Bunbury Mahogany (f. 1979), which took the male championship at the Royal Welsh in 1984 and 1986.

Brockwell Cobweb's sire was Harford Starlight (f. 1944), an overgrown Welsh Mountain pony, which Mrs Binnie had found in a pigsty on a farm in the Elan

Brockwell Joannie How, named after a hill near her breeder's farm.

Valley, where he was being used as a shepherding pony. Although poor in condition he proved to be a bargain at just £25. He had been recommended to her by Theron Wilding-Davies, the breeder of Cobweb's dam, Fayre Ladybird. On paper Starlight was beautifully bred, with the best of Crabbet Mountain Pony breeding. However, Mrs Binnie herself cast doubt on his pedigree when she wrote of him in an article in the 1992 'Gold Journal' of the Scottish and Northern Welsh Pony and Cob Association, 'He [Harford Starlight] was bred by Lady Wentworth and I believe (although registered Welsh) his parentage was in some doubt – he was a super mover and charming character, both characteristics he passed on to his children and grandchildren).'

Irrespective of her doubts, Harford Starlight proved a perfect cross for the Foundation mares that were gathered at her Brockwell Stud. To Brockwell Rainbow, an FS mare of unknown breeding, he produced the highly successful show mare, Brockwell Misty Morning, and to Criban Ester, Brockwell Joannie How (named after a hill overlooking the farm), the start of the famous 'J' line at the stud which included her daughter by Cobweb, Japonica, sold to the Weston Stud as a yearling. Coincidentally, Japonica's half-sister by Downland Beechwood, Jemima, was put to Brockwell Prince Charming and produced the chestnut colt Jaguar, which helped form the Stoak Stud, at a time when the Johnston family took over their new property at Llangollen, formerly home of the

Weston Stud. Harford Starlight died of a twisted gut, but his daughters would provide Mrs Binnie with a perfect cross for her beloved Cobweb, thus providing her with a foundation for her stud as other stallions were brought in for subsequent generations. These included Downland Beechwood, Senlac Choirmaster, Mistral Starling, Bunbury Thyme and Gunthwaite Fleur de Clocke.

Just as Brockwell Cobweb's arrival in 1959 had proved a major event for the Stud, so too did the arrival of two mares purchased at the Coed Coch dispersal sale. Mrs Binnie's first purchase at the sale was the 13-year-old brood mare Coed Coch Pendefiges, another by Tanybwlch Berwyn but out of Tanybwlch Penwen by Cairo; registered as an FS1, she was also registered in the Part-bred section of the Arab Horse Society Stud Book, as indeed she had a great deal of Arab breeding in her pedigree. Pendefiges was previously owned by Mrs and Mrs E Griffith and won at the Royal of England with them in 1949 and 1950. She bred a succession of colts at Coed Coch including her foal at foot, Padog (by Coed Coch Blaen Lleuad), which sold on the day for 22 guineas; Padog later became a very famous pony under saddle for several families, including the Daffurns from Evesham. For Mrs Binnie she bred several fillies, the most important of which was the one she was carrying by Rhyd-y-Felin Selwyn, Brockwell Penguin (f. 1960). She proved to be an excellent cross for Cobweb, whose 1968 filly Brockwell Puss was one of Mrs Binnie's all-time favourites. Among her most successful progeny was the stallion Brockwell Prince Charming by Keston Royal Occasion – he stood at stud at Rotherwood and later at Millcroft, where his dam had already found a new home following a successful time at Brockwell.

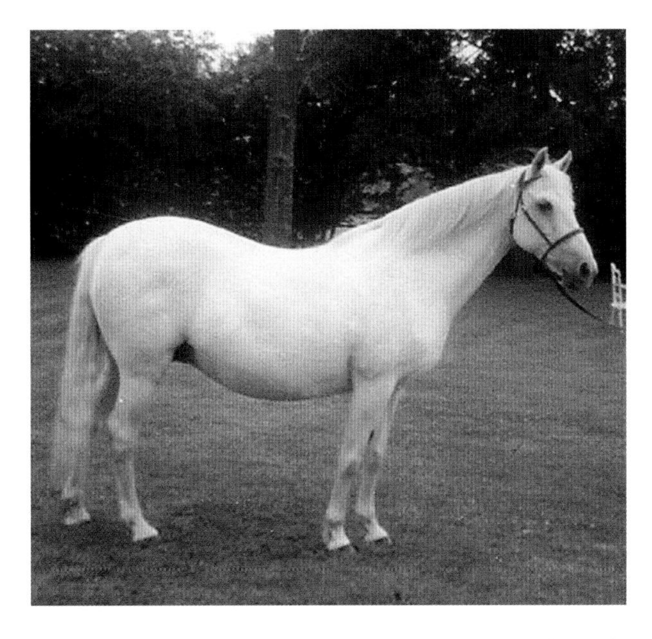

Coed Coch Bugeiles sold for 210 guineas to Vivian Eckley at the 1959 Sale and later became a foundation mare at Rotherwood Stud.

The second purchase, for 170 guineas, was the senior mare at Coed Coch, Berwyn Beauty, the 17-year-old daughter of Tanybwlch Berwyn x Dinarth Wonderlight. Covered by the three-year-old colt Rhyd-y-Felin Selwyn (Coed Coch Blaen Lleuad x Rhyd-y-Felin Seren Wyb by Tanybwlch Berwyn), her son Brockwell Berwyn (f. 1960) was full of Tanybwlch Berwyn blood. Rhyd-y-Felin Selwyn was sold that day to Sweden for 320 guineas. In 1964 Beauty produced a top-class filly by Cobweb named Brockwell Beauty Queen, who was the dam of the 1972 Royal Welsh Champion, Mrs Margaret Williams' chestnut five-year-old stallion Brockwell Chuckle (f. 1967) by Lydstep Ronald. Mrs M C Williams

Mrs Margaret Williams with her influential stallion Nefydd Autumn's Chuckle.

Nefydd Antur winning the Section B championship at Glanusk 2008.

had bought him in 1971 at Fayre Oaks, a purchase that would prove to have the greatest influence on her Nefydd Stud situated in Clwyd in North Wales.

Mrs Williams grew up in Birkenhead near Liverpool and had always ridden as a child around the parks. Her family was evacuated during World War II to Cilcain in Flintshire, where the only mode of transport was horse or bicycle. She developed a great love for horses and went on to be a land girl, working on the neighbouring farm to her future husband, Bill, who travelled Shire horses in the district. They married in 1949 and in 1951 bought a mountain farm on the Denbigh Moors, with a little black pony called Polly thrown in for luck; with the help of Mrs Griffith (of Plasnewydd) she was accepted as Foundation Stock in the Welsh Stud Book. Polly was trotted behind the Austin 7 the ten miles from the farm to Coed Coch for six seasons, and a foal was produced every year, three by Coed Coch Proffwyd and three by Coed Coch Blaen Lleuad. To the latter she produced a bay filly, Stella (f. 1959), which was sold to Lady Margaret Myddelton as a lead-rein pony for her grandchildren. The next year she produced another filly to Blaen Lleuad, Sherry, which became a good child's pony before joining the stud – through Sherry, Polly's bloodline has persisted in the stud for six generations.

As the stud grew, mares were added, including Downland Autumn (f. 1962 Downland Dauphin x

Downland Springtime), purchased for 200 guineas at Bangor-on-Dee in 1965 from Peter Ward. Autumn proved to be an ideal mate for Brockwell Chuckle, producing among others Nefydd Autumn's Chuckle, foaled 1973 – the same year that his sire was champion at Glanusk, winning both in hand and under saddle. Autumn's Chuckle was very successful in the show ring and was later sold to Israel. Two of his winning progeny include the Royal Welsh female champion, Talhaern Eirlys y Pasq, and Hafodyresgob Buzby, sire of the champion stallion Eyarth Rio. Another stallion by Brockwell Chuckle, Nefydd Super Sir (f. 1992), and Super Sir's son, Nefydd Buccaneer (f. 1996), have been the champion dressage ponies of Spain ridden by children. Buccaneer has won six times and performed at the World Equestrian Games held in Jerez. Another of Brockwell Chuckle's daughters was Eyarth Antonia, which has been a major influence in the stud; her son by Eyarth Tayma, Nefydd Antur (f. 2003), has proved to be successful under saddle and in hand. In 2008 he was champion at Glanusk, like his grand sire 34 years earlier; a filly from the same line, Nefydd Awel Ebrill by Rhoson Adonis, was also in the ribbons at Glanusk in 2012. The stud is now run by Margaret Edwards (née Williams).

As we have seen, Mrs Binnie's purchases at the Coed Coch Sale in 1959 had far-reaching effects on many studs, and so did those of others that day. The success of Pendefiges for Mrs Binnie was replicated by her full sisters, Coed Coch Pilia and Coed Coch Pluen, which were also offered for sale at Abergele in 1959. Pluen was the most expensive of the three when she sold for 270 guineas (second-highest price for a Section B female at the sale) to Mrs Crisp (neé Gooch) for her Kirby Cane Stud in Norfolk. Catalogued at a height of 13.2½ hands, at twelve years old she was classically Coed Coch bred and offered her purchaser the height she required for her stud, which aimed to produce a top-class riding pony of 13.2.

As early as 1964, Mrs Crisp defined the purpose of her breeding in an article in the WPCS *Journal*:

> My aim is to breed a 13.2 Child's Riding Pony, who can take its place in the 13.2 Ridden Classes and yet retain its Welsh characteristics. It obviously must be a finer type than some of the present Section B. I feel the demand is for this pony, and for those who say, why then bother to breed pure? the answer (anyway mine) is because there is no pedigree breed of pony of this height, other than Welsh Section B.

When Barbara Alexandra Gooch married Raymond John Steffe Crisp in April 1936, two of the most prominent families of East Anglia came together. Her father, Sir Thomas Vere Sherlock Gooch of Benacre Hall, Suffolk, was a Justice of the Peace and High Sheriff of Suffolk in 1911. The Benacre Estate has been in the Gooch Family for almost 300 years and extends to some 7,000 acres. Her father-in-law, John Robert Crisp of Kirby Cane Hall, was also a Justice of the Peace for both Suffolk and Norfolk, actively engaged in the local community of Beccles and head of the firm of John Crisp & Sons, owners of one of the largest maltings in the country. Although now out of their ownership, the family still lends its name to the Crisp Malting Group, the leading privately-owned maltster in the UK, operating in both England and Scotland. Together they were great hunting

Kirby Cane Shuttlecock. This photograph shows the influence of his Mountain Pony breeding.

families, with expansive estates on low-lying fertile land. Kirby Cane Hall was a large Georgian-style house dating from the 17th century situated about fourteen miles from Norwich between the villages of Beccles and Bungay, and would provide Mrs Crisp a distinctive prefix for her stud of Welsh Ponies.

Having been brought up with hunting ponies as a child and following the tradition with her own children, Mrs Crisp had no intention of starting a stud, but a visit to the 1948 Royal Show at Shrewsbury would change all that. Having spent two days immersed in non-horsey activities, she took herself away from her party of friends to watch pony classes on the third day of the show, whereupon she totally fell in love with the Welsh ponies on show. Her catalogue revealed that they were mostly bred by Miss Brodrick and sired by Tanybwlch Berwyn, so she immediately tracked down Miss Brodrick, who was a steward in the collecting ring of the Children's Riding Pony classes. Never one to miss out on a sales opportunity, Miss Brodrick's invitation to stay at Coed Coch and view the ponies was quickly accepted and

Kirby Cane Generous.

Mary Reveley's Royal Highland Show Champion, Kirby Cane Vogue.

unwittingly the first steps towards forming the Kirby Cane Stud had been taken.

The trip turned into a shopping foray, with two Welsh Mountain fillies and a mare, Kilhendre Celtic Greylight II (f. 1946) with foal at foot, destined for Norfolk as a foundation for the stud which initially set out to breed Section A. Miss Brodrick recommended that a Section B colt would be a good choice and sold Mrs Crisp a young colt bred by Lady Wentworth, Wentworth Penglyn, whose dam, Tanybwlch Penllyn (Tanybwlch Berwyn out of Tanybwlch Penwen) was registered as Foundation Stock. Although he was registered in the Part I of the Welsh Riding Pony section of the Stud Book, he was born at a time when Section B stallions were very low in number and the WPCS Council was considering the entry of a select number of colts, otherwise ineligible for full registration, to enter the Stud Book after inspection. All was going well until 1949, when the Council had a change of heart, thus leaving Penglyn offspring ineligible for full registration. Miss Brodrick agreed to take him back, but before he returned to Coed Coch he covered a few visiting mares including a Thoroughbred, Myrra, owned by Mrs Yeomans, a well-known personality in equine circles at the time. Their filly was duly entered in the Foundation Stock Register as Kirby Cane Vaudeville (f. 1954), which in turn produced the beautiful mare, Kirby Cane Vogue (f. 1962) by Kirby Cane Shuttlecock (f. 1954). Having been sold as a five-year-old at the Fayre Oaks Sale in 1967 for 480 guineas, Vogue went on to a highly-successful showing career for Mary Reveley, who bred Welsh ponies long before she became a famous racehorse trainer.

Desperately in need of a stallion and besotted with Coed Coch breeding, Mrs Crisp returned to Miss Brodrick, who had nothing available, so resorted to searching for the breeding she liked from other breeders. The Royal Show catalogue revealed that Mrs Chadwick from Dwyran, Angelsey, had just what she wanted and her homebred Bryntirion Rowan (f. 1947) duly joined the Kirby Cane Stud – much to the delight of his new owner, who had bought him blind. As it happens, she could not have chosen better, as he had the best of Section A bloodlines, being by the great Coed Coch Glyndwr out of Coed Coch Pansi (f. 1942 Coed Coch Glyndwr x Tanybwlch Prancio). He particularly impressed Mrs Crisp as he was similarly bred to Coed Coch Powys, which she had seen and loved in the riding pony classes. Although Mrs Crisp would have preferred to purchase a Section B stallion, Rowan proved to be an outstanding sire and a cornerstone of the stud's success in the years to come, as he provided a perfect outcross for the Tanybwlch Berwyn breeding that would soon be introduced.

The first foals bearing the Kirby Cane prefix arrived in 1951, although it was surely inconceivable that Rowan on Celtic Greylight II would provide Mrs Crisp with one of her proudest moments in her career as a breeder. Their daughter, Kirby Cane Greensleeves (f. 1951), the very first filly born at the stud, was chosen as a riding pony for HRH Prince Charles. With a double cross of Bleddfa Shooting Star on both sides of her pedigree, it was little wonder that she bred top-class riding ponies, although interestingly it would be her cross with Downland Chevalier some 15 years later that would provide Mrs Crisp with one of her most famous mares, Kirby Cane Generous (f. 1966). Rowan also proved an excellent choice for the 13 hands Section B mare Coed Coch Seron (f. 1947 Tanybwlch Berwyn x Coed Coch Serog by CC Glyndwr), which had generously been exchanged by Miss Brodrick for Wentworth Penglyn, who returned to Coed Coch. Without doubt, Seron proved to be the outstanding matron among the foundation mares, providing her new owner with a wide selection of colts and fillies over the years to a selection of stallions, which would take the Kirby Cane prefix to studs throughout Britain and continental Europe. This included Kirby Cane Shuttlecock (f. 954) and Kirby Cane Scholar (f. 1965 by Downland Drummer Boy), male champion at the Royal Welsh in 1969 for Miss Wheatcroft.

The breeding of Mountain ponies at Kirby Cane would end almost as soon as it began due to the unsuitability of the rich land, which seemed to make the ponies grow over height or dangerously close to laminitis much of the time. Nevertheless, the foundation mares chosen would prove their weight in gold as they enhanced the bloodlines in the stud through a large number of sons and daughters and their offspring. Others of note included Mostyn Partridge (f. 1939), a mare brimming over with Starlight breeding through her sire, Caerberis Kindle (f. 1929), and dam, Grove Ladybird (f. 1921), many times a champion and sold at the 1927 Grove dispersal sale for 50 guineas to Lord Mostyn. Partridge produced a wonderful line of ponies bearing bird names for the stud and, within two generations, the influential sire Kirby Cane Pilgrim (f. 1966) by Kirby Cane Plunder (f. 1963). Another notable pony was Downland Grasshopper (a daughter of the Cuff children's Craven Good Friday), which was carrying Kirby Cane Gopher (f. 1956) to Dauphin's

HM The Queen with Princess Anne and her Section B Kirby Cane Greensleeves.

grand sire, Downland Serchog. Gopher was also the dam of another good stallion, Kirby Cane Gauntlet (f. 1959) and the Vardra Stud's big winner, Kirby Cane Golden Rod (f. 1967 by Kirby Cane Plunder).

One breeder who took advantage of these particular female lines was prominent WPCS Council member, Treasurer and former President, Peter Ward, who purchased Kirby Cane Gossamer (f. 1960 out of Gopher) and Kirby Cane Pigeon (f. 1959 out of Mostyn Partridge) for his Brynore Stud at Ellesmere in Shropshire. He favoured Criban Victor and used him extensively – in 1964 Victor's crossing with Gossamer and Pigeon produced two outstanding mares born, Brynore Tarantella and Brynore Petronella respectively. As youngsters and later as mares they won extensively at major shows.

The Kirby Cane Stud became very self-sufficient in stallions in the years to come, but initially there was a need for an up-to-height section B stallion, so Mrs Crisp jumped at the opportunity to take Coed Coch Berwynfa when he was offered on lease for three years by Miss Brodrick. Among his offspring was a colt to another of the initial Section A mares, Bowdler Glasen (f. 1932) – this was Kirby Cane Gamecock (f. 1958). Meanwhile, to Coed Coch Seron he produced a filly, Kirby Cane Songbird (f. 1958), which, put to Kirby Cane Juggler (f. 1960), produced Kirby Cane Statecraft (f. 1963), the sire of Kirby Cane Grayling (f. 1967), resident stallion at the Roseisle Stud in Scotland. Juggler came from a double source within the Welsh as his dam, Revel Jetlass (f. 1942), was an FS mare while her sire was Gredington Mynedydd by Coed Coch Planed.

It may have been due to the distance of her stud from other mainstream Section B breeders that Mrs Crisp tended to become self-sufficient in homebred stallions, or perhaps it was an inherited trait, coming from a family of livestock breeders. Whatever the case, Mrs Crisp had a talent for successfully mixing her bloodlines to great effect, sometimes bringing in mares from established studs to augment her own. She acquired the Griffith's Twinkle IX by Gay Presto which, through her daughter by Coed Coch Samswn called Kirby Cane Spangle, produced the Berwynfa filly, Tinsel (f. 1958); Brenhines (FS1 f. 1950), by Coed Coch Siabod, bred by the Dowager Lady Kenyon, started the 'B' line at the stud; and Criban Heather Bell, so successful at Downland, produced a fillies with 'Bell' in their names.

Looking at the pedigree of the Kirby Cane ponies, it was obvious that it was to Coed Coch Seron that Mrs Crisp continually looked for her next generation of stallions – and no wonder, as she proved to be such a success for the stud. Over the years, right up to the twilight of the stud, her influence through the use of outside stallions was felt. Bryntirion Rowan had been a great success, producing Shuttlecock and Spindlewood (f. 1953), dam of Kirby Cane Smuggler (f. 1963). Trefesgob Benedict brought in Criban Victor to produce Sorcery (f. 1963), the dam of Spellbound (f. 1966 by Kirby Cane Plunder), whose son by Paith Astronaut brought his dilute colour into the stud with the distinctive dun, Kirby Cane Sundog (f. 1971). Similarly, colour came into the stud via a young stallion, Wortin Cornflakes, whose sire, Belvoir Zoroaster, would make a name for himself within the breed. Put to Snapdragon (f. 1964), the Seron daughter by Downland Drummer Boy, she produced another dun colt, Kirby Cane Spruce (f. 1974), which was used extensively in the stud.

Throughout nearly 30 years of breeding Section Bs, Mrs Crisp did much to publicise the breed throughout her own area of East Anglia and beyond. She showed her ponies most successfully at major shows and provided other breeders with a reservoir of top-class breeding to satisfy their own breeding requirements. Ponies from Kirby Cane also were most successful in saddle classes, although the cross-bred show pony was soon to take over from the Welsh ponies so successful after the war. She was a regular consigner of ponies at the Fayre Oaks Sale and achieved good prices from home and overseas buyers, who were quick to settle on her bloodlines in order to boost the gene pool. Mrs Crisp would be the first to credit Miss Brodrick with her success as a breeder, and they became great friends over the years.

Therefore it came as no surprise that Mrs Crisp looked to Coed Coch for another foundation mare for her stud on the occasion of Miss Brodrick's major reduction sale in 1959. Of the three full sisters on offer, her choice was Coed Coch Pluen, then 12 years old, full of the breeding she knew, three times a winner at the Royal Show and born the same year as Coed Coch Seron. Although one of the highest prices of the day at 270 guineas, luck was on Mrs Crisp's side when the foal by Coed Coch Blaen Lleuad she was carrying at the time of the sale proved to be a filly, which she called Kirby Cane Plume, a great breeder in future years and dam of Kirby Cane Plunder, born in 1963 by Kirby Cane Gauntlet. Plunder would have an enormous influence on her own stud – as he would on those of others, among them Anne, Duchess of Rutland's Belvoir Stud, who leased him in 1970. Her choice of Pluen had been a wise one.

I suspect that Lady Margaret Myddelton, the purchaser of the last of the three full sisters, the 15-year-old Coed Coch Pilia, would have been a bit disappointed with her purchase as the best of her progeny were produced at Coed Coch and not at Chirk, where she had founded a stud of Section Bs in the early 1950s. The prefix Chirk is not to be confused with Chirk Castle, which was a prefix selected by Lord Howard de Walden (see Chapter III), who we know leased Chirk Castle for a 30-year period. Indeed it remains one of the best examples of a Norman stronghold in existence in Britain today, its origins going back to the late 13th century and serving as a fortress in North Wales for King Edward I. It was bought by Thomas Myddelton in 1595 and has remained in the family since; in 2004 it became the responsibility of the National Trust although Lady Myddelton's grandson Guy and his family reserve a home there.

Lady Margaret (neé Petty-FitzMaurice) came from a very distinguished family, her maternal grandfather being the 4th Earl of Minto, Viceroy of India from 1905 to 1910; her father, the Earl of Lansdowne, died when she was young and her mother remarried, this time to Lord Astor of Hever, an American by birth but an English gentleman through and through. A soldier and politician of note, he will be best remembered for his ownership of *The Times.* Lady Margaret was brought up at Hever Castle in Kent, along with her half-brother, Gavin Astor, whose wife, Lady Irene Astor, shared her sister-in-law's love of ponies and later founded her own stud at Hever.

Lady Margaret married Lt Col Ririd Myddelton, who, like her, was keen on fox-hunting – she always rode to hounds side-saddle, keeping a hunting box in Leicestershire, where she hunted most weeks during the winter. Her collection of a dozen side saddles on racks in the tack room at Chirk Castle was a sight to behold. She and her husband moved into Chirk Castle in 1946, whereupon she undertook a major refurbishment of the garden, which became well-known in its own right. She also became an accomplished water-colourist, specialising in flowers. However, in keeping with her friends in North Wales, she decided to embark upon the breeding of ponies suitable for children to ride, so the Welsh Section B was an obvious choice. It was to Lord Kenyon's successful Gredington Stud that she looked for her foundation mare and found it in the three-year-old filly Gredington Bronwen (f. 1946), a daughter of Tanybwlch Berwyn and Silver, the dam of his best mare Gredington Milfyd. Her resultant foal, born in 1950 and the first to carry the Chirk prefix, was a colt by Coed Coch Siobad and her next, by the bay, Craven Debo, was a filly named Chirk Deborah. Following the same successful path taken at both Coed Coch and Gredington, Lady Margaret then sent Bronwen to Criban Victor, resulting in a filly, Chirk Heather (f. 1955). Her choice could not have been better and now, by doubling up the Victor blood, she put Heather to his son, Coed Coch Blaen Lleuad, producing the two colts that would make her stud truly famous. Chirk Caradoc (f. 1958) was retained at the stud while his full brother, Chirk Crogan (f. 1959) was first used, then sold to the Weston Stud, where his influence was considerable. Unfortunately Heather died of strangles aged five, after producing two outstanding offspring.

In an interesting move, Lady Myddelton put the two grandsons back on their grandmother, Gredington Bronwen, to produce fillies which gave their breeder much success in the show ring. Crogan to Bronwen produced Ceinwen (f. 1964) and Caradoc to Bronwen produced Brangwyn (f. 1962) and Curigwen (f. 1965). Unusually for breed classes then and now, Brangwen was shown very successfully as a gelding, taking the Section B championship at the Bath and West as well as wins in riding pony breeding classes at both the National Pony and Ponies of Britain Shows in 1964. Curigwen became one of Lady Margaret's most successful show winners, with the championship at the Royal Show in 1967 among her many wins. This victory reflected Lady Margaret's keen interest in the show ring – the team of Chirk ponies appeared extensively throughout the major shows across Britain, initially guided by the very capable head groom, Arthur Bowling (who fell foul of the law in 1971) and latterly produced by Joan Hanmer, an experienced horsewoman with an abiding passion for the Chirk ponies. Lady Margaret was a great supporter of the National Pony Society and Ponies of Britain Shows, where her ponies gained some of their major successes. Chirk Caradoc himself was a good winner – in 1964 he was champion at the National Pony Society, and the same year he won under saddle there as well as at Glanusk, Northleach and the Ponies of Britain Scottish Show at Kelso. Three years later he was a winner at the Ponies of Britain Stallion Show at Ascot. Up to 1974 Chirk Caradoc took the Ponies of Britain Mountain and Moorland Progeny Award no fewer than eight times, and was three times winner of the Riding Pony Group Award at the annual National Pony Society Show.

The Chirk ponies never claimed the overall championship at the Royal Welsh, although Lady Margaret twice took the Youngstock title, in 1978 with the two-year-old filly Chirk Francolin (Llandecwyn

Ffefryn x Brockwell Cobweb mare, Langford Sanderling), and in 1980 with her half-sister Chirk Cornbunting by Caradoc. This was not the case at the Royal Show of England, where she took championships with regularity, such as Chirk Shirley's win in 1965. By Caradoc out of one of the foundation mares at the stud, Shan (Section A), Shirley (f. 1963) was a big winner in both breed and riding pony classes, with championships at the Ponies of Britain and National Pony to her credit.[1] Chirk Shirley did very well in youngstock classes within the riding pony section, and later went on to qualify for the 12.2 hands class at the Horse of the Year Show. Her son by Downland Chevalier, Chirk Knight Templar (f. 1971), was first used as a stallion at the stud before going on to enjoy a highly successful career in Working Hunter Pony classes. Chirk Caradoc bred a good number of smaller ponies which excelled in children's ridden pony classes, such as the Royal International Champion Chirk Seren Bach, Chirk Aderyn, Chirk Delightful, Drayton Caraway (dam of the 1992 HOYS champion Drayton Penny Royal) and the Horse of the Year Show winners Nantcol Cariadus, Nantcol Arbennig and Firby Fleur de Lys – whose dam was a lucky purchase for her breeder, Mrs Jeremy Ropner from Yorkshire, who bought Sinton Perl in foal to Caradoc from Lady Margaret at the 1967 Fayre Oaks Sale for 430 guineas.

From the outset, Lady Margaret purchased well-bred foundation mares for her stud and made every effort to provide Caradoc with the best bloodlines. At the 1961 Gredington Sale, she purchased the Coed Coch Berwynfa daughter, Desert Queen, whose dam, Desert Flame, was by Incoronax. She made little impact on the stud, although her daughter by Caradoc, Chirk Coronet, had a major impact on the Section Bs in the Netherlands through ponies bred at the Roman and Steehorst studs. Gredington Iris (f. 1954), a full sister to Daliad by Criban Victor out of Silver, was added to the stud. One of her most successful purchases was Cusop Sunshade, bought at Fayre Oaks in 1969 for 500 guineas; her daughter Chirk Sundance (f. 1972), by Lydstep Barn Dance, won extensively and bred the good saddle winner, Chirk Saraband (f. by Knight Templar). When Sunshade returned to the Cusop Stud (sold by Lady Margaret to Vivian Eckley at the 1973 Sale for 600 guineas) she bred such well-known ponies as Cusop Steward (f. 1982) and Sunset (f. 1980), top price at the 1983 Fayre Oaks Sale at 1,250 guineas. Lady Margaret also brought in stallions as outcrosses for her Caradoc mares, such as Shawbury Bittermint (Downland Dauphin x Oare Brenda by Coed Coch Glyndwr), Llandecwyn Ffefryn (Ardgrange Debonair x Weston Crista Bell by Chirk Crogan), Beaulieu Buzzard (a double cross to the Chevalier son, Belvoir Zechin) and Hever Ulyssues (Downland Mandarin x Hever Marinka by Brockwell Cobweb).

Despite every effort by his breeder to maximise the excellence of her much-loved stallion, many experts would comment that Chirk Caradoc bred better riding ponies than Welsh ponies, unlike his brother, Crogan, who is found in the pedigrees of many top-class Welsh

---

1 This was at a time when, unlike today, ponies were shown with plaited manes and tails; during the interim period they would be shown with manes down and tails plaited, and now the only plait that may be found is the small long plait behind the ear, which shows off a clean gullet; apart from this, the Welsh Section B is shown in a more natural state than previously.

ponies; in reality, the brothers were very different in looks and type and this was reflected in their progeny. Perhaps the colouring of Caradoc was partly responsible, as he was a 'hard' bay with no white on his legs and only a small star on his forehead – his stock looked very like him, showing his correct conformation and riding quality more akin to the Thoroughbred, although there was none in his breeding. This proved no drawback and was arguably an asset when Chirk Caradoc was used as a riding pony sire; he crossed extremely well on Riding Pony mares and none more so than Lady Margaret's small Thoroughbred, Kitty's Fancy, to whom he bred some of the most successful ponies found in the annals of the British Riding Pony. Chirk Catmint, Caviar and Cattleya were all big winners in hand for their breeder and their progeny went on to major honours, both in hand and under saddle. Another of his Welsh Part-bred offspring worthy of special mention must be Rotherwood Peepshow, Mrs Mansfield-Parnell's outstanding 14.2 hands show pony mare, a champion under saddle and in hand, and an exceptional breeder whose children and grandchildren have taken the limelight at the Horse of the Year Show as well as every major show in the country.

Lady Margaret was a charming personality – every inch an English Lady, and immensely popular with the equestrian community and hunting, riding pony and Welsh breeders alike. She was a respected judge for the WPCS, a Council member and its President from 1972 to 1973. Equally as charming and popular, but less prominent in pony circles, was her sister-in-law, Lady Irene Astor of Hever, whose public service during her lifetime was exceptional by any standards. She was the youngest daughter of Field Marshal Douglas Haig, 1st Earl Haig of Bemersyde (the Haigs' home in the Scottish Borders). Haig had an exceptional military career, having fought in the Nile Expedition in 1898 and the Boer War, and serving as Chief of Staff in India between 1909 and 1911, an Aide-de-Camp General to King George V in 1914 and commander of the British Expeditionary Force in France from 1915 to the end of World War I. His family name is associated with Haig's whisky, while his own name is associated with the Earl Haig Fund, responsible for the Poppy Appeal that annually raises funds for war veterans.

Lady Irene's father died when she was eight, but his memory undoubtedly inspired her to voluntary work and she worked for the Red Cross during World War II. In 1945 she married Gavin Astor, the son of John Jacob Astor, 1st Baron Astor of Hever. It was Gavin Astor's father who had acquired the controlling shares in *The Times* in 1922 and who married Lady Margaret's mother. Following the death of her husband in 1984, Lady Irene devoted much of her life to charity work; she was awarded the Red Cross's Queen's Badge of Honour, was a President of the Kent branch of the Red Cross and, as chairman of the Royal National Institute for the Blind, raised more than £14 million towards the Sunshine Fund for Blind Children over a 40-year period.

Before succeeding to the title on the death of his father in 1971, in 1962 Gavin Astor and his wife had moved from Wickenden Manor in Sussex to Hever Castle in Kent, well known to members of the Royal Family, who were frequent visitors. This friendship extends to the present day – a great-granddaughter of Lady Astor was a flower girl at the wedding of Prince William to Katherine Middleton in 2012. The move to Hever explains why

Above: Chirk Caradoc pictured in the stable yard at Chirk Castle, handled by Joan Hanmer.

Opposite: Lady Myddelton's youngstock winner Chirk Curigwen, which shows a great likeness to her sire Chirk Caradoc.

Lady Astor's Welsh ponies were first registered as Wickenden, only to be followed by Hever after her move there; it is not widely known that she had two prefixes before this, Tillywick and Sotra, which she never used. This applied to the Section Bs, while the Mountain ponies had another prefix – Tillypronie, the name of the Astor's Scottish estate on Royal Deeside, which formerly belonged to Queen Victoria's physician, Sir James Clark (he was instrumental in introducing Queen Victoria to Deeside and Balmoral, when she and Prince Albert were looking for a Scottish home). The large granite house sits among heather hills and dates from 1867, when Queen

Victoria laid the foundation stone – it has been the Astor family's Scottish home since 1952.

By contrast, Hever Castle is located near Edenbridge, some 30 miles east of London; at the turn of the 16th century, it was owned by the Boleyn family and became the property of King Henry VIII after the death of his second wife, Anne Boleyn. The building went through various stages of refurbishment during its long history but by the turn of the 20th century had fallen into disrepair; it was purchased in 1903 and restored by the American millionaire William Waldorf Astor, who used it as a family residence for almost a century. Sold in 1982, it has become a popular function venue.

Encouraged by her sister-in-law, Lady Astor followed her lead at the 1959 Coed Coch Sale and in some ways did better with her purchases. While Lady Margaret bought Coed Coch Pilia, Lady Astor chose her daughters – at 170 guineas, Coed Coch Pwsi (f. 1959 by Royal Reveller) covered by Rhyd-y-Felin Selwyn and, at 220 guineas, Coed Coch Perfagl (f. 1956 by Coed Coch Blaen Lleuad) in foal to Coed Coch Berwynfa. The catalogue states that the latter was eligible for registration as a Part-bred Arab and had been a winner in the show ring. In addition to the Pilia daughters, Lady Astor also purchased for 180 guineas Coed Coch Bettrys (f. 1956 Coed Coch Blaen Lleuad x Bess Heulog by Tanybwlch Berwyn) covered by Berwynfa. In addition to these mares she also bought Coed Coch Bonny Girl (f. 1942), another daughter of Tanybwlch Berwyn. Like others before her, Lady Astor had equipped her new stud with an adequate supply of Tanybwlch Berwyn breeding on which to build its future. Unlike Lady Margaret, who also acquired the Berwyn blood for her Chirk Stud and then based much of the future around her homebred stallion, Caradoc, it was to her mare lines that Lady Astor built up her Hever Stud to the success it enjoyed over a 30-year period.

Without question, Perfagl was the outstanding breeder among her Coed Coch purchases, although Pwsi gave her a useful female line through the big show winner Piazza, by Solway Master Bronze; Bonny Girl did the same with Hever Boleyn (f. 1964) by Downland Nursery Rhyme (by Downland Roundelay), and Bettrys produced Wickenden Partridge (f. 1960), by Coed Coch Berwynfa, another good female carried in utero from the sale. Snowball, Lady Astor's children's riding pony purchased from Welsh breeder Miss Anne Lumsden, was passed for the Foundation Stock Register and interestingly was responsible for the FS 1 mares, Wickenden Cloudy (f. 1960) and Misty (f. 1959), the only mares with Foundation Stock breeding which Lady Astor chose to retain in her stud of Section Bs.

Never one to make height an issue, Lady Astor kept a very open mind in her use of stallions, as demonstrated by the best of the colts bred out of Perfagl, all of which went on to be useful additions to the Welsh Stud Book. They included Wickenden Osprey (f. 1960 by Trefesgob Benedict), Wickenden Platignum (f. 1963) and Hever Imperial (f. 1970), both by Solway Master Bronze, and Hever Fiesta (f. 1967 by Chirk Caradoc). Osprey was sold to the Belvoir Stud and Platignum won extensively for Lady Astor – he claimed the Section B championship at Northleach in 1968, and the same year was Reserve Male Champion at the Royal Welsh. Arguably the best of Perfagl's show winners, Hever Fiesta excelled as a yearling, winning at the Royal Welsh, Royal of England, Ponies of Britain and Kent County, her local show. It was at Kent County in 1972 that he was judged Supreme and took the coveted Lloyds Bank In Hand Qualifier for the Horse of the Year Show. He was produced to perfection by the Hever groom, Heather Silk, who was devoted both to her ponies and to Lady Astor, who died in 2001.

Hever Piazza with Lady Astor in 1964.

Perfagl's success as a matron was not restricted to her colts, as her female line remained strong throughout the life of the stud. Her first filly, Wickenden Pheasant (f. 1961) by Trefesgob Benedict (Criban Victor out of Berenice) made a name both for the stud in the show ring and a brood mare. Pheasant's 1968 filly by Brockwell Berwyn, Hever Grania, bred a good colt, Hever Lyric (f. 1973 by Solway Master Bronze), which was reserve champion at the Royal Show in 1978 and was later sold to Australia; following her sale to Miss Margo Rees Grania bred the Royal Welsh champion, Boston Bodecia (f. 1984) by Downland Mohawk. Pheasant's 1969 filly by Chirk Caradoc, Hever Heartsease, was kept for stud duties at home, as was her Master Bronze filly, Katriona, whose son Hever Omega (f. 1976) by Keston Royal Occasion was kept entire. The Perfagl daughters, Jade (f. 1971) and Ophelia (f. 1976), by Chirk Caradoc and Keston Royal Occasion respectively, kept the bloodline alive in the stud.

None of these fillies could compare with Lady Astor's personal favourite, Hever Guinevere, the 1968 chestnut daughter of Solway Master Bronze; the dam of 16 foals, Guinevere seemed to breed well to every stallion to which she was mated, endorsing the cross of her sire on this particular mare line which worked exceptionally well. To the Downland Chevalier son, Downland Mandarin, Guinevere produced the stallions Hever Noble (f. 1975) and Quiver (f. 1978). Both were good winners in the show ring, although Quiver excelled in taking championships, among them Surrey County and Royal Windsor in 1981, when he was also second at the Royal Welsh. To Rotherwood Goldfinch by Keston Royal Occasion, Guinevere produced the roan colt Hever Xan. Her fillies included the Mandarin daughters Hever Unity (f. 1982) and Violet (f. 1983), and, while living at Tillypronie, to Paddock Gemini she produced a filly, Hever Willa (f. 1984). One of Guinevere's most successful daughters was Marinka by Lydstep Barn Dance, whose son, Ulysses (f. 1982), was another Hever stallion of note.

The Coed Coch purchases of 1959 by Mrs Binnie, Mrs Crisp, Lady Myddelton and Lady Astor did them proud, but they were not alone in their successes – as we will see in the next chapter. The Coed Coch blood would travel to other parts of Britain, including the Wye Valley, where Vivian Eckley and his wife Pat had already established their very successful Cusop Stud of British riding ponies.

Chapter X

Cusop, Sinton, Elphicks, Reeves, Lechlade, Twyford

Cusop Banknote shown by Vivian Eckley at the Royal Welsh in 1976, where he took the Section B championship.

Vivian Eckley with his 1968 Royal Welsh Champion Cusop Hoity Toity.

Ladies didn't have it all their own way at the 1959 Coed Coch Sale. It was Vivian Eckley who spotted a colt that would enhance his Cusop Stud and a filly which made her mark on the show ring. The latter, at 210 guineas, was Coed Coch Bugeiles, a yearling by Coed Coch Blaen Lleuad out of Bet's Heulog, a mare carrying a double cross to Tanybwlch Berwyn. Both her sire and dam were sold to the United States that day, Blaen Lleuad going to a Mr Bonnie for 320 guineas (equal the top-priced Section B) and Bet's Heulog for 180 guineas to a Mr Chambers. Bugeiles' two-year-old full brother, Coed Coch Buddai, was sold for 65 guineas; in time he became one of the most successful 12.2hh riding ponies in Britain.

More importantly for Eckley was his purchase, for 55 guineas, of lot 90 a yearling colt by Coed Coch Blaen Lleuad – Coed Coch Pawl, whose breeding potential others had obviously dismissed as he was ineligible for full registration in the Welsh Stud Book. Pawl was described in the sale catalogue as a Welsh Part-bred Section B and 'a beautiful mover. Likely to make a high class 13.2 riding pony.' He was ineligible for stallion registration due to his FS1 dam, Mrs Crisp's purchase Coed Coch Pluen (by Tanybwlch Berwyn out of Tanybwlch Penwen by Cairo). It was his riding quality that obviously appealed to Eckley, who took advantage of a relaxation in the rules for stallion registration in a desperate attempt by the WPCS to introduce some new stallions into the Stud Book. After inspection, he was accepted as a Section B and won at the Glanusk Show in 1961; he was used briefly at Cusop before his sale at Fayre Oaks Sale in 1962, when he sold for the top price of 380 guineas to Mr Philips of Cromer for his Paith Stud.

Most people associate the Cusop Stud with the best of British riding ponies, largely through its highly influential stallion, Bwlch Valentino – a legend within the breed who, along with his sons and grandsons, dominated the breeding of children's riding ponies. The Stud was initially engaged in the breeding of Welsh Mountain ponies which lived on Cusop Hill, part of the Black Mountains on the border between England and Wales – the commercial value of the ponies lay in their use as pit ponies for the mining valleys, which were only a short drive away. Shire horses were also bred for use on the farm, which was situated in the village of Cusop, a few miles from Hay-on-Wye in Herefordshire, with its famous old house of Llydyadyway (which means 'entrance to the hills'), home to Vivian and Pattie Eckley and their four children. Vivian Eckley's father had purchased the house and farm in 1917 when Vivian was only two years old; it is currently farmed by John Eckley, along with his wife and sister Suzanne, who still keeps the Cusop Pony Stud.

Eckley loved his Welsh ponies and bought many at local sales which he put to registered stallions for entry in the Welsh Stud Book. He took full advantage of the Foundation Stock Register and registered almost 100 mares therein during the period 1952 to 1974. Llydyadyway was well placed, with easy access to many of the big studs, including Criban owned by his great friend Llewellyn Richards, who influenced him greatly in those early days – not because of the breeding policy at Criban but because they had much in common, with a preference for ponies capable of a day's hunting or shepherding. Richards was responsible for Valentino going to Cusop and for several of the mares which would feature in the breeding of the Welsh and riding ponies found there. Cusop Architect was the first Welsh pony registered with a Cusop prefix in 1952. He sired the promising colt, Cusop Call Boy, the sire of Mrs Borthwick's Cusop Sheriff and his full brother Cusop Sentry out of Coed Coch Brenhines Sheba. Sheba had been Lord Kenyon's top show mare before her sale in 1952 to Mrs Cottrell, who in turn sold her to Eckley; Call Boy was later gelded in 1959 to become a top 12.2 hands show pony for Mrs Carleton-Smith of Market Harborough.

Coed Coch Pawl winning at Glanusk in 1960.

Coed Coch Pawl immediately had an impact on the stud – when put to Coed Coch Bugeiles he produced Cusop Burglar (f. 1962), sire of the 1972 Royal Show Male Champion, Cusop Citizen (f. 1966), the senior stallion at the Williams's Crawel Stud. Citizen's dam was half-sister to Call Boy out of Cusop Crystal by Revel Newsreel, a highly influential bay Section A stallion at Cusop. It was another daughter of Newsreel, Cusop Hazel (f. 1960), which put Pawl on the map when she produced to him in 1963 Cusop Hoity Toity, a solid bay colt with no white. It was a reflection of the upgraded breeding of the Section B through the Foundation Stock Register that Cusop Hoity Toity had Arab on both sides of his pedigree – Sahara from his sire and Skowronek through the Rangoon colt Munis on his dam's side. As a two-year-old Hoity Toity took the Male Section B Championship at the Royal Welsh, a feat he repeated two years later in 1967; on both occasions he failed to take the overall award, beaten by the fillies Belvoir Tosca (f. 1964) and Lydstep Rosetta (f. 1964). Like Chirk Caradoc, his major breeding success as a sire of ridden ponies including the 13.2 hand Show Pony of the Year, Cusop Sequence, and the Royal International winner, Cusop Escalade.

This began a remarkable run of success for Eckley, whose stallions took the Male Championship for Section Bs at the Royal Welsh on no fewer than eight occasions and overall champions four times over a 20-year period. Next up was Cusop Banknote, an extremely attractive grey born in 1973, which took the Royal Welsh Male Championship four times (1974, 1975, 1976 and 1979) and Overall Champion in 1976 and 1979. He was by Downland Cavalcade, a colt initially sold by Mrs Cuff to Hugh Edwards (Sarnau Stud) by the Downland Dauphin stallion, Shawbury Bittermint, out of Downland Cameo out of Criban Heather Bell. (Criban Heather Bell was the grand dam of Criban Heather by Bwlch Valentino, a mare which founded the very famous 'H' riding pony line at Cusop.) Banknote's dam was Cusop Blush II (f. 1962), a daughter of Coed Coch Pawl and Criban Belle (a granddaughter of Silverdale Loyalty, and Heather Bell's half-sister by the Welsh stallion Bolgoed Squire). Eckley bought Belle from William Thomas, known locally as 'Will Coity Bach', who had been given her by his neighbour, Llewellyn Richards, for retirement.

Criban Belle was a remarkable mare and dam of the Eckleys' highly-successful 12.2 hands ridden show pony, Criban Activity, a great breeder herself for them and their daughter, Jocelyn, who used her as a foundation for her own Courtway Stud. She was also dam of Criban Biddy Bronze, whose sire, Criban Gay Snip, was out of Criban Whalebone, the dam of Criban Victor. Biddy Bronze was nothing like Victor – in fact, quite the opposite, as she was quite petite and full of quality. By the time she was 15 years old Belle had been begun a well-earned retirement from the rugged Criban Hills. However, when Richards spotted the chestnut filly she produced to Gay Snip in 1950, he quickly bought it and gave her a Criban prefix; this was Biddy Bronze. She was small but full of quality, something that attracted Miss Elspeth Ferguson, a well-known figure in riding pony circles from Worcester. In due course she was sold to Mrs Reiss (later Lady Reiss), whose daughter Virginia Booth-Jones was in need of a good pony for the 12.2 hands show pony classes. This family was among the best producers of ponies in the country, so it was no surprise that Biddy Bronze set off on a glittering career under saddle, ridden by Jabeena Maslin, before her retirement to stud in 1957, when she was sent to the renowned Coed Coch Glyndwr, which by this time stood with Miss Marguerite de Beaumont at her Shalbourne Stud in Wiltshire.

There was a filly born in 1958, Solway Summertime, and, following a return to Glyndwr in 1958, the resultant colt born the next year was Solway Master Bronze, a bright chestnut with white socks, 12.3 hands in height, pretty and full of Welsh character. He was appreciated not only

Downland Gold Leaf, Champion at Northleach in 1983.

in the show ring but also by the prominent breeders of the day, who were desperately in need of a pony of his type and breeding to enhance the bloodlines which had been established. His full sister became a foundation of Lady Reiss' Solway Stud of British riding ponies, which would be among the best in the country; despite the success of Master Bronze, his breeder would not pursue the breeding of purebred Welsh ponies in the years to come.

His Criban background did not tempt Vivian Eckley to seek the Master Bronze bloodline for his Cusop mares, preferring to return to Mrs Cuff when he found his next stallion at the Fayre Oaks Sale in 1979. Having sold Downland Cavalcade to Mrs Carolyn Bachman for her newly-formed Carolinas Stud in Hampshire, he purchased the sale's top price of 2,000 guineas, Downland Gold Leaf, a two-year-old by Downland Chevalier out of Mere Fire Myth, also dam of Twylands Firecracker. Her breeding was well known to Eckley as her sire was Cusop Hoity Toity and her dam, Bwlch Firebird (f. 1947), an FS mare whose sire was full of Grove breeding and whose dam, Bwlch Goldflake, was also the dam of his famous riding pony stallion, Bwlch Valentino. Gold Leaf was shown extensively and was Overall Section B Champion at the Royal Welsh in 1981 and 1985.

Of the three stallions, Banknote was the one which most satisfied his breeders' quest for the ideal Section B – so much so that the Foreword to the Cusop Stud Major Reduction Sale in 1980 read: 'Although perhaps overshadowed by the Riding Ponies, this Stud has made notable contributions to quality yet typical Welsh Section B Ponies, especially Cusop Banknote, arguably the great white hope of this Section.'

History would show that Cusop Banknote did not live up to his breeders' expectations, although it was obvious that there had been much pressure within the WPCS during the 1960s and 1970s to fix the type within the Section B.

Mrs Crisp's view on the Section B was shared by many of the aspiring breeders of the time, whose model of a Welsh riding pony, particularly a 'finer type' did not conform to that of 'the establishment', whose own model had emerged some 50 years earlier from the hunting field and the 'Pony' shows in London. Had there been a classification for the Show Hunter Pony as a well as the

Working Hunter Pony, as developed by the end of the 20th century, perhaps the modern Section B would have taken a different direction as the quest for bone and substance would have been one shared by new and old breeders alike. Since this would not happen for another 40 years, it was understandable that Section B breeders aimed for ponies capable of holding their own in the 'Show' pony classes that emerged after World War II, when the trend had moved away from the small Thoroughbred hunter type prevalent before this time. The quality cross-bred pony posed the greatest threat to the modern Section B.

The term 'Welsh Ponies of Riding Type' had more or less disappeared by the early 1960s as more and more breeders identified the type of pony by its Stud Book section, namely Welsh Section B. It had come a long way in a relatively short period of time. The Stud Book definition had remained the same for many years and was not about to change: still unaltered, the 2010 Stud Book reads:

> THE WELSH PONY
> Not exceeding 13 hands 2 inches high
>
> Section B of the Stud Book
> The general description of ponies in Section 'A' of the Stud Book is applicable to those in Section 'B', but more particularly the Section 'B' pony shall be described as a riding pony, with quality, riding action, adequate bone and substance, hardiness and constitution and with pony character.

Given that the definition of the Section A was (and is) one that could be applied to any well-made horse or pony with the exception of the description of the head, tail set and action, it was little wonder that there was such dissent among Welsh enthusiasts of the developing Section B. Without the political will to describe attributes such as 'riding action', 'adequate bone and substance' and 'constitution' objectively, they would always be open to interpretation, as remains the case to the present day – although not at such a controversial level.

The division between the 'old' and 'new' Welsh Pony breeders heralded a flurry of articles on the subject of 'type' in WPCS *Journals* during the early 1960s. In 1962, Mrs Pennell and Llewellyn Richards outlined the origins of the Welsh Pony over its 100-year history of use on hill farms to dispel a belief at the time that it was a 'new innovation created solely to meet the demand for the Children's Show Pony of to-day'. While the authors made little attempt to define in detail the type of pony required in the Section B, the following year Eddie Griffith most certainly did – in an article the likes of which had not been written before nor, more importantly, since. It is an insight into his intelligence, knowledge and clarity of vision, all the while anchored in sound practice, that Griffith was able to bring together a definitive article on the subject as relevant then as it is 50 years later. With an emphasis on purpose rather than type and unaware that in the years to come some people would wish to keep ponies for in hand showing alone, he suggests:

> ... breeds that fail to move with the times relapse into obscurity. We, and our ponies, live in a highly competitive world. We do not keep ponies merely to look at or to show in hand (reasonable people will not devote their energy to such useless and unprofitable ends); we keep them because we hope

to use or sell the 'end product'. The 'end product' of Section B is, of course, the child's pony. And therein lie both an opportunity and a challenge; opportunity because there is a wonderful market to exploit; challenge because we must be able to compete with top class ponies of mixed blood. In the long run, people will buy our ponies not because they are Welsh but because they are good, and our task, therefore, is to make the name 'Welsh Pony, Section B' a hallmark of quality.

Whatever methods are employed, the first essential is that breeders should have a definite objective in mind, and a definite plan as to how that objective should be achieved, and, therefore, the purpose of this article is not to lay down the law, but rather to take stock of our many assets and to plead for careful thought so that they can be used to full advantage.

Politically, there was also a move afoot to educate breeders on the subject of the Welsh Pony and a Section B Conference was organised by Mrs Pennell at her home in Gloucestershire in 1962. The 'Brains Trust' of Miss Brodrick, Mrs Borthwick and Llewellyn Richards was aided and abetted by Mrs Yeomans, herself a breeder of Welsh ponies (Bywiog Stud) and a well-known judge and commentator. The participants that day were treated to the best of ponies, which included the most popular exhibit among the 30 ponies forward – the mare Downland Lavender (Coed Coch Sidi out of Criban Loyal Lass), bred by Mrs Cuff but owned by Miss Brodrick, who showed her with such success in both Welsh and Riding Pony classes. Of the stallions, Coed Coch Berwynfa was preferred to Criban Victor, a 'close second' according to Mrs Yeomans, who lamented the fact that Berwynfa's height (over by a quarter of an inch) prevented his appearance in the show ring.

In her report in the 1963 *Journal*, Mrs Yeomans obviously chose to forget the influence of the Arab on the Tanybwlch blood in Coed Coch and the Thoroughbred in Criban when she wrote:

As in the Hunter ring and many others, judges will always vary in their ideas in ideal types, and how very dull it would be if they are always exactly the same! A truly made horse or pony that may lack that little bit quality will, in the hands of some, be placed over the bright little flat-catcher that is full of attraction (especially from the ringside) but does not always bear close inspection in vital ways …

At the end of the afternoon, after four hours or more concentration, the general opinion was that the young ponies showed a great deal improvement over those seen a few years back … nice fillies, but we all know there is a lack of young stallions. These will appear, for we have sufficient nice mares to produce them. It would be a thousand pities if outside blood was introduced in a panic to hurry matters on – loss of type, big heads, big feet, cow hocks, all sorts of things could turn up in future generations.

However, there was some consolation for the 'modern' breeder in her judge's report from the 1966 Royal Welsh

Show, when Lady Margaret Myddelton commented:

> Seeing the wonderful and well-filled classes, there can be no doubt that the Section B has arrived – though what exactly has arrived is still not clear. They came in all shapes and sizes, but amongst them there were many lovely animals which seemed to me to combine the qualities for which I think we should aim, namely, substance combined with quality, and true Welsh character.

Prices received for Section Bs rose steadily during the early 1960s from home and overseas buyers (some would say to the detriment of the Welsh Pony), sparking comment in the equestrian press. Mrs Pennell penned an article in the 1968 WPCS *Journal* which attempted to explain the difference in opinion while at the same time urging that caution be exerted by modern breeders:

> Soon after the last War [WWII], the demand for children's ponies was very much on the increase, and good prices given for anything good enough to win in the show ring ... A move was then made to open the Book [Welsh Stud Book] to a few carefully selected stallions such as Criban Loyalist, who although himself only 13hh was by a Polo Pony stallion but out of a fully Registered Welsh mare. There was such an uproar by a certain section of members that the idea had to be abandoned. It was then decided that a safer and more acceptable policy would be to grade up through some 13.2hh mares to be entered as Foundation Stock. Although these mares were only put in on inspection, I fear some were passed that never should have been, and before the Book was closed the damage was done. ...
>
> In spite of all these upsets and arguments Section B caught on like wildfire, both here and overseas. Prices rose astronomically. So much so that in the last few years, any pony claiming to be Section B, often sold far above its true value for the demand greatly exceeded the supply....
>
> It seems that most Section B breeders are now trying very hard to put their house in order, and I am sure they will succeed, especially if they will only follow the excellent description laid down in the Stud Book. We must aim at breeding a pony suitable for present day requirements, with good limbs and all the attributes of a riding pony, but with the essential qualities of the breed.

By her own admission a relative newcomer to the Section B when she wrote in the 1969 *Journal*, Teresa Smalley urged for uniformity among breeders and judges alike. Reflecting on a more selective market appearing for the Section B, she explained:

> When the boom first started many of the best ponies were sold abroad. We can readily understand why we are now faced with a situation where there is still a great dearth of really good stock and a very large number of middling animals that may or may not in time, breed something of value. Where do we go from here?

> Time is what we need – time in which to achieve a greater uniformity of type within the section. Unless we can do this, the already somewhat unsteady market could, in time, collapse altogether, and the Section B pony [be] relegated to the obscurity of ten to fifteen years ago.

The discussion on type within Section B raged into the early 1970s, in some ways exacerbated by an article written Elwyn Hartley Edwards for the 1973 Welsh *Journal* provocatively entitled, 'The Sins of Section B'. It might be cynical to suggest that his great friendship with both Mrs Pennell and Llewellyn Richards may have influenced his view somewhat, but as a respected equestrian journalist and expert horseman, there is no doubt that his words carried some 'clout'. Having introduced the notion of how 'over-breeding' in the Welsh Mountain Pony had brought about a 'lowering of standards', Edwards brings himself to the crux of his piece on the Section B. After spending much time over the summer months by the ringside, he found himself both concerned and amazed. Edwards takes up the story:

> Fortunately, the inexorable law of supply and demand provides a salutary remedy for the over-production of all commodities, and ponies are no exception to the rule. What, in my view, is a matter of greater concern is the state of the 'Welsh Pony not exceeding 13.2hh, Section B of the Stud Book'.
>
> Here, it is not a question of over-breeding, since it is probable that one of the heaviest demands today is for just such a pony as that described in the Stud Book under Section B. And there's the rub. How many Section B ponies conform to the Stud Book description? ...
>
> I confess to being amazed. Amazed, that is, that so straightforward a description as that appearing in the Stud Book, relative to Section B, should be open to so wide a variety of interpretation by both exhibitors and judges – for the latter must bear much of the blame for the blatant disregard of the Society's definition that is often evident in Section B classes.
>
> Space prohibits me from reminding the transgressors of the definition in full. Suffice that 'the general description of ponies in Section A of the Stud Book is applicable to those in Section B, but more particularly the Section B shall be described as a ridding pony, with quality, riding action, adequate bone and substance, hardiness and constitution and with pony character ...
>
> A goodly number of our Section B breeders are in need of one helluva shake-up – it is up to the judges to see they get it. If they don't, we shall see the victory of the nondescript, the triumph of the irresponsible ignorants, and we may as well close the pages of the Stud Book on the section B forever.

John Mountain, whose wife Alison owned the Twyford Stud, mainly consisting of Welsh Section A,

C and D with only a few Section Bs, waded into the argument the following year when his article, 'The Two Bs' appeared in the 1974 *Journal.* It began as controversially as it ended when he wrote:

> The outstanding talking point one year at Fayre Oaks sale was the similarity in type between many of the top-priced Welsh Ponies (Sec B) and many of the top-priced Part-breds. Indeed, many of the Part-breds showed more Welsh characteristic than some of the registered Welsh Ponies. There is a widely held theory which explains how this odd state of affairs has come about, but as it is based on pure speculation and not on proven facts, I do not propose to spell it out. The fact remains that there are now two distinct and dissimilar types of registered Welsh Ponies. For the purpose of discussion they need labels.

Mountain goes on to define them as 'true' type and 'refined' type. He continues:

> I see no reason why those who prefer the refined type should not continue to breed, register, exhibit and sell for satisfactory prices the ponies of their choice. I am, however, sure that it is the duty of all judges, who have accepted an invitation to serve on the Society panel of judges, to put up the true type and down the refined type ...
>
> It is not the duty of a judge to place the pony which he himself prefers. He is acting for the Society on whose panel he has agreed to serve and that Society has laid down a certain specification. If the pony he personally prefers does not conform more or less to that specification he must put it down ...
>
> If there are any judges not prepared to put up those ponies which most closely resemble the Society's official description, they should resign from the panel ...
>
> I do not wish to run down the refined type because I am sure there is an honourable place for them in the pony world, but I also believe that because they look like Part-breds they will be genetically incapable of passing on the hard constitution, good bone and pony character of the true type.

While there is no evidence of the breeders of the 'refined' type hitting back, or at best, defending their breeding policies, there can be little doubt that Mrs Cuff was among the targets of criticism. Her advertisement in the 1963 Welsh *Journal* said it all when she stated, 'THIS STUD SPECIALISES in Section B Ponies that can win in any company in children's riding classes'. As if working in tandem with her fellow breeder, on the directly opposite page, Mrs Crisp made her own very obvious and clear statement when she wrote, 'The policy of this stud is to produce a top-class Riding Pony of 13.2 with a really good temperament and all Welsh character'.

One of the problems facing the breeders of the Welsh Riding Pony Section B was that of height, and perhaps therein lay the problem of type – as the height went up to the maximum of 13.2 hands, the perception of 'true Welsh' type diminished. There was no question that there was a demand for up-to-height ponies, as observed in the 1974 WPCS *Journal*. In the section 'Trades and Trends', with reference to the 1973 Fayre Oaks Sale, Michael Wyatt not only comments that nearly 50% of Section As sold overseas measured over 12 hands but also suggests that both Sections B and C need to be bigger. He wrote:

> Part-breds have gone fast towards filling the demand for bigger ponies up to 14.2 but this lucrative market remains closed to Sec B; perhaps there is a case for this section together with Sec C to be reframed towards the demand.

After all, this was exactly the thinking of the Americans, who set the regulation heights in the United States for Section A at 12.2 hands and Section B at 14.2 hands.

A number of breeders chose to plough their own furrow and breed for type irrespective of the demand for the bigger ponies. Among them was Mrs Kathleen Bullock, who was another buyer at the 1959 Coed Coch Sale – she paid 80 guineas for the two-year-old filly Coed Coch Pefr, by Coed Coch Blaen Lleuad out of the prolific show winner and dam of winners, Coed Coch Penllwyd. Pefr, another full sister to Coed Coch Padog, proved to be a wonderful buy for Mrs Bullock. She provided her owner with two stallions which would be a great asset to her Sintoncourt Stud situated on the outskirts of Worcester. The stud name began simply 'Sinton', a prefix shared with neighbour and great friend Mrs Jean Houghton, whose husband had served in the Royal Navy as a medical officer alongside Commander Bullock. Mrs Bullock herself held the position of Chief Commander of the Auxiliary Territorial Service (ATS), the women's branch of the Armed Services during the World War II, for which she was awarded the OBE.

Initially Mrs Bullock purchased a few Mountain ponies to graze the pasture at Sintoncourt, but she recognised the demand for the Section B and shifted her attention to breeding them instead (she also bred Jersey cattle). Her venture into the Welsh Riding Pony started with the very best when she purchased the young mare Cusop Glamour (f. 1955), in foal to the bay Section A stallion, Revel Newsreel. The resultant filly, Sinton Moving Charm (f. 1959), not only gave her instant success in the show ring by claiming the Royal Welsh title in 1968, but also ensured that the stud would have a place in history as Moving Charm was selected by Albany Fine Arts (a porcelain factory in Worcester) as the subject for a limited edition study of a mare and foal sculpted by David Lovegrove. Glamour was an outstanding success in the show ring and bred the following year the Section B stallion, Sinton Gyration (f. 1960 by Bolgoed Automation).

Gyration was eventually sold to Mrs Poldervaart in the Netherlands, but not before he had sired Ceulan Largo (f. 1963 out of Dyrin Larina), which became the foundation of the Section B in New Zealand after export there in 1973. Gyration provided Mrs Bullock with some nice mares to carry on his name and two sons out of her foundation mares. In 1964, Kirby Cane Song-Belle (a daughter of Criban Heather Bell) produced Sinton Bellhop by Gredington Mynedydd, which became a resident and very influential stallion at the stud. The following year Coed Coch Pefr foaled a colt, Sinton Whirligig (f. 1965), a substantial grey purchased by Mrs Betty Knowles and her daughter Frances (later to become

Sinton Moving Charm, Royal Welsh Champion 1968 modelled by David Lovegrove for Albany Fine China.

Mrs John Carter), who had started breeding Section Bs at their Millcroft Stud near Dawlish in Devon. Produced by Miss Ferguson from her Rosevean yard near Pershore, Whirligig won extensively at major shows prior to his sale to Australia.

Pefr proved to be a valuable breeder for Mrs Bullock. She bred the aforementioned Perl (dam of Firby Fleur de Lys), as well as another influential stallion, Sintoncourt Perilustre (f. 1967) by Reeves Golden Lustre, Mrs Reeves' famous stallion which came to live at Sintoncourt in semi-retirement aged 20. The records show that Mrs Bullock also found a home for several well-known older mares by Tanybwlch Berwyn which included Berenice and Coed Coch Silian, the former Royal Welsh champion. Mrs Bullock favoured the 'old' breeding and brought Gredington Obringa, another offspring of Planed, into the stud as well as several Criban mares including Criban Chiffon (f. 1959), whose daughter by Gredington Mynedydd born in 1965, Sinton Bridal Gown, bred some of the loveliest small ponies in the Stud Book including Roseisle Bridesmaid, the winner of the first leading rein class ever staged at the Horse of the Year Show.

Sinton Whirligig, champion in hand when produced from Rosevean.

It is noteworthy that Mrs Bullock found no use for Solway Master Bronze, a stallion which attracted a great deal of attention since his purchase as a foal by Miss Miriam Reader from Lady Reiss in 1959, regarded as a turning point in the history of the Welsh Stud Book. Along with Downland Dauphin, Chirk Crogan and Brockwell Cobweb, Master Bronze was seen to breathe life into the Welsh Section B. Dr Davies in *One Hundred Glorious Years*, he states:

> 1959 can be regarded as the 'turning point' in the fortunes of the Welsh Pony Section B. The various well-intentioned schemes which had been devised to introduce new blood in the section by allowing stallions out of FS1 mares etc into the Stud Book had produced slight expediency and a few useful outcrosses, but not sufficient to satisfy the growing need for ponies of this type which were able to hold their own as children's riding ponies against all comers.

It was Miss Reader's brother who named the colt 'Red' when he first saw him, and the name stuck for the remainder of his life at her Elphicks Stud in Sussex, a name taken from her family's farm in Kent where they grew hops and fruit. Aware that Criban Biddy Bronze had produced a beautiful filly in 1958 and also that she was once more in foal to Glyndwr, Miss Reader wrote to Lady Reiss to secure the next foal, which she duly did. Although he had been extensively shown with his illustrious dam, Solway Master Bronze proved a bit of a handful for his new owner in the show ring, who had no previous experience of a colt. He was an instant success despite this – Lady Reiss helped show him at his first event when he was second to Criban Victor in the only class for Welsh Section Bs at the 1960 Ponies of Britain Stallion Show at Ascot. He did better in the riding pony section, where he won the Muir Trophy as best yearling and the Simsbury Cup for the champion youngstock. Bill Beaney and David Reynolds took over as showmen as 'Red' became too strong for his owner. He enjoyed a very successful show career, including the championship at the Royal Welsh no fewer than three occasions (in 1961, 1962 and 1963) – the only pony other than Criban Victor to have taken the championship in three consecutive years. In 1962 he won both in hand and under saddle at the National Pony Society Show.

Solway Master Bronze was the bright red chestnut colour described by Miss Reader's brother, with four white socks and an attractive blaze; as his breeding suggests, he was full of Welsh character and relatively small in height. He could not have had better parents, whose individual attributes brought the very best qualities to their son – it was no wonder that he was very popular as a sire as he stamped his stock as his own. With over 500 offspring attributed to him while he was still only 12 years old, he had an enormous influence on the Welsh Stud Book.

Miss Reader would be the first to say that he was the making of her stud, although she had been breeding ponies for a few years before his purchase. Having broken a shoulder as a two-year-old, the yellow dun filly Harvest Moon was registered by Miss Reader within the Foundation Register as Section B and covered by the popular roan stallion Reeves Golden Lustre. She produced the cream filly Honey Moon in 1956 and in 1958, Elphicks Harvest Gold, the first with her own prefix. The palomino Section A stallion Marsh Crusader was used for the next two years until the first of the Master Bronze foals arrived in 1963. Honey Moon immediately produced top-class fillies by him such as Elphicks Half Moon (f. 1967) and Crescent Moon (f. 1970) and in 1962 the most famous of all, Elphicks Honey B, which went on to become one of the best 12.2 hands ponies in the country before returning to stud. Harvest Gold would also breed a host of lovely ponies by Bronze including the fillies Elphicks Burnished Gold, Golden Grain and Golden Gift, a winner in hand and later under saddle, including the Royal Welsh when ridden by Jennifer Daffurn. In her 1971 WPCS *Journal* advertisement Miss Reader wrote that all her mares had been winners under saddle other than Honey Moon; she loved showing mares and foals, which she did with great success at the shows in and around the south of England. No victory was enjoyed more than that of Reeves Fairy Lustre, which took the championship at the Royal Welsh for Miss Reader in 1973 under the watchful eye of Mrs Teresa Smalley.

It was no wonder that Mrs Smalley liked the type of Fairy Lustre, as she had already purchased Reeves Turquoise (f. 1969) by Godolphin Pendragon out of Reeves Silver Thread – a daughter of Ceulan Silver Lustre like Fairy Lustre herself. Both mares were bred by Mrs Dorothy Gilbert, whose love of Welsh ponies stemmed from the need for ponies for the Reeves Riding School which she ran at Penn in Buckinghamshire. Her first 'Welsh' purchase from E S Davies from Talybont was the colt Ceulan Revoke (f. 1942 out of Ceulan Silverleaf), which she registered as Reeves Ceulan Revoke when she registered her own prefix. In 1947 she managed to persuade Davies to sell a half-sister out of Silverleaf called Ceulan Silver Lustre (f. 1938), which was registered as Foundation Stock as she was by the Arab stallion, Incoronax. The sale included her son Reeves Golden Lustre, born in 1945 by Ceulan Revolt, her daughter Reeves Crystal (f. 1946) and foal at foot, Nantgarw. At the time Golden Lustre was ineligible for full registration since he was out of an FS mare, so he was registered as a Part-bred – due to the dearth of Section B stallions, he was one of those selected after inspection to be registered within the main register in 1960.

Apparently it was a quite spectacle to see the small herd of ponies, which had lived their lives unhaltered on the hills, run through the streets of Penn from the

Miss Reader's famous stallion Reeves Golden Lustre at Springbourne, where he stood at stud.

railway sidings to the stables. To their credit, they very quickly acclimatised to their new life and soon made excellent riding ponies – and as such were in much demand in the Home Counties. Ceulan Silver Lustre, aged ten, took to saddle as if she had done it all her life and she won extensively for the next two years, including four National Pony Society Silver Medals as well as eight championships for a young Geoffrey Carter from Maidenhead; in 1950 they were reserve for the Country Life Cup at the Royal Welsh Show. She also won extensively in hand; one of her more notable victories came in 1954 at the Ponies of Britain Show, where Mrs Spooner staged classes for Foundation Stock mares, and where Llewellyn Richards famously placed Silverleaf above one of Miss Brodrick's mares. A glittering show career came to an end with a win at Bucks County in 1960 – when Silver Lustre was aged 22.

Her success under saddle was matched, if not bettered, by her success as a breeder. Her son, Reeves Golden Lustre, attracted much interest at the time – especially from the riding pony breeders, who were desperately looking for an outcross for their finely bred mares. In 1954, Mr Deptford's brilliant winners Pretty Polly and Firefly, as well as Mr Blythe's Glide On, were all booked in to him. His reputation as a sire quickly grew, as his purebred stock started to do well in the show ring, including one of his sons, Springbourne Golden Flute (f. 1966), which was Overall Section B Champion at the Royal Welsh in 1968.

The stud's reputation was much enhanced by the many fillies out of Silver Lustre, which proved outstanding. Luckily for her owner, she bred a large number of them which proved themselves under saddle and in hand. Put to Kirby Cane Shuttlecock (which stood at stud nearby at Bracknell with Mrs Yeomans), she produced Reeves Sapphire in 1959 and her full sister,

Reeves Fairy Lustre, in 1961. The former was sold to a near neighbour, Mrs Jennifer Williams, wife of the well-known showjumping commentator Dorian Williams, whose Pendley Stud had become a major force in Riding Pony circles; she showed her very successfully in brood mare classes. None of the family proved more successful than Reeves Fairy Lustre, which was purchased privately as a four-month-old foal by Miss Reader for £500. It was a wise move, as she proved a marvellous show filly from the outset, culminating with her win at the Royal Welsh.

One of Ceulan Silver Lustre's last foals was Reeves Celadon (f. 1964), by Solway Master Bronze, which had sired a very successful leading rein pony named Master Cinnamon, which was out of the Silver Lustre daughter, Reeves Cinnamon. Indeed, Solway Master Bronze bred many top-class ponies for others breeders, including Mrs Janet Meyer's Clyphada Periwinkle (f. 1964), dam of Keston Royal Occasion, and Miss Featherstone's Senlac Fleurette (f. 1965 out of Hafod Fleur), a prolific winner of children's ridden classes. Master Bronze's stock was so successful in the show ring that he won the Ponies of Britain Mountain and Moorland Progeny Award in 1966 and 1967 and the WPCS Sire ratings in 1969 and 1970. When Miss Reader's deteriorating eyesight forced her to retire from breeding ponies, it was a real heartbreak to part with Master Bronze, but in 1975 she asked the Honourable Maxine Ponsonby if she would like to buy him for her Lechlade Stud situated in the village of that name in the Cotswolds.

The Ponsonby family farmed extensive lands at Little Faringdon near Lechlade. Coming from an enthusiastic fox-hunting family who seldom missed a day with the VWH, her husband, Tom Ponsonby, was appointed Deputy Lieutenant of Gloucestershire in 1972 and was High Sheriff of the county in 1978: most importantly, he was very supportive in his wife's pony interests. (In 2002 their son Rupert succeeded his uncle to become the 7th Baron de Mauley. In March 2005 he was the first new peer to gain a seat in the House of Lords, after his by-election victory for a Conservative hereditary peers' seat following the 1999 House of Lords Act. In 2012 he was appointed Parliamentary Under-Secretary at the Department for Environment, Food and Rural Affairs. Lord de Mauley has an interest in hunting and in racing, in which has been both owner and jockey.)

Mrs Ponsonby had an interest in ponies from an early age although she says her parents were 'Londoners'. Her godmother was Miss Gladys Yule, a renowned breeder of Arab horses, who encouraged her to ride and appreciate fine horses, something that would guide her through her days as a well-respected judge and breeder of riding ponies. Lechlade became a name synonymous with the best of British riding ponies during the latter part of the 20th century, when they were shown both in hand and under saddle – Lechlade Violet and Lechlade Melissa were two of her exceptional ponies. While they took centre stage at Lechlade, there was room for a few Welsh ponies, including the extremely pretty Section A mare, Pencoedcae Lily, purchased in 1961 with a filly foal at foot, Foxhunter Golau. Bred by the Olympic showjumper and Welsh pony breeder, Lt Col Sir Harry Llewellyn, Golau was by his noted stallion, Coed Coch Brenin Arthur, purchased by Llewellyn at the 1959 Coed Coch Sale for 250 guineas and to whom Lily produced the first foal to be registered with the Lechlade prefix, in the WPCS Stud Book in 1962.

Lechlade Scarlet Pimpernel shown as a yearling at Malvern by David Reynolds.

Impressed by his beautiful head, Mrs Ponsonby used Brenin Arthur in a move to improve the slightly plain head of Tanwen (f. 1948) by Tanybwlch Berwyn, a Foundation Stock mare bequeathed to her in 1964. Tanwen evidently carried the type of breeding so prevalent in the early days of the Section B and her cross with Coed Coch Brenin Arthur was either a stroke of luck or of genius as the resultant filly, Lechlade Arum, born 1965, would play a very important part in the future of the Section B. Arum was a dun carrying the dilute gene from Brenin Arthur, whose sire, Coed Coch Madog, often bred colour and his dam, Snowdon Arian II, was a cream. Interestingly this colour would carry on through several generations of champions in the years ahead, culminating in the dun 1998 Royal Welsh Male Champion, Wortley Fisher King, whose sire, Cottrell Artiste, was a great-grandson of Lechlade Arum.

Having a love for pretty, quality ponies, Arum was sent to Solway Master Bronze in 1967, producing the chestnut colt Lechlade Scarlet Pimpernel, an influential stallion while standing at the Belvoir Stud and then the Congarinni Stud in Australia after his export there in 1979.

A previously unpublished painting by Tom Carr of Mrs Ponsonby's foundation mare Tanwen.

Her next foal was a filly by Downland Chevalier called Lechlade Angelica (f. 1971), one of the few mares which successfully joined the Downland Stud, where she was a favourite of Mrs Cuff. Following the successful pattern of Mohawk on Chevalier mares, the recipe once more came good when a filly, Downland Almond, was foaled in 1975. Almond became one of the principal producers at Mrs Johns-Powell's Cottrell Stud in Glamorgan, which would come to be one of the most influential studs in the Stud Book as it moved towards the 21st century. The arrival of Master Bronze at Lechlade in 1975 made the choice of stallions for Arum easy and she bred a series of lovely fillies to him over the years. In addition, her half-sister, Lechlade Sunflower (f. 1970 Solway Master Bronze x Tanwen) was shown extensively in brood mare classes by Mrs Ponsonby and took the championship at the Three Counties Show, among other wins.

Although Mrs Ponsonby brought together a small band of mares for Master Bronze, Arum remained his most important partner at Lechlade, although other top-class mares were sent to him until his death in April 1983. He seemed to fulfil the needs of those breeders who were

Solway Master Bronze at a personality parade at Malvern.

unhappy at the time about the type emerging within Section B, as he had everything they were looking for in a 'typical' Welsh pony. Not only was he regarded as an ideal sire for the up-to-height ponies with Downland breeding, but he was also regarded as suitable for those with Coed Coch blood and was used extensively for both. He was particularly useful for breeders living in the south of England, such as Alison Mountain, who lived near Miss Reader and was keen to take forward the bloodlines of the small number of Section B mares she had gathered at her Twyford Stud. Her views on type by this time were well known, so Master Bronze was the ideal choice.

Mrs Mountain was the first Editor of the WPCS's annual *Journal*, a Council member for many years, Council Chairman and President of the Society. Like her father, Arthur McNaught (Clan Stud), she was an established breeder of Welsh Mountain ponies (Twyford Stud) following the purchase of her first ponies at the end of World War II. Her father bought the best of breeding when Mrs Armstrong-Jones (mother of Princess Margaret's husband, Lord Snowdon), dispersed her herd of Mountain ponies, most of which she had acquired from Lady Wentworth.

McNaught also purchased from Lady Wentworth the mare Tanybwlch Penllyn (by Tanybwlch Berwyn), which, despite her height of 13 hands, was eligible to breed both Section A and B. She had a great influence on the Section A in his stud as within two generations he

Rhyd-y-Felin Selwyn in Sweden in 1980.

had bred from her the Royal Welsh champions Clan Pip and Clan Peggy.

Penllyn's daughter, Clan Prue, was a foundation for the Section Bs at Twyford along with Reeves Crystal (Ceulan Revelry x Ceulan Silverleaf), Chirk Andante (Chirk Caradoc x Kipton Allegro), Bowdell Quest (Chirk Caradoc x Kirby Cane Goosefeather), Rhyd-y-Felin Seren Wyb (Tanybwlch Berwyn x Coed Coch Sirius) and Kirby Cane Songbird (Coed Coch Berwyn x Coed Coch Seron). Among the stallions she used other than Master Bronze were Brockwell Berwyn, Coed Coch Pedestr and Pendock Peregrine. She also brought back from Sweden Rhyd-y-Felin Selwyn, which she used before selling him to retirement in the Isle of Wight with Ted Bucket. Selwyn's most famous Twyford offspring was Twyford Sparkle (f. 1971), which was the last of the Section Bs to be sold when they were dispersed in 1976 with Twyford Signal at foot. Sparkle had a great impact on ponies in neighbouring Denmark when she produced MollegaardsSpartacus.

Mrs Mountain's interests in the Section B were limited and short–lived, as she could see that the ponies she bred were not fashionable during the 1960s. As she wrote in an article about the Twyford Stud for the WPCS's *Golden Anniversary Journal* in 2007:

> We did not have the section B ponies for very long. Nobody wanted the sort we were breeding. At

Twyford Sparkle, a champion in Denmark and the dam of Twyford Signal and Mollegaards Spartacus.

that time the fashion was for very fine ponies with no shoulders and which scuffed the ground as they moved. Our mares were of good size with good bone and feet and were built to do a job of work ... It is ironic, but gives me great satisfaction, that the wheel of fashion turned again and that the ponies we bred and the type we wanted are now proving themselves in the pedigrees of successful section Bs today.

The comments made by Mrs Mountain said much for the progress that she had witnessed over almost 30 years since selling her Section Bs. It was brought about in no short measure by the last of the notable stallions born in 1959, Chirk Crogan, and the Weston Stud, his home during a critical time in his life and that of the still-developing Welsh Riding Pony.

Twyford Signal, a champion both in hand and under saddle when shown by Mr and Mrs Bigley.

Chapter XI

Coed Coch (Williams-Wynn), Weston

Weston Mary Ann, Royal Welsh Champion in 1975, seen here following her export to Australia where she continued her winning ways.

Lt Col Williams-Wynn with 'the Boys' at Coed Coch as they appeared in the WPCS Journal in 1975.

Just as the Coed Coch Sale of 1959 put down a marker in the history of the Welsh Riding Pony, so would two sales which took place within three remarkable days in September 1978, both set in North Wales and both offering a standard of Section Bs that had no previous equal. The first was the more remarkable because it marked the final dispersal of the Coed Coch Stud following the death a year earlier of Lt Col Edward Williams-Wynn OBE. There were 77 Section Bs offered at the 1978 sale, all but three born at Coed Coch. The other was a major reduction sale of the Weston ponies belonging to Jack Edwards and his family, who had built up a highly-acclaimed stud of both mountain ponies and Section Bs over a 20-year period at Llandyn Hall, Llangollen. Little did they know that the following year would witness their own final sale.

Following Miss Brodrick's death in 1962 'The Colonel', as Williams-Wynn was affectionately known, had embraced the breeding of Welsh ponies at Coed Coch, as he had the running of the estate which he had inherited. His previous experience was largely Thoroughbreds and fox-hunting – he was the second son of Sir Watkin Williams-Wynn, a Master of renown with the Wynnstay, a hunt of repute and one associated with his family over centuries. The Colonel was a quiet, unassuming man with a distinguished military career – he was held in high regard, like Miss Brodrick, by the staff at Coed Coch, and employed his former army batman, Willie Davies, as a chauffeur when he was demobbed. Davies soon embraced the pony interest of his employer, and became part of the show team along with the Jones 'Boys'. One of their proudest days under their new employer came in August 1969, when Williams-Wynn organised an open day at Coed Coch to mark the investiture of the Prince of Wales at Caernarvon Castle.

From modest beginnings of one stallion, Coed Coch Berwynfa, and three mares – Coed Coch Penllwyd, Downland Lavender and Vesta – Williams-Wynn had rebuilt the herd, with considerable show success along the way. Ironically Berwynfa (along with the great Mountain Pony stallion, Coed Coch Madog) died only a few weeks before the 1978 sale; although successful in his own right, he did not have the best of show records. But two of the mares had achieved much, especially in the Section B championships at the Royal Welsh. Vesta herself was never champion there but her full brother and sister, Valiant and Verity, had been for their breeder, Mrs Griffith, in the early 1950s. Downland Lavender, by the Coed Coch sire, Coed Coch Sidi, out of the Foundation Stock mare, Criban Loyal Lass (by the polo pony, Silverdale Loyalty) had been A L Williams's female champion in 1960, one of the years when Criban Victor was judged Overall Champion.

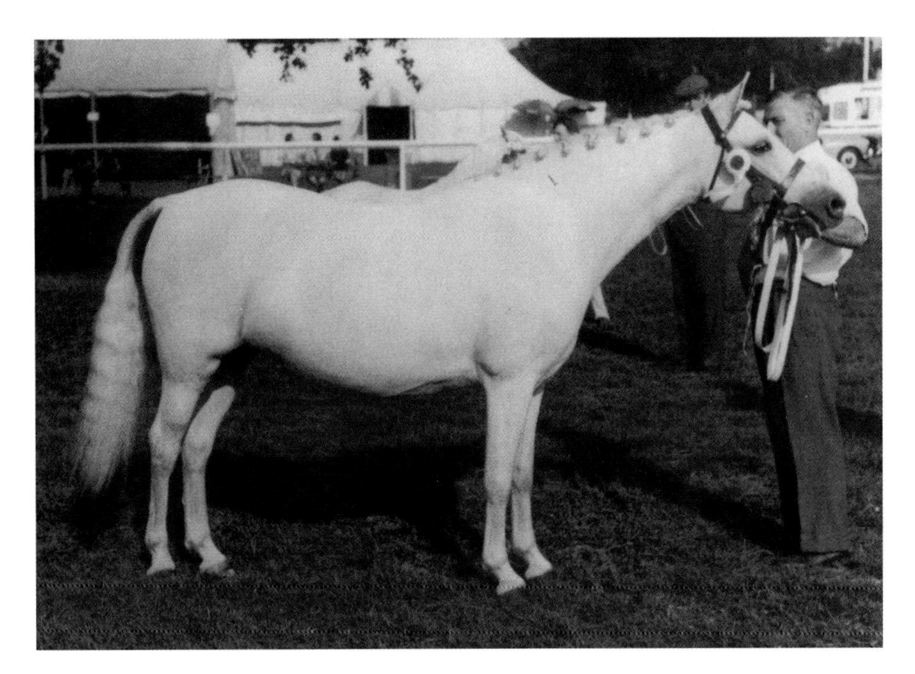

Coed Coch Dawn shown by Shem Jones in 1974.

Last but not least was Victor's daughter, Coed Coch Penllwyd (out of Pendefiges), which had been champion at the Royal Welsh in 1964, two years before her full sister, Coed Coch Priciau, which on retirement was given to Ann Bale-Williams for her Baledon Stud.

Showing mares was a favourite of the Coed Coch 'boys' and they excelled, each year producing one better than the last. They had campaigned the aforementioned with great success but afterwards their offspring did equally as well, starting with Coed Coch Llawring, a Berwynfa daughter out of Lavender. Over two seasons (1969–70) she accounted for championships at the Bath and West, Great Yorkshire and Royal Shows and took the coveted Lord Arthur Cecil Cup at the National Pony Society in 1969 when judged champion Mountain and Moorland Brood Mare. Her full sister, Llywy, took up where her sister left off when she was champion at the Three Counties Show as well as two North Wales favourites – Anglesey and Flint and Denbigh, where competition was stiff for all the North Walian breeders. Both Llawrig and Llywy were sold to Dr Wulf from Germany during the 1978 sale for 2,000 guineas and 1,400 guineas respectively.

Penllwyd's daughter by Coed Coch Blaen Lleuad, Coed Coch Peoni, had proved to be a good breeder for the stud. Her daughter by Berwynfa, Penwn, was the next mare to be shown with success, winning consistently as a youngster and later as a brood mare. Her sister, Gala, was not shown to the same extent but would prove her worth for her next owners, Bernard and Maureen Butterworth, who had gone to the sale specifically to buy her; anxious that she might go over their budget but keen to secure the bloodlines, they bought at 1,400 guineas Penwn (which was the previous lot to Gala)

Coed Coch Berwynfa at Warrington Show in 1961.

and then Gala herself for 3,800 guineas. The following two lots also went to one buyer, this time Mrs Olive Weston, who was building up her Section B interest at her Seaholm Stud in Lincolnshire. She bought for the bargain price of 1,300 guineas the next prominent show mare, Coed Coch Dawn, another daughter of Berwynfa out of Penllwyd; Dawn had been Overall Champion at the Royal Welsh in 1974 and Reserve Female Champion in 1975. Mrs Weston also purchased Coed Coch Nina, a Lavender granddaughter out of Coed Coch Llawrig. Nina's sister by Cusop Banknote went to New Zealand at 2,000 guineas, while a brother by Solway Master Bronze headed for the Netherlands.

It was something of a mystery that Coed Coch Alarch made the second-highest price for Section Bs at the sale, for she was only lightly shown as a yearling and quite small in height. Perhaps her chestnut colouring accounted for this – she was one of the few daughters of Coed Coch Berwynfa which was not grey. Another possible explanation is that her famous dam, Pelydrog, was arguably the most successful show mare of all time and carried the very best Mountain Pony breeding. Either way, she made 3,600 guineas when sold to Mrs Akehurst, who sold her in 1983 to Mrs Owen for her Owendale Stud in Australia. Top price among the males was the senior stallion, Coed Coch Targed, very much the image of the old Coed Coch stallions. He had been very successful in the show ring from an early age, taking championships at leading shows such as the Great Yorkshire, West Midlands Stallion, Ponies of Britain, National Pony Society, Royal Show and many more. His show successes did not attract an exceptional price or attract the home buyers, and he left Wales for the Netherlands for a maiden bid of 3,000 guineas.

The 1978 Coed Coch Sale attracted a huge crowd of enthusiasts keen to witness the greatest sale in the history of the WPCS, and buyers keen to secure part of the Coed Coch Stud for their own. An account of the final chapter in the Coed Coch story would not be complete without mention of the exceptional prices paid for two Section A stallions at the sale. Coed Coch Saled sold for 14,000 guineas to Mrs Gadsden for her Bengad Stud in Gloucestershire; and the world record price in auction for a Mountain Pony, 21,000 guineas, for Coed Coch Bari destined for the Nattai Stud in Australia following his sale to Miss Brodrick's great friend, Lady Creswick.

E S Davies wrote that Miss Brodrick was 'one of the great stalwarts of the breed during the bleak times when we kept our breeds going, often for no financial rewards'.

The sale may have heralded the end of the Coed Coch Stud, but not the Coed Coch story. Just as in 1959 with the sale of so many good mares, several of the purchasers maximised on the bloodlines they had secured for their own studs. In the final paragraph to the foreword to the 1978 sale catalogue, E G E Griffith summed up the occasion thus: 'So ends the story of a great herd; of two gifted and dedicated breeders, Miss Brodrick and "Colonel Edward"; of a skilled and devoted staff. Theirs has been a great achievement; they have helped make history.'

In the overall history of livestock breeding in Britain there have been men and women who possessed a talent for breeding good animals, ranging from Silkie hens to Shire horses. They possessed a good eye for correct, sound conformation and movement and an instinct of how to bring together the attributes of individuals in a way that improves one generation for the next. It also involved a sound knowledge and understanding of pedigrees, an understanding of how they might merge and, most importantly, a sixth sense of what is required of the breed for the future. Among the many talented breeders of Welsh ponies in the annals of the WPCS, few would dispute that John ('Jack') Edwards of the Weston Stud stands out as one of these exceptionally talented livestock breeders.

Edwards originally bred and trained Shire horses for dray work for the railways and breweries, and showed them with success. In 1958 he moved from the dairy farm of Weston near Oswestry in Shropshire, to Llandyn Hall situated near Llangollen, a small town in North Wales famous for its outdoor activities based around the neighbouring hills and the River Dee. Llangollen is also home to the International Musical Eisteddfod, a prestigious cultural gathering for enthusiasts throughout Wales and the rest of the world. The 500-acre farm, which rose to over 300 metres, was in a poor state of repair when the Edwards family first moved there, but its beautiful old house and extensive hill land was substantially improved over the 21-year period that they lived there. Llandyn Hall became famous not only for ponies, which it suited, but also for Murray Grey cattle.

Edwards' wife, Dilys Gethin, was an accomplished horsewoman from a well-known family from Tregynon near Newtown. Her father bred ponies for the pits, shires for transport and riding horses; her brother was a prominent breeder and showman of hunters. Initially, ponies were purchased for their children, Carol and Janet, who were keen riders and very successful in children's riding pony classes; even when they moved to adult classes, their famous show hack Olympian competed with the best. One of the first ponies to arrive at Weston

Farm was a palomino mare of unknown breeding named Primrose, born in 1948, later to be registered in the Foundation Stock Register as Section A. She proved to be a perfect child's pony, regularly travelling to the Gethin family home at Tyn-y-Bryn, from which she hunted with the David Davies Hunt. With a future pony for the girls in mind, Primrose was put to the larger Section B, Coed Coch Siabod, prior to embarking upon a ridden career herself. Weston Lavender Blue, a filly, the very first to be born with a Weston Prefix, was born in 1951, setting Edwards on a pathway to success in Section B that even he would have found difficult to believe. The record books show that Lavender Blue was the family's first winner at the Royal Welsh, when she won three ridden class prizes in 1954, ridden by Carol. Within three decades they bred five Royal Welsh Female Champions, of which two stood Overall Champion Section B.

Weston Lavender Blue at the Great Yorkshire Show in 1963.

Edwards was second at the Royal Welsh in Llandrindrod Wells in 1951 with the two-year-old Shire gelding, Crimwell What's Wanted. Little did he realise that day that it would be with ponies that he and his family would return in later years. Thus, born into showing, the Edwards girls moved from the ridden rings into the in-hand rings with great ease; their skill of both producing for the ring and handling ponies in it was a great asset to their parents, and much success was gained by this breeding/showing family. Colour had been an attraction to Edwards from the beginning and several of the girls' ponies were either cream, palomino or dun with Arab breeding, including the Connemara mare, Clonkeehan May Morning by the well-known Arab, Naseel, and Sunset Caribbean Princess by Algosha. Princess' dam, Honeysuckle Rose, had been champion palomino at the British Timken Show, where Princess was also foal champion. Honeysuckle Rose, by the Thoroughbred Cosmopolitan Jack, was one of the mares passed after inspection for the FS Register to enter the Weston Stud. Edwards bred some top-class riding ponies alongside his Welsh over the years as the trade in quality children's ponies grew. Two of the stallions used for this purpose were the Welsh Part-breds, Rossall Gazelle, a chestnut stallion used during the 1970s, and Small Land Otto by Downland Romance, purchased as a yearling at Fayre Oaks and a big winner both in hand and later under saddle. Perhaps it was the marketplace that persuaded him to move into Welsh riding ponies in due course.

Edwards' first foray into the Welsh breeds came in 1953, when he purchased the pretty Mountain Pony yearling, Revel Fair Maid (FS2 mare by Ceulan Revolt), which carried her own cream colouring through her dam, Revel Fair Lady from her sire, Cwm Cream of Eppynt. Interestingly the colour would carry for six generations to Weston Picture, born in 1978, a palomino which was Section B Female Champion at the Royal Welsh in 1979. With an early interest in Section A, Edwards used the best stallions available in his area, favouring offspring originating from Coed Coch Glyndwr in their immediate back pedigree.

In the case of Revel Fair Maid, he produced two extremely influential lines from an early start with Mountain Pony breeding. To Coed Coch Planed, she produced Weston Fair Lady in 1956, which in turn produced two 'queens', firstly Weston May Queen in 1962 by Gredington Iolo and then Weston Pearly Queen (f. 1964) by Gredington Simwnt. These mares made an early name for the Weston Stud in the show ring –they soon embarked upon breeding some of the best Section Bs for the Edwards family, but not before Pearly Queen was put to the pretty-headed Section A Brierwood Blue Boy to produce Weston Pearly Princess (f. 1965) and Weston Pearly Necklace (f. 1969).

Edwards used Criban Victor on a few mares, with particular success on the FS mare, Miss Crimpy Peek-a-Boo (f. 1955), a daughter of the good small Thoroughbred stallion Potato, out of Criban Brenda. Peek-a-Boo was very successful in the show ring, with arguably her best result coming at the 1966 Ponies of Britain Show at Peterborough, where she was judged Reserve Overall Brood Mare Champion. Her 1965 daughter by Victor was Weston Romany, an excellent producer for the stud and one selected to travel to Australia when the stud moved there in 1979.

There was an obvious need for a Section B stallion at home, and it was at this time that Edwards made his

Chirk Crogan winning under saddle at the Ponies of Britain Scottish Show in 1965.

most important move within Section B breeding when he bought the up-to-height colt, Chirk Crogan, to add some height when crossing with his Section A mares. By Coed Coch Blaen Lleuad out of Chirk Heather, he was shown successfully in hand and under saddle, where he remained unbeaten. Crogan carried a double cross to Criban Victor as well as Tanybwlch Berwyn on either side of his pedigree, which brought both size and substance to the Weston mares. All the time the Section A blood maintained Welsh character, which the family will tell you was largely responsible for the type being fixed within the stud and hence responsible for its success. He was an instant success on the Primrose line when he produced the cream dun filly, Weston Airbell (f. 1965) out of Weston Heather Bell, the product of Coed Coch Planed on Primrose. Airbell won as a yearling at the Royal Welsh in 1966, when her class second was Weston Romany.

It was the arrival of the first of the Crogan fillies that prompted Edwards to invest once more in a stallion for his expending Section B stud. This time he selected a foal, the grey Gorsty Firefly (f. 1965), yet again by Criban Victor but, following his instinct on the value of the Mountain Pony bloodlines, he secured a stallion whose dam, Gorsty Dusk, carried Glyndwr on both sides of her pedigree through his grandson Coed Coch Madog. Here was a case where a breeder invested heavily on two selected lines – a risk for some, but not for Edwards, who demonstrated the excellence which can come from line breeding; of all the qualities he maintained, it was the Welsh breed characteristics along with substance that remained intact. Firefly was only shown as a yearling, when he was a big winner including Champion Mountain and Moorland and Reserve Supreme at the Ponies of Britain Stallion Show at Ascot.

Crogan built up an array of top-class mares over the next few years and proved an ideal cross for some of the other mares that had been bought in as the stud stepped up its production of Section Bs while all the time maintaining a strong Section A herd. Put to Weston May Queen, Crogan produced Weston Sugar Plum (f. 1966), the dam of Weston Chilo (f. 1970) by Gorsty Firefly. Chilo had an enormous influence on the 1974 crop of foals at Llandyn Hall, including the 1975 overall Royal Welsh Champion, Weston Mary Ann (f. 1974) out of Weston Romany. Chilo was an outstanding success at Mark Bullen's Imperial Stud in Australia following his export there in 1974. Another Crogan colt, Weston Gigli (f. 1969), came from a different source but it comes as no surprise that there was yet more Glyndwr blood in his dam, the 1969 Royal Welsh Champion, Revel Glimpse by Kirby Cane Shuttlecock (grand sire Glyndwr) and out of Revel Gorse also by a son of Glyndwr. Gigli on Weston Pearly Princess produced the 1976 Royal Welsh Champion, Weston Rosebud (f. 1975), while Chilo produced out of her full sister Pearly Necklace, the stallion Weston Neptune (f. 1974), the sire of Weston Burgundy (f. 1979) and the 1979 Royal Welsh Champion, Weston Picture (f. 1978), both out of Llysun Blue Mist. Her son by Gigli was the much-acclaimed Weston Charmer (f. 1973), while her Chilo daughter was Weston Twiggy (f. 1976), a foundation mare at the Williamsons' important Eyarth Stud. Gigli's outstanding filly, Weston Carousel (f. 1976 out of Weston Romany) was one of the consignment destined to Australia in 1979.

Three Brockwell mares were among these, including Brockwell Penelope, which bred top-class stock by Crogan, among them the show mare Weston Vogue (f. 1965) and the stallions Weston Spider (f. 1966) and Paul (f. 1967). Brockwell Flirt, another daughter of Brockwell Cobweb, brought more Blaen Lleuad blood via her dam Criban Fair Faith to produce Weston Fair Faith, whose crossing with Gorsty Firefly could not have been more effective. Mention must be made of one of the most sought-after lines within the Weston Stud – that of Weston Choice (f. 1969), a granddaughter of Primrose, who had first been crossed with Coed Coch Madog to produce the filly Weston Princess Mandy (f. 1958) who, when put to Gorsty Firefly, produced the 1972 Royal Welsh Champion, Weston Choice. Incredibly, three of her sons – Olympian (f. 1977 by Neptune), named after their celebrated show hack, and Chippendale (f. 1978) and Cherrywood (f. 1979), both by Gigli – would head to Australia, where they brought all the successful pony families of Weston established at Llandyn Hall.

Lucrative prices came the way of Edwards during the time he was breeding ponies at Llangollen, and his regular drafts – initially at Gredington followed by Bangor-on-Dee and Fayre Oaks – brought good prices and an outlet for the riding ponies which augmented the

Left: Weston Chippendale, purchased as a foal for 1,700 guineas at the 1978 Weston Sale by Mark Bullen for his Imperial Stud in Australia, where he set up a great partnership with the other imported stallion Weston Chilo.

Below: Gorsty Firefly, a prolific breeder for the Weston Stud.

income of the Welsh ponies at Llandyn Hall. The three sales held at home will be best remembered, as prices were recorded for the Weston ponies like no others for Section Bs previously. The first came in September 1968, with the Foreword stating:

> They represent a fair cross section of the Stud, which is being reduced because of the accumulation of quality female stock sired by Chirk Crogan, augmented by this year's filly foals. Previous Sale Drafts have been selected with a view to retaining the best at home – now purchasers have the opportunity of selecting their choice from these ponies.

Chirk Crogan was offered but unsold that day; it

Weston Choice, Champion Female Royal Welsh in 1972.

was two years later that he was bought for 850 guineas at Fayre Oaks for the Seaholm Stud. Weston Airbell did find a buyer that day, however, when Mrs Meyer secured the Ponies of Britain champion at the top price of 1,200 guineas for her Keston Stud. Weston Sugar Plum was catalogued but not forward – which was as well for Edwards, as she produced Weston Chilo two years later. The major sale and the big process came six years later, when 64 lots were offered – the Section A mare Weston Mink Muff stole the show when Norma Book bought her for Australia at 2,300 guineas, the highest price at that time recorded at auction for a Welsh pony. The Section Bs also sold well – top price was Brockwell Japonica at 1,000 guineas to the Butterworths' Paddock Stud in Yorkshire; Gorsty Firefly, having had his time at Weston, went to Mrs Gadsden's Bengad Stud; and the big winner, Weston Vogue, went to the Thompsons' Cnapton Stud.

As we have witnessed from the Coed Coch sales, stud reductions and dispersals provided breeders an opportunity to buy stock otherwise not offered for sale publicly and the 1978 Weston Sale was no exception. Overseas bids came fast and furious for the early lots, with Mark Bullen – by this time a great fan of the Weston ponies – topping the sale with 4,000 guineas for Weston Choice, followed up with 1,700 guineas for her colt foal, Weston Chippendale. Weston Gigli by Chirk Crogan had been a successful sire at the stud and was now offered – he headed to another Australian, Mary McDonald, at 3,000 guineas, the same price as his dam, Weston Glimpse, which remained in Wales but headed south to Glamorgan. She was knocked down to Mrs Pat Johns-Powell for her Cottrell Stud, the home of her dam, Revel Glimpse. Section Bs continued to dominate the sale, with David Lawrence paying 3,600 guineas for Weston Neptune (unsold at 100 guineas in 1974) and Mrs Mansfield-Parnell taking Weston Lark (f. 1966 out of Weston Heather Bell) and her filly foal, Lapwing, home to her Rotherwood Stud for a combined price of 4,000 guineas.

The final sale at Weston, a year later in 1979, secured a few good prices including Weston Sunday Guest (f. 1978), a colt by Weston Gigli out of Weston Odett, which sold for 1,700 guineas to Eric Dudley for his Kirreway Stud in Australia. Alan Wilding-Davies bought the good breeder Weston Romany for Australia, but she never left the country. The black yearling filly Weston Louisa, a half-sister to the Royal Welsh Youngstock Champion, Weston Consort, was top price on the day at 2,000 guineas and has since proved her worth at the Cadlan Valley Stud in West Wales. The bargain of the day was

Weston Gigli (left) with Carol Jones and Weston Chilo with Janet Evans stand in front of the rostrum at the 1974 Weston Sale.

a filly, rejected by the buyers and to some extent by her breeder on account of her colour. Weston Twiggy, a blue-eyed cream by Weston Chilo out of Llysun Blue Mist, was spotted by the Williamson family from Clwyd, who bought her for 520 guineas along with the Chilo daughter, Weston Crystal (f. 1976) for 600 guineas. Both mares would go on to breed well for their Eyarth Stud – Twiggy in particular, which would keep the Weston prefix prominent right up to the 21st century.

This sale marked the end of an era for the stud's Section B enterprise in Britain, as Mr and Mrs Edwards headed to the other side of the world on account of Jack Edwards' deteriorating health. Having visited Australia to buy Murray Grey cattle, he and his wife decided to 'up sticks' and head to a new life and a healthier climate in Victoria. Their daughter Carol, with her livestock auctioneer husband John Jones, remained at home, where she has continued with the Weston prefix, although confined nowadays to Section A. Janet, with her veterinarian husband Dick Evans, went with her parents to Australia along with 29 ponies (of which nine Section Bs) to found the Weston Park Stud. During a visit to Wales many years later Janet purchased as a foal, a grandson of Weston Twiggy called Eyarth Sama – he was successfully shown prior to his export to Australia, where he has been a big success. This was followed up with the importation of the Dutch-bred stallion, Steehorst Kyro, another grandson of Twiggy.

In keeping with many farm livestock breeders, Edwards set great store in the quality of the head, as he noted in an article in 'Horse and Hound' in 1967. He commented: 'An over-large head in any stock means rough bone throughout, for the head structure carries right through the skeleton. It is the head that denotes character, and if you have a poor head you generally have a poor animal.'

Most Welsh Pony enthusiasts would certainly associate the Weston ponies with good heads, although this belies the fact that Edwards also got it right in other aspects of his breeding policy. He liked good conformation with bone and substance as well as free movement – while never compromising height. Most importantly, the ponies he and his family bred were Welsh in character through and through and it was probably this more than anything else that secured their popularity with the many breeders who would emerge during the latter part of the 20th century.

By this stage, unlike in the recent past, there was now a wealth of stallions and bloodlines available – largely thanks to the Foundation Stock Register – to take the Welsh Pony forward into the next century. All the breeders previously mentioned would have their place in that process as the current type of pony emerged within Section B of the Stud Book and the show ring became a principal focus of a strong home market.

Jack Edwards enjoys a joke with well-known judge Eric Worthington.

Chapter XII

Springbourne, Belvoir, Pendock, Mynd, Polaris

Water colour of Pendock Prince Pinnochio by Lionel Edwards.

Winning yearling Menai Shooting Star in 1962.

The Weston ponies were not alone in having a high level of Welsh Mountain pony in their breeding, as there was a trend within breeders of Section A to 'upgrade' to Section B as they witnessed the Section flourish and the potential to show them in breed classes grow. As we have seen, they responded to a demand for quality children's ponies as well as demand from overseas. In recent times, the stud name Springbourne has been associated with Mountain ponies – and good ones at that, for they have been extensive show winners at major shows and sought after by breeders at home and abroad. The stud was started by Miss Lorna Gibson, a Scot living in Wiltshire, who married in 1965 David Reynolds, one of the best-known showmen in Wales. He and his brother Cai were born into the world of Welsh ponies as their grandfather, John Reynolds, registered them as early as 1903 while living in Merthyr Tydfil and their father, J E Reynolds, who lived at Dowlais in the Welsh Valleys, shared that interest (he was a WPCS Council member from 1929 to 1932).

Pony mad since childhood, Miss Gibson started with Welsh Mountain ponies including Revel Fawn, one of two fillies purchased at Fayre Oaks in 1958. Fawn was a 'fairly ordinary hill pony' according to her new owner, but the saving grace was her colour – she was palomino like her grandsire, Cwm Cream of Eppynt. The following year, having decided that she needed something bigger to ride, she decided to travel for the first time to North Wales and to the Coed Coch Sale in 1959, where she hoped to purchase a Section B. She fell in love with Berwyn Beauty, but her funds couldn't match those of Mrs Binnie, so she decided to go for second best, her yearling filly, Coed Coch Berin (f. 1958 by Coed Coch Sandde), which she bought for the 150 guineas she had in her pocket. Along with Revel Fawn, she became a foundation mare for Section Bs at Springbourne.

By 1961 the search for a Section B stallion had begun, leading Lorna back to Wales, where she found a beautiful black colt foal, Menai Shooting Star, at Willie Jones's Menai Stud. Jones would not sell him, but sent him on lease on the condition that he would be shown. Shooting Star had a successful show season, winning the yearling colt class at the National Pony Society in 1962, and bred some good ponies while at Springbourne. Fawn and Berin bred well for their owner – especially Berin, whose first two fillies went to the Netherlands and Sweden. Her third foal by Menai Shooting Star, called Brocade (f. 1963), was a great producer. Eight of Brocade's 13 offspring were sold abroad, including Springbourne Bounty (f. 1967 by Solway Master Bronze). In 1971 Bounty sired Burstye Kythnos, reserve male champion at the Royal Welsh in 1972 for his breeder, Mrs Anne Knowles from Sussex;

he went on to a highly successful career in performance classes, with wins including the 1982 National Pony Society Working Hunter Pony of the Year.

The first of the Springbourne Section Bs to hit the headlines was Springbourne Golden Flute (f. 1966), a son of Revel Fawn and palomino like his mother. He was by Reeves Golden Lustre, which was standing at Stud at Springbourne in 1965 although his fertility by this time was not so good and Fawn was one of only two mares in foal to him that year. Although on the small side at 12.2 hands, Golden Flute's quality and pretty Welsh head carried him through; he won the Royal Welsh as a yearling as well as being Reserve Male Champion in 1967, and then went one better the following year when he beat Sinton Moving Charm for the championship. The result provided the Reynolds (now married) with a great double, as they produced Treharne Tomboy for Col Rosser-John to win the Section A championship in 1968.

Coed Coch Berin.

Golden Flute would end his days in the Netherlands, but not directly from his breeders, who sold him to Mrs Ridgeway at Fayre Oaks in 1969 for 1,200 guineas.

This was not their only reason for celebration, as their homebred Section B colt, Springbourne Blueberry (f. 1967) took the yearling class; he returned in 1971 as a four-year-old to take the stallion class. Blueberry was a son of Coed Coch Berin, substantial in size and build, something that defied his breeding as he was by the Section A stallion, Brierwood Blue Boy by Revel Pattern. Although it was a great success, the cross was not planned, as Berin had visited Solway Master Bronze that year but turned at six weeks – whereupon they decided to cover her with Blue Boy, who had been sent by his breeder, Captain Brierley, for showing. Blueberry proved to be a good stallion at Springbourne but had the additional attribute of being a wonderful ride. He was regularly hunted and on his only show ring appearance under saddle was placed second at the National Pony Society Show. Mrs Soster bought him for Australia, where he was unbeaten under saddle. Springbourne Brocade was exported to Jannys McDonald's Glenmore Stud in Australia in 1980, the year before the last foal was registered at Springbourne.

Another breeder with roots in Section A was Anne, Duchess of Rutland, who began her Welsh pony interests while married to the 10th Duke of Rutland at their home of Belvoir Castle, a stately home set in 15,000 acres near Grantham. Their marriage in 1946 at St Margaret's, Westminster, attracted great crowds as the fashion model and equestrian, Anne Cumming Bell from Yorkshire, married the most eligible bachelor in Britain – a former Grenadier Guard who had inherited his title and fortune

Above: Springbourne Blueberry, Royal Welsh winner 1971.

Left: Springbourne Golden Flute, Champion at NPS and Royal Welsh in 1968.

when aged 20. The Duchess adopted the castle's name for her stud, which was started in 1949 and made up of Mountain ponies; she moved to Herefordshire in 1964 and registered her last pony in 1982, when the stud was dispersed.

She was attracted to palomino or cream-coloured ponies with attractive heads – characteristics that would be associated with the Belvoir ponies throughout the stud's history. Criban and Cui bloodlines from the Richards brothers, Llewellyn and Dick, found favour with the Duchess, whose instinct had taken her, perhaps unwittingly, towards Coed Coch Glyndwr; the foundation ponies she selected were full of his breeding, especially through his son, Criban Winston. Among the stallions she used, it was the influence of the cream stallion Dyrin Goldflake (f. 1949) that would influence both the Welsh character and colour of the Belvoir ponies in both Section A and Section B. Bred by well-respected breeder, judge and Council member, Gwyn Price, who farmed near Sennybridge on the Eppynt, Goldflake was line-bred to Glyndwr through his dam, Criban Vanity, and his sire, Criban Cockade, the son of Criban Socks. Goldflake would have a great influence on the foundation of the Section B ponies at the stud through his son, Belvoir Gervais (f. 1955) and grandson, Belvoir Talisman (f. 1958).

Gervais was full of Section A breeding, although a small Section B himself – his palomino dam by Criban Winston was Jasmine of the Golden Fleece (f. 1944), which was bred by Miss de Beaumont, who latterly owned Coed Coch Glyndwr at her Shalbourne Stud in Wiltshire. Talisman was by Cui Florin, a son of Dyrin Goldflake, while his dam, Belvoir Trinket (f. 1954), was by Goblin, a dun Section A colt by Glyndwr, again bred by Miss de Beaumont. Trinket's dam, Meifod Tlws (f. 1951), was one of the foundation mares at the Stud, and herself a double cross of Glyndwr. To Dyrin Goldflake she produced in 1955 a filly, Belvoir Tulip (later exported to Canada), which, put to Coed Coch Berwynfa, produced the beautiful grey mare Belvoir Thalia (f. 1962), highly successful for the Duchess in brood mares classes. Thalia not only took championships at major shows (although never the Royal Welsh, where she was second on three occasions and third twice), but she was also a successful breeder – especially when put to the Duchess's young stallion, Belvoir Zoroaster (f. 1965); of note is her son, Belvoir Turks Cap (f. 1971), which was exported to Germany, and the two sisters that followed, Hollyhock

Dyrin Goldflake, resident stallion at the Belvoir Stud.

Anne, Duchess of Rutland holding Belvoir Taffeta (left) and Belvoir Hollyhock.

and Hyacinth. Hollyhock was a favourite of the Duchess and was sold to Dr and Mrs Maze for the Congarinni Stud in Australia.

Zoroaster came from a different line, but through a pattern of breeding which attracted his breeder and reflected the import of Arab through the Foundation Stock Scheme. The FS mare Fairy Gold (f. 1945), a chestnut by the Arab stallion Indian Grey, had been admired by Mrs Mountain on a judging trip to the West Country and subsequently joined the stud at Twyford, where she bred exceptional stock by Solway Master Bronze and the like. Her 1953 filly by the Mountain Pony stallion Cui Beau was called Belvoir Zinnia, which bred for the Duchess and later retired to Scotland, where Robert Graham had started breeding Section Bs modelled on the Belvoir ponies. He also purchased Gervais, which was produced under saddle to take the championship at the 1967 Ponies of Britain Scottish Show at Kelso;

An oil painting of Belvoir Zoroaster by Richard Dupont, commissioned by Anne, Duchess of Rutland.

together Gervais and Zinnia produced for Graham the 1970 Royal Highland Champion, Balone Electra (f. 1966). At Belvoir, Zinnia bred well to Gervais, producing among others Zenobia (f. 1960), which was sent to West Wales, where she visited Downland Chevalier. She produced the exceptional colts, Belvoir Zoroaster in 1965 and Zechin in 1969. Zoroaster was shown extensively in hand, winning among other top awards the youngstock championship and reserve supreme at the Ponies of Britain Stallion Show in 1968, the same year that he was Champion at Glanusk and a winner at Northleach. After making his mark at the stud as a sire, he was sold to the Netherlands.

Another mare of Arab breeding to make her mark on the Belvoir ponies was Belvoir Tangerine (f. 1945), a cream-coloured mare bred by the Honourable Mrs Vaughan Williams. Tangerine, by the Arab, Azim, crossed very well with the small-height Belvoir stallions with all their Mountain pony blood; both Belvoir Gervais and Talisman were used on her with great success, thus providing their breeder with a host of lovely females such as Tansy (f. 1962), Tiger Lily (f. 1963) and Tosca (f. 1964), which was Champion at the Royal Welsh in 1965 for Mrs Binnie. Tansy put to Zoroaster produced Belvoir Tiarella (f. 1970), the dam of the Royal Welsh winner, Belvoir Endymion (f.,1974) by Lechlade Scarlet Pimpernel.

The Duchess of Rutland was always on the lookout for an outcross which would maintain her pretty pony type as well as maintain the colour. Drawn to Solway Master Bronze, whose stock were prominent in her area, she purchased Lechlade Scarlet Pimpernel after being enchanted by him one day when he was shown with his dam, Arum. As Rae Cashmore, Assistant Publicity Officer of the Australian Welsh Pony & Cob Society, wrote in her society's 1982 *Journal*, 'Put to daughters of Belvoir Zoroaster, Lechlade Scarlet Pimpernel got exactly the Welsh ponies expected of him – ponies with beautiful heads, lovely action, quality and "Welshness". With the other mares of Belvoir he combined to produce outstanding stock.'

Dr and Mrs Maze were fully committed to the Belvoir ponies and added to their stud the mares Belvoir Taffeta (f. 1973) and Belvoir Brocade (f. 1978), and the stallions Belvoir Zechin and Lechlade Scarlet Pimpernel, which sadly died after a relatively short stay at Congarinni. In order to make up the loss, the stud secured another top-class stallion to replace his sire – Chamberlayne Don Juan (f. 1975), whose dam, Chamberlayne Nell Gwynne, carried both Downland Chevalier and Coed Coch Blaen

Lleuad through her sire, Tanlan Julius Caesar. Another of Scarlet Pimpernel's sons to be exported to the southern hemisphere was the striking dun stallion Belvoir Jasper, who was successfully shown in hand and under saddle before his sale to New Zealand.

Prior to the stud's dispersal in 1982, the Belvoir ponies were in great demand both at home and overseas – a great many found homes in Europe as well further afield, as we have seen. At various stages in the stud's history, outside stallions were brought in as an outcross – such as Brockwell Berwyn and Pendock Plunder, the sire of Belvoir Taffeta. Plunder was bred by the Duchess's neighbour and great friend, Miss Rosemary Philipson-Stow, whose mother had started their very successful Pendock Stud at Priors Court and later Barlands in the village of Pendock near Malvern. Her father, Major Guyan Philipson-Stow, was a son of Sir Frederic Philipson-Stow, a Cape Town lawyer for the diamond firm de Beers and a leading personality during the early years of South African politics. Major Philipson-Stow bred Red Poll cattle which, like the ponies, continue to be bred at Pendock. Miss Philipson-Stow has been a prominent member of the WPCS, including its President in 1994–1995, Council member, Chairman and well-respected judge on all four panels.

In 1934 Miss Philipson-Stow's godmother gave her a pretty liver chestnut filly foal called Peggy by Revel Chief, bred by John Griffiths (father of Emrys) of the Revel Stud. Peggy proved to be a brilliant first pony for the Philipson-Stow children and, when outgrown, she was sent off to the Craven Stud in 1939. The intention was to cover her with Grove Sprightly, but on the advice Tom Jones Evans she went instead to Craven Cyrus by the Arab King Cyrus. The result was a chestnut colt named Pendock Prince Pinnochio, registered in Section A by Mrs Philipson-Stow and Miss Elizabeth Morley, who broke the ponies to ride and with whom she was initially in partnership. The success of Peggy in the show ring sent Mrs Philipson-Stow back to the Revel, where she bought two fillies foaled in 1937, which she registered as Pendock June and Pendock Bunty, both by Revelation, a son of Criban Chief like Peggy's sire, Revel Chief; June was 12.2 hands in height and quite thick-set of build.

When Pinnochio covered Revel June in 1942, little did his breeders realise that they had embarked upon a breeding policy which would take them to the top in Welsh Section B, although it would be for Section A ponies that Pendock would be best known in the years to come. The mating gave them a bay filly, Pendock Bambi, which was initially used to breed Mountain ponies, which were particularly successful under saddle. Pendock Sabre's success under saddle at White City in 1954 should have been sufficient to point Mrs Philipson-Stow towards breeding Section B from Bambi, as he was by Tanybwlch Berwyn. After a succession of colts, Bambi eventually produced a bay filly in 1959 by Criban Victor, which was aptly named Patience, and then another by him called Pandora (f. 1961).

Like all the ponies bred at Pendock, there was – and remains – an emphasis on suitability for hunting and riding in general, so it was no wonder that Patience should embark on a successful career under saddle as well as in hand. She went hunting, and won not only the Working Hunter Pony Class at the Badminton Horse Trials in 1964 but also the Section B brood mare class at the Royal Welsh in 1967 under Dick Richards (Criban R) – and her foal,

Pendock Patience taking part in a Pony Club One Day Event in 1964.

Pendock Prudence (f. 1967), by the famous stallion Coed Coch Berwynfa out of Pendock Patience, became a corner stone of the section B ponies bred by Miss Philipson-Stow.

Pendock Prudence, was second that day. By Coed Coch Berwynfa, Prudence brought the famous cross with Criban Victor to the fore once more and provided her breeder with a springboard for breeding some of the most successful Section Bs of the day.

The success of this stud lay in a very strong female line stemming from the Section A. Bambi's second filly, Pandora was successful in her own right, although less prolific than her sister, Patience, which produced a number of colts which were exported. One that remained at home and excelled in ridden classes was the bay Pendock Plunder (f. 1975), by his namesake

Kirby Cane Plunder. Plunder's sire was Kirby Cane Gauntlet (another dark bay), whose dam was the good breeder Kirby Cane Plume, the daughter of Coed Coch Blaen Lleuad and Coed Coch Pluen. Gauntlet became the resident stallion at Pendock and crossed really well on Prudence: a filly from the breeding named Poise was kept in the stud, while others were sold overseas. These included Priscilla (f. 1976) to Germany and Pia (f. 1975) to New Zealand. It was to New Zealand that Patience's colt by Kirby Cane Plunder, Pendock Pirate (f. 1969) travelled, while Prudence's colt Pioneer (f. 1974), also by Kirby Cane Plunder, went to Sweden.

The Kirby Cane breeding obviously suited the Pendock mares – and no wonder, as they were full of the 'old' breeding on which Mrs Crisp had started her own stud. The latest stallion used was Kirby Cane Golden Rod, a stallion which stood at stud nearby at Hereford and belonged to Tom Williams of the Vardra Stud, who showed him with great success – a champion at Glanusk in 1971, he was three times champion at Lampeter. Prettier and possessing more quality than some of the Kirby Cane ponies, when put to Prudence he produced in 1975 the filly Pendock Prunella, the dam of the big winner, Pendock Penny Royal (f. 1984 by Pendock Plunder). Penny Royal produced one good foal after the other including, in 1996, Pendock Penny Lane, whose sire, Duntarvie Nominator, carried yet more Kirby Cane breeding through his dam, Longnewton Sundance (by Kirby Cane Sundance). Nominator's sire was Duntarvie Crusader, a son of the champion ridden stallion Longnewton Maestro, himself a son of Keston Royal Occasion. Bringing the line right up to date, Penny Lane produced in 2002 a free-moving chestnut filly, Pendock Pennyworth by Lindisfarne Guardsman. Pennyworth has been a major contender in ridden classes, qualifying annually for the Horse of the Year and Royal International Horse shows and Champion at the latter in 2010.

Lindisfarne Guardsman was bred in Northumberland on the mainland overlooking the Holy Island of Lindisfarne by Sheila Henderson, who has continued to breed a small number of Welsh ponies based on bloodlines assembled by her near neighbour, Mrs Teresa Smalley, whose Mynd Stud provided the Lindisfarne Stud with a number of foundation mares when it was dispersed in 1982. Mrs Smalley was almost bred into Welsh ponies as her father, Julian, was the younger son of Charles Coltman Rogers, a founder member of the WPCS who bred Welsh ponies and cobs at the family's Stanage Park in Radnorshire. Like her family, Mrs Smalley's first steps into Welsh breeding were taken with Mountain ponies, to be followed for a short time with a few cobs, although her main interest settled on Section B.

Although she showed her ponies all over Britain, Mrs Smalley had a remarkable run of success at the Royal Highland, where her homebred ponies stood either champion or reserve between 1972 and 1977. The first of them was Mynd Coral, a daughter of Reeves Coral (out of Reeves Crystal) and Gredington Mynedydd, Lord Kenyon's Coed Coch Planed colt. Coral was reserve champion to Kirby Cane Vogue in 1972. Her half-sister, Mynd Cowrie by Downland Romance, went one better the next year when she claimed the championship. Cowrie was a successful mare at the Mynd Stud before joining the Lindisfarne Stud in 1982, the year in which Mrs Smalley registered her last foals. Interestingly, there was little

Downland blood to be found directly at Mynd other than that of Mynd Gala, a pretty chestnut mare by Downland Chevalier out of Mrs Borthwick's FS1 mare Starlight, a Fayre Oaks purchase in 1967. Another trace of Downland came through a stallion which stood at Mynd for a few seasons – the attractive grey colt Collenna Cnicht, bred by Mrs Smalley's great friend Ann Colbatch-Clark. He was by Gredington Mynedydd and out of Collenna Ladylove, sired by Downland Dauphin and out of the Coed Coch Berwynfa mare Wyrhale Cariad.

1974 brought further Royal Highland success for the Mynd Stud when Merriemouse was judged Champion. She was by Gredington Mynedydd and out of Criban (R) Field Mouse, which was a foundation mare at the stud along with her Section A dam, Criban (R) White Mouse. These mares were bred by Dick Richards and his wife, Sheila, who was a great friend of Mrs Smalley and her husband, Basil, who farmed at West Kyloe, Berwick-upon-Tweed. By this time, the resident stallion at the stud was one of the last colts bred by Miss Brodrick, Coed Coch Pedestr, a grey son of Coed Coch Berwynfa and Coed Coch Pannwl (Lady Margaret Myddelton's purchase in 1959). Pedestr was full brother to Lady Astor's Coed Coch Perfagl. He exerted quite an influence on the stud; to Field Mouse he produced in 1967a filly, Mynd Tittlemouse. She in turn produced the 1975 Reserve Royal Highland Champion, the yearling Mynd Titus by Collenna Cnicht. For the following two years another of Pedestr's daughters led the ranks, taking reserve in 1976 and Champion in 1977. Mynd Nesta came from another foundation line established by Mrs Smalley – still based on Criban lines, as her dam was Bwlch Nereid (by Kirby Cane Pirate out of Criban Nella), purchased at Fayre Oaks in 1965.

Above: Coed Coch Pedestr at Mynd Stud, Northumberland.

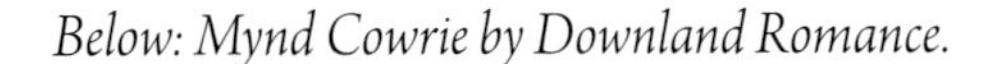

Below: Mynd Cowrie by Downland Romance.

Like others of her generation, Mrs Smalley enjoyed much respect within the Welsh Pony community. She was a Council member of the Welsh Pony and Cob Society, serving as its President from 1985 to 1986, as well as a founding member of the first Welsh Pony and Cob Breeders Association, started in 1962,. Her breeding policy bridged the gap between the old bloodlines of Coed Coch and Criban while adding that touch of Downland, which was seen to be a way ahead for many of the breeders entering the latter part of the 20th century. The same could be said for a true stalwart of 'old' breeding, Rosamund Greaves, 11th Countess of Dysart, whose Polaris Stud became one the largest strongholds of Coed Coch bloodlines after the latter stud had been dispersed. She was a great personal friend of Miss Brodrick and of Lady Wentworth, another enthusiast of the Coed Coch ponies with whom she shared a love of Arab horses. If anyone can be credited with keeping the Coed Coch breeding policy – especially in the Mountain ponies – going until her death in 2003, it was Lady Dysart.

The Dysart title originated in 1643, when William Murray, a great lifelong friend of King Charles I – and his whipping boy as Prince of Wales – was granted the earldom, as well as the title of Lord Huntingtower. Lady Dysart succeeded her mother, who had married Major Owain Greaves of the Royal Horse Guards (Blues), whose uncle, R M Greaves, bred Welsh ponies at The Wern, Portmadog. A founder member of the WPCS, R M Greaves was its Vice President from 1911 to 1912; his niece was President in 1979. Greaves also bred Welsh Black cattle, another of Lady Dysart's passions in life next to her ponies. Having spent much of her life in Scotland, the very popular Lady Dysart boasted her Scottish ancestry with great pride, while always extolling the virtues of her Welsh roots, which accounted for her love of Welsh ponies and cattle.

Lady Dysart lived in Scotland for a large part of her life, initially at Stobo Castle – where she was buried in 2003 alongside her parents in the small graveyard next to the hill – then at Beaufort Castle near Inverness. It was here in 1962 that she started her stud, which consisted largely of Mountain ponies. Her stud name came from the North Star (Polaris), a traveller's guide and inspiration for generations. Aware of the long distances to shows where the Polaris ponies were enjoying considerable success, Lady Dysart moved first to North Devon and finally to Bryngarth, with its 1,000 acres of hill land near Abergavenny. She loved nothing better than to show people around her stud of ponies, which at times numbered over 200.

As Lady Dysart's interest in the Coed Coch bloodlines grew, so too did the Polaris Stud, regularly augmented by ponies from the sales, where she was a regular buyer. She made several purchases at the 1963 Coed Coch draft offered following Miss Brodrick's death, and again in 1979 at the Coed Coch dispersal sale, where she bought Coed Coch Tarian, the top-priced Section A mare at 4,200 guineas. It was while at Bryngarth that she decided to concentrate her efforts on breeding Section B, as demand at the time was healthy. She turned once more to the Coed Coch bloodlines and in 1975 purchased a mare, Coed Coch Gwenfron (Coed Coch Berwynfa x Coed Coch Pannwl), a full sister to Perfagl. She was purchased from her niece, Mrs Diane Haak, who arguably bred the best out of her when Uplands Genevieve by Downland Tanglewood was foaled in 1974. Genevieve joined the Polaris Stud and was lightly shown by Lady Dysart, winning among others the Supreme at the Northleach Show in 1984, produced by Len Bigley and judged by Dr Davies.

Lady Dysart gathered together at Bryngarth a herd of mares of mixed breeding, ranging from her initial buy, Downland Jasmine (Downland Chevalier x Lydstep Jasmine), to Baledon Bronze Poppy (Solway Master Bronze x Gredington Blodyn by Coed Coch Berwynfa), while maintaining Coed Coch breeding as and when she could acquire it. One of her foundation mares was Hazelwood Verbena, whose sire was Ardgrange Heritage (Downland Dauphin x Chirk Deborah) and whose dam was Weston Veronica (Coed Coch Salsbri x Weston Peke Victoria by Criban Victor). She bred well for the Countess, and progeny out of her were in demand at the stud's dispersal sale in 2004, which was held at the Royal Welsh showground. Polaris Euros secured the top price of the day of 3,300 guineas, while his sire Horsegate Spark sold to the Thornberry Stud for 2,000 guineas. Second top price at the sale went to another Verbena offspring, the 11-year-old bay mare, Polaris Elfrida, which sold for 2,800 guineas.

It could be argued that the best of her offspring on offer was the grey 18–year-old daughter, Polaris Elsie, one of Lady Dysart's favourites; she was by Paddock Mistral, the product of the Coed Coch ponies Lygon and Penwn, both purchased by the Butterworths at the 1978 sale. Another bred by them, Paddock White Lightning, line-bred to Coed Coch Gala through his sire Paddock Camargue and dam Paddock Garland, also stood at the Polaris Stud and was later sold on the 2004 sale. Although Elsie was one of the top prices at 1,400 guineas, it was for her son, Polaris Elmer, that she and the Polaris prefix will go down in WPCS history. Born in 1994 by the Chirk-bred stallion, Polaris Fagus, Elmer had taken the show ring by storm at the same venue only a few months before, when Dewi Evans (Hilin Stud) selected him as Male Champion of the Welsh Section Bs at the Centenary Royal Welsh Show, presided over by Her Majesty the Queen. Owned by Mr and Mrs Bethell and shown by their son, David, this up-to-height big mover impressed at Llanelwedd – none would have been more proud than his breeder, who was surely looking down on him that day.

In his report of the Polaris dispersal sale in the 2005 WPCS *Journal*, the Editor, David Blair wrote:

> The Countess invested forty years in establishing a type of pony she was extremely proud to see in her fields. The interest at the sale reflected the appreciation other breeders had for her ponies. With the ponies now dispersed into many different studs, it will be interesting to see their influence in the years to come.

It will indeed be interesting as, during her time breeding Section Bs, Lady Dysart bred more than many of her day – registering 144 fillies and 117 colts. She was the last of her generation with an appreciation for the work of Miss Brodrick, although there were others with a determination to maintain Coed Coch bloodlines into the 21st century. Ruth Thacker has been one of them. She already had as her stud stallion Coed Coch Pele (f. 1978), a colt by Coed Coch Tysilio (by Lydstep Royalty x Coed Coch Priciau); his dam was Coed Coch Peleu, a Section A granddaughter of Tanybwlch Prancio. Pele

was sold as a foal at the 1978 Coed Coch Sale for 100 guineas, and for Mrs Thacker this big-moving grey built up a great show reputation in the South East, where she lived and bred, among others, Farchynys Rhianwen, Mrs Mountain's Youngstock and Reserve Female Champion at the Royal Welsh in 1996. To complement her Coed Coch bloodlines, Mrs Thacker added Seaholm Dawn, which she purchased for 1,300 guineas at the Polaris Sale. The 17-year-old's dam, Coed Coch Dawn, had been sold to Mrs Weston at the Coed Coch dispersal sale.

Polaris Elmer in full flight with handler David Bethell at Swansea Show. Elmer was Male Champion at the Royal Welsh in 2004.

The breeders who have become custodians of the Coed Coch bloodlines in Britain are Mrs and Mrs Bethell of the Cross Foxes Stud, Llandysul, Carmarthenshire, mentioned previously as the owners of Polaris Elmer and now of the stallions Polaris Durward, a son of Polaris Euros, and Polaris Dorcas and Polaris Farmer's Boy by Polaris Dial out of a daughter of Coed Coch Dawn. With a family connection to Lt Col Williams-Wynn and a personal friend the Coed Coch 'Boys', Davinia Bethell felt compelled to carry on the work of Coed Coch and said as much when she wrote, 'By 1978 I had made the decision that I would continue these lines and keep them as pure as possible, for as along as possible'.

She started in Section B with the foundation stallion, Coed Coch Onid (born 1974), which was by Coed Coch Berwynfa, followed by Coed Coch Nestor (born 1968), a grandson of Berwynfa through his sire, Coed Coch Gawain. The stud's foundation mares included Coed Coch Aden and Coed Coch Gwenfron, both by Berwynfa, Coed Coch Nicola by Targed and Coed Coch Patsy by Tysilio. The stud's commitment to Lady Dysart is no less great, with seven Polaris mares at Cross Foxes as well as Elmer's dam and his full sister. Again, Mrs Bethell's words express the stud's commitment to Lady Dysart and the type of pony she tried to breed: 'Her lines were carefully picked to breed proper Section Bs able to do a job of work. These were her words, but ones that I wholeheartedly agree with.'

Burstye Kythnos, Royal Welsh winner in hand and a famous champion in ridden classes.

There is no doubt that the Cross Foxes Stud has rejected the popular show ring model of the Section B for one that is more robust and workmanlike, and in this respect it stands alone in its aspirations. Consequently its owners have done everything possible to preserve the old bloodlines in order to achieve this. It will be interesting to learn if breeders turn to this remarkable gene pool in the years ahead. That does not mean to say that the Coed Coch bloodlines were rejected completely by other breeders, as we shall discover in the following chapters. Far from it, a recipe was found simultaneously by several breeders that would produce an ideal cocktail to meet the demands of the late 20th century.

Chapter XIII

Pennwood, Lydstep, Breccles, Ardgrange, Rhoson, Rosedale, Cennen, Sianwood, Tetworth

Rhoson Shem takes the championship at Glanusk in 2004.

Downland Finch, the Downland Mohawk daughter which was reserve Section B champion at the Royal Show in 1988 when shown by Len Bigley for Stanley Griffith.

Just as there was support for the Coed Coch bloodlines in North Wales, so too was there a following for Mrs Cuff's contribution to the Stud Book through her Downland breeding in the South. Several breeders made choices based on genetic preferences and had the funds to support the travel costs to either or both; however, for many it was a case of cost effectiveness and logistics. It may also have been a case of liking that to which you are accustomed and, during the early days of her stud, Mrs Cuff showed extensively in the South Wales area as well as further afield, so enthusiasts regularly saw the type of pony she was breeding – and chose to follow her example.

One of the first to come under the Downland spell was Clive Morse, who lived close by Mrs Cuff while she lived near his hometown of Llandovery. As a boy he rode his little mare, Bess, up to Mrs Cuff's farm at Llanddeusant to be covered by the 14.2 hands part-Arab, Downland Dominie, and then rode her home again. Bess' colts were not registered but sold at the local market at Llanybydder. However all that changed in 1968, when he bought at Hay-on-Wye Sale a yearling Section B filly, Sinton Fairy Bell (f. 1967), by the Gredington Mynedydd stallion, Sinton Bellhop, and out of Sinton Chablis (f. 1963), again by Mynedydd; he later bought Sinton Claribel (f. 1965) and Sintoncourt Airbell (f. 1969). At much the same time he also purchased Springbourne Bunting (f. 1969), a big strong grey filly by Springbourne Golden Flute. These mares had riding qualities – not as good as those of the Downland ponies that he had admired so much as a lad, but he felt he couldn't afford them at the time.

His first foals were registered with his Pennwood prefix in 1971; Fairy Bell had a colt by Downland Chevalier, while her dam had fillies by him in both 1971 and 1972. Once more smitten by the Downland ponies, Morse embarked upon a quest for ponies of Mrs Cuff's breeding and became a great friend and client over the ensuing years. At the 1971 Fayre Oaks Sale, he was able to buy the 14-year-old mare Downland Dawn Chorus, by Ardgrange Llun Gwyn out of the renowned Downland Misty Morning. She proved to be a bargain at 400 guineas, since the following year she produced a beautiful bay colt foal by Downland Chevalier which was named Pennwood Mujib (f. 1972). Destined for stud duties with Morse, he was first shown by the top professional, Colin Rose, who enjoyed many good results with him at the early stallion shows at Ascot and Malvern.

Having made a start, he went on to add to his stud Downland Rosewood (f. 1971), second top price at the 1974 Fayre Oaks Sale – his strategy of bidding at the opening asking price of 1,700 guineas by well-known auctioneer, Michael Wyatt, wrong-footed others also keen on the mare. By Downland Chevalier out of Downland

Camelia, Rosewood was one of his favourites, and bred what he wanted in the form of a bay colt born in 1978 by Mujib called Pennwood El-Dorado. He later became a resident stallion at both the Weston and Cottrell studs. Just as Mrs Cuff found that her own breeding crossed well on itself, her breeding away from home at Pennwood did exactly the same. Mujib was a great success on the Downland mares – Delphine produced to him Indiana (f. 1979), the dam of Milan, a winner at Ascot as well as coming second at the Royal Welsh and Royal shows as a yearling. Other mares at the stud at this time were Downland Carmen, Demelza, Titlark and Downland Finch, the Downland Mohawk daughter which was reserve Section B champion at the Royal Show in 1988 when shown by Len Bigley for Stanley Griffith – whose Merryment Stud in Cheshire was also based on Downland lines. Griffith bought Downland Titlark from Morse and she also won the Royal Show before being exported to Australia. Several of Morse's stock found themselves 'Down Under', including the fillies Santa Fe (f. 1977) out of Rosewood, Mukima out of Dawn Chorus and the highly successful colt, Pennwood Bodine (f. 1977) out of the Tanlan Julius Caesar mare, Shimpling Madonna – whose dam was by Downland Dauphin. Full brothers to Bodine, Emperor (f. 1978) and Monsanto (f. 1980) were sold on as stallions, the former to Donald Cooke's Bureside Stud and the latter to Mr and Mrs Rennocks to breed riding ponies.

There were several show ring successes for the Pennwood ponies over the years. Pennwood Milan, the 1985 colt out of Pennwood Indiana (out of Downland Delphine), won extensively in colt classes before he was sold to David Williams's Moelgarnedd Stud at Bala. Milan was the sire of Eyarth Rio. Later came Pennwood Nicosia, champion at Lampeter as a yearling in 1998 and a winner at the Royal as a two-year-old; she was by Merryment Moondreamer, a colt by El-Dorado which Finch was carrying when sold to Griffith.

Business commitments along with his personal circumstances and health problems curtailed Morse's continuing breeding programme, and twice he has sold out his herd only to come back again. By 2012, Wiltshire sheep had become a more manageable option, with no plans to return to Welsh pony breeding. Like many breeders, while retaining a taste for beautiful ponies, his ideal type of Section B had turned away from the larger quality riding types and moved more towards a pony showing more Welsh characteristics.

Following her move to Cardiganshire, Mrs Cuff found a niche market for her stallions and a keen interest in her foals as the Downland name spread and flourished. One stud which took advantage of the bloodlines on offer was that of Lydstep, based nearby at Tenby in south-west Wales and owned by Mrs Patricia Hutt and her sister, Miss Jo Pringle. They were brought up on a small gentleman's estate near the village of Buckland St Mary in Somerset, where their father, Lt Col John C Pringle, had retired following an Army career with the Royal Engineers. He had seen action in Mesopotamia and India, where he played polo and hunted, activities he maintained on his retirement; his two daughters had ponies and enjoyed hunting at an early age, although Mrs Hutt was never as interested as her younger sister.

Miss Pringle trained in physiotherapy at St Thomas' in London; her professional skills and equestrian interests merged when she took an active role in the Riding for the

Disabled Association and very quickly she recognised a link between horse riding and physical rehabilitation. In the early 1970s she became one of the foundation members of a clinical interest group which introduced a formally-recognised examination based on the Horses in Rehabilitation, which remains a professional qualification. In 1999 she was awarded the MBE for her work with disabled riders. In the late 1940s her sister had come to live at Lydstep following her husband's army posting to Manorbier. Miss Pringle opted to join her sister there and found a job locally; together she and her sister pondered what to do with the field adjoining the house, which was crying out for apony or two.

It was during the early 1950s that they decided that a Welsh pony might provide the fun and interest they envisaged, and in 1952a sale at the Cui Stud beckoned. They bought two colts, which set them off on their pony journey, but more importantly they met Llewellyn Richards, who would become their mentor and great friend. They showed Cui Red Robin with some success and Richards was so impressed that he sent them a filly to show which was of a better standard than the geldings and worthy of their production skills, which he admired. The filly, by Bwlch Valentino, was Criban Viola (f. 1956), a grey which they showed successfully and broke to saddle before she embarked on a highly successful ridden career with the well-known Bullen family. On retirement from the show ring, Viola joined Mrs Egerton's renowned Treharne Stud and was responsible for the full sisters by Shalbourne Camelot, Treharne Angela (f. 1969), youngstock champion at the Three Counties Show and Treharne Stephanie (f. 1970), overall Section B at the Royal.

Lydstep Lady's Slipper, Royal Welsh Champion and Lloyds Bank finalist at HOYS in 1977.

Buoyed by the success of Viola, the sisters purchased Criban Lily, born in 1955 by Bwlch Valentino, and another Valentino mare, Cusop Polly Garter. They attended the 1959 Coed Coch Sale, where the ponies they preferred sold over their budget; guided by Richards, they settled on two mares which brought Miss Brodrick's Coed Coch blood into their stud – Melai Ruth (f. 1955) by Coed Coch Seren Aur at 180 guineas, and her 1959 foal by Coed Coch Blaen Lleuad, Melai Ruthene, at 110 guineas. Of the two, it was Ruthene which made her mark, by producing two significant colts – Lydstep Ronald (f. 1964) by Downland Romance and Lydstep Royalty (f. 1965) by Downland Chevalier. Following her sale in 1966 to the Tanlan Stud, the foal she was carrying by Chevalier turned out to be the influential stallion Tanlan Julius Caesar, which Tanlan sold in 1968 at Fayre Oaks for 600 guineas to Mrs Gowing for her Shimpling Stud.

Lydstep Blondie with her new owner Mrs Archdale (left) and Mrs Hutt, her vendor, at 1969 Fayre Oaks Sale.

The journey to North Wales brought with it an introduction to other Section B enthusiasts who proved to be most helpful – none more so than Eddie Griffith, who agreed to sell them Pleiad, a daughter of Coed Coch Planed and his wife's FS mare, Twinkle by the Thoroughbred, Gay Presto. Pleiad was in foal to Criban Victor and produced his daughter, Lydstep Lyric in 1961; Lyric was one of the mainstays of the stud and crossed well with the Downland stallions. To Chevalier she produced two outstanding daughters, Lydstep Toffee Apple (f. 1966), the dam of the influential stallion Radmont Tarquin, and Tarantella (f. 1967), a beautiful filly which sadly succumbed to grass sickness. In 1961 Criban Lily produced a filly to Coed Coch Berwynfa called Lydstep Jasmine (f. 1961), a winner as a three-year-old at the Royal Welsh. Given the success that Jasmine gave Mrs Cuff when she joined Downland, perhaps it was a rash move – especially since Lily's beautiful daughter, Lydstep Rosetta (f. 1964) by Downland Dauphin died at a very early age, like her sire and Tarantella, from grass sickness. The three-year-old Rosetta gave her breeders their greatest showing success at the Royal Welsh when she was judged overall champion in 1967 under Dick Richards. The show would be the site of another great Lydstep victory when Lady's Slipper (f. 1967), Lily's daughter by Downland Chevalier, was judged champion Section B, as well as qualifying for the Fredericks In Hand Final at the Horse of the Year Show where she came a creditable third in 1977. Lady's Slipper was a great winner for her new

Lydstep Ginger Girl, Section B Youngstock Champion Royal Welsh 1985, shown by Mrs Pringle.

owners, Mrs Betty Knowles and her daughter Frances for their Millcroft Stud in Devon.

Among the various mares selected for their stud, Criban Lily and Pleiad were described by Miss Pringle as their 'golden girls', for they produced so many good ponies for them – although Cusop Polly Garter gave them a good line through her Royalty daughter, Lydstep Fairytale (f. 1969). Pleiad arguably did better as she crossed very well with Downland Chevalier, his son Lydstep Royalty and Downland Gondolier. In 1962, she produced a cream filly, Lydstep Blondie, which was loaned to Mrs Binnie, for whom she bred Brockwell Chipmunk (f. 1965 by Brockwell). The following year she produced another Brockwell Cobweb colt named Lydstep Barn Dance and, the year after that, Lydstep Barman by Downland Chevalier. Blondie eventually bred a filly in 1968 called Lydstep Bridesmaid by Coed Coch Maentwrog, which had been swapped for the season with Lydstep Royalty, which stood at Coed Coch. Lyric produced, almost ten years later, a filly by Downland Gondolier named Prairie Flower (f. 1975), which was sold although her presence continued to be felt at Lydstep through Bridesmaid's daughter, Lydstep Little Gem (f. 1976) by Downland Mohawk. An outcross for Gem was sought, and found in one of Miss Pringle's favourites, Abercrychan Spectator (Pennwood Mujib x Synod Champagne); three full sisters resulted, including Lydstep Ginger Girl (f. 1982), which was youngstock champion at the Royal Welsh in 1985 and a winner there for new owners in 1990.

Ginger Girl was almost the last of the ponies bred at Lydstep, a stud founded by breeders with a desire to breed 'down' to Welsh from Arab and Thoroughbred foundations. Miss Pringle admitted to having the pony enthusiasm, and her sister a 'very good eye'; together, their talents benefited others who will be discussed later. Interestingly, although they were well placed to use Mrs Cuff's stallions, they were never tempted to buy one of her fillies for their stud – which was not the case for some other breeders, like Morse, who selected them as foundations for their studs.

Lydstep Prairie Flower became a principal brood mare for the Douthwaite Stud and a champion when shown by Len Bigley. To his stallion Twyford Signal she bred the well-known stallion Douthwaite Signwriter and show mare Douthwaite Sign. The Douthwaite ponies were bred in County Durham by antiques dealer Alan Jackson and his wife, Betty – he liked quality things, including ponies. In addition to Prairie Flower he owned the aptly-named Downland Glamour, a small quality mare by Downland Mohawk.

Lydstep Blondie was sold for top price at the 1969 Fayre Oaks sale for 1,350 guineas to Major and Mrs Archdale, who lived at the Elizabethan manor house of Breccles Hall in Norfolk. The Archdales bred Welsh Section Bs extensively using the fashionable Downland stallions as top line on the pedigrees of their foundation mares, which included the homebred Dauphine (f. 1965) by Dauphin, Mayday (f. 1965) by Romance, Cherie (f. 1967) by Chevalier, Filigree (f. 1969) by Chevalier and Fleur (f. 1970) by Lydstep Royalty (by Chevalier).

Based so far east in the country at a time when road communication was less agreeable than today, like many breeders in East Anglia she largely restricted her showing to the county shows of Norfolk, Suffolk and Essex as well as the Royal of England, and the East of England and Ponies of Britain shows at Peterborough. Her record with homebred ponies was excellent, with her best result coming at the Royal in 1974 when Breccles Filigree took the Section B championship and her foal, Roundelay by the Gower-bred stallion, Cefn Choirmaster, won his class, one of many wins that season. He was a future stallion at the stud, along with Breccles Quicksilver, which later stood at the Priestwood Stud before being sold to the Continent. Tetworth Master Mercury (Solway Master Bronze x Cusop Glamour) was resident stallion at Breccles Hall for several years and Downland Mandarin was leased from Mrs Webster in 1978.

Like the Breccles ponies, it was the use of Downland stallions initially that played a large part in the Ardgrange Stud, which was situated some 30 miles up the coast from Downland near Llanrhystyd, Cardiganshire (now Ceredigion). The stud had been founded by a Yorkshireman, Harry Chambers, whose men's tailoring store in Aberystwyth was well known, both during his own time and after his death when his daughter Daydre took over. He was introduced to Welsh ponies through his daughter, who was keen on ponies as a child and a regular winner in the show ring with her ridden ponies, produced by Mrs Joan Owen from Llandysul. Chambers took second place in the youngstock class in 1953 with Gredington Ffyddlon, a three-year-old Section B by Criban Victor out of Coed Coch Brenhines Sheba; in the same class that day, Lady Myddelton came fourth with Chirk Deborah (f. 1951), a filly by Craven Debo out of Gredington Bronwen by Tanybwlch Berwyn.

Obviously taken with the Chirk filly, Chambers secured her for his emerging stud, where she held court for the rest of her life.

Chirk Deborah bred some exceptional colts, a number of which were retained as stallions at Ardgrange – including her first, Ardgrange Pimpernel (f. 1955 by Criban Victor), Llun Gwyn (f. 1959) and Astronaut (f. 1961) – both by Coed Coch Planed – Heritage (f. 1964) by Downland Dauphin, Debonair (f. 1966) by Downland Chevalier and Dihafel (f. 1968) by Downland Romance. Ardgrange Dihafel (Welsh for 'incomparable'), was shown successfully in 1969, when he won at the Royal and at Lampeter (where he returned to win in 1970), but swept all before him a few weeks later when he took the championship among the Section Bs at Glanusk. He was a favourite among Deborah's sons, and left the stud only once in his life when he spent a season at the Rotherwood Stud, where he sired the Royal Welsh Champion, Rotherwood Lorikeet. Deborah was also represented in the stud by two daughters, Demelza (f. 975) and her older sister Ardgrange Derwena (f. 1967) by Downland Chevalier. The latter was second to the eventual champion, Rotherwood Honeysuckle, at the Royal Welsh in 1970. The next year she had a useful colt, Ardgrange Drygionus, by Quern Fijian, the resident stallion at Will Jones's Towy Valley Stud, which was best known for its top-class riding ponies. However her colt by Dihafel, Ardgrange Difreg, born 1975, was selected by Mrs Carter (née Knowles) for her Millcroft Stud in Devon; not only was he a good show winner for her, but additionally a great stock-getter.

Prior to owning Chirk Deborah, Trixie – of unknown breeding but registered in the Foundation Register of

Ardgrange Dihafel.

Ardgrange Derwena.

the Welsh Stud Book – proved to be a good choice as a foundation mare at Ardgrange since she produced good foals including her first to carry the Ardgrange prefix, a filly named Persuasion (f. 1961 by Ardgrange Llun Gwyn). She had the distinction of beating Coed Coch Priciau in the youngstock class at the Royal Welsh under Mr Eckley as a yearling in 1962, and then standing reserve female champion to Gredington Milfyd; Solway Master Bronze was overall champion that day. Trixie's daughter by Ardgrange Debonair, Pefren (f. 1969), proved to be a successful matron at the stud producing, among others, Ardgrange Pili-Pala (f. 1973 by Dihafel). Pila-Pala's daughter, Prydferth (f. 1989) was by Ardgrange Cristalle (second at the Royal Welsh in 1982), who in turn was by Downland Gold Leaf out of the Dihafel mare, Ardgrange Catryn (f. 1976); Dihafel was used exclusively on Catryn's dam, Downland Cameo, which was sent to Ardgrange in retirement by Mrs Cuff, who was by this time a great friend. After the death of her father, Miss Chambers kept the stud going on the Pila Pala and Catryn lines she had established with him, and added a few Downland mares including Downland Meadow Pippit (Downland Mohawk x Downland Moonwalk) and the stallion Downland Hamlet (f. 2002) by Downland Arcady out of Downland Elf.

Another prominent West Wales stud based on Downland lines was Rhoson, owned by John Davies, a teacher of agriculture who lived with his wife Glenys on a small farm near Blaenffos in Pembrokeshire, where they kept a Welsh pony for their three boys to ride. Although he came from a respected family whose interests lay in Welsh cobs, he himself erred initially towards Section Cs and considered breeding a few, although this changed following a meeting with a former schoolmaster, Gwynfil Rees (of Gwelfro Cobs), who he met while on a trip to Aberystwyth. Their conversation got round to Welsh cobs and the schoolmaster promised to take Davies round a few studs by way of kindling his interest in Section C. Instead, the following day Davies received a phone call from his former teacher asking him if he knew Mrs Cuff, who lived fairly close by, and if so, could he arrange to take him there as he had never been. That visit to Downland – and particularly the sight of Downland Mohawk – would make such an impression on Davies that from then on he knew it would be on the Section B that he would set any ambition. He was impressed not only with the quality and balanced outlook of Mrs Cuff's ponies but also by their excellent temperament, qualities which he and his wife would look for in their own breeding stock. In Davies' own words, 'Everything about them was beautiful'.

With the purchase of a Section B in mind, Davies and his wife found their first registered Welsh pony at the 1970 Fayre Oaks Sale: Claydon Bluechip (f. 1967) was beautifully bred by Weston Winston (Coed Coch Berwynfa x Coed Coch Poerlys) out of a mare by Coed Coch Planed. This breeding was far removed from that upon which the stud would later embark, but – as we have learned previously – the Berwynfa bloodline worked well with Downland and it was little wonder that Bluechip proved to be an excellent foundation mare for the stud, producing many fillies, several of which were kept on. Interestingly, the first registered with the Rhoson prefix was Awel Mai (f. 1972) by Lydstep Royalty; she would give Davies his first homebred stallion, Rhoson Mandinka (f. 1977) by Downland Manchino.

Rhoson Anja as a yearling.

Downland mares would feature strongly in the years to come – such as Downland Baled (f. 1975) and Downland Rhamant (f. 1976), the latter bought as a foal, an up-to-height daughter of Downland Romance and Downland Rapture. Her 1981 foal by Mandinka, Rhoson Taranaki (named after the hometown of the New Zealand rugby team captain), was surprisingly small but crossed well with the Bluechip descendents, especially Mirain (f. 1979), whose sire, Cippyn Man o' War (f. 1975), lived only eight miles away. Bred by Mary Abraham, he was by the Downland Mohawk son, Downland Warcry, and out of Cippyn Morfran by Chevalier out of a Berwynfa mare. Davies was so keen on the Rhamant breeding that he fell for and purchased from Mrs Cuff her half-brother, Downland Rembrandt (f. 1986), a dark liver chestnut, which he sold to the Glenhaven Stud in the United States in 1997 – but not before he left his mark at Rhoson. Line-bred to his half-sister, he produced a filly called Rhoson Tlws (f. 1992). She was a favourite in the stud and produced two outstanding offspring, Rhoson Sipsi (f. 1996) by their champion colt, Cottrell Fabergé, and the grey colt Rhoson Shem (f. 1999) by Eyarth Sama. Sipsi bred several useful winners, including Rhoson Mandolin (f. 2008), which was second at the 2010 Royal Welsh for Rhoson enthusiasts Mr and Mrs Doughty, and Rhoson Dorti (f. 2009), which was reserve Supreme at the Royal Welsh Winter Fair in 2010. Their sire was the Dutch-bred grey, Steehorst Kyro, Dewi Evans's selection for youngstock champion at the 2004 Royal Welsh.

Shem had an outstanding show career when shown in hand – he was unbeaten as a yearling in 2000 and, along with the other Rhoson yearlings Anja and Persawr, made up the winning progeny group at the Royal Welsh for Eyarth Sama, which by then was resident in Australia. Shem was judged overall champion at the Royal Welsh in 2006, when he also qualified for the Cuddy In Hand Finals at the Horse of the Year Show. With the type of temperament his breeders were aiming to produce, he proved to be a great success under saddle, qualifying for the Horse of the Year Show on three occasions, twice champion at the National Welsh and Welsh Part-bred Show and best of breed at Olympia when produced and ridden by Jonathon Stevens.

Sipsi was a winner at the Royal Welsh like Shem and was selected by Davies and his wife to continue the Rhoson line when they dramatically reduced their stud for health reasons. The other female they kept was Rhoson Annalena, a chestnut daughter of Millcroft Ghost and Rhoson Anja, which proved to be a very good breeder for Davies. It was hard to part with her when the stud reduced to two mares – she made 2,800 guineas at the High Flyer Sale at Fayre Oaks when sold to Sweden. Anja's three-year-old filly by Kyro, Rhoson Anoushka, has been retained at Heniarth and was shown successfully at the Royal Welsh, where she won in 2012.

The stud gathered a dozen Downland mares over its history and they paid handsomely for the faith shown by the Davies family for its support. Downland Rosewood – the best of the mares at Pennwood, according to her previous owner, Clive Morse – continued to breed well at Rhoson. Her son Rhoson Gaugin (f. 1990) by Hever Olympic was chosen by the Buttfields for their Section B stud in Perth, Australia. Meanwhile Downland Carmen's daughter of the same name, Rhoson Carmen (f. 1995), made the top price recorded at auction at the time when she sold for 4,000 guineas at Fayre Oaks in 2000, the year the sale had to be postponed for three weeks due to the fuel crisis in Britain. Ponies from the stud have been popular with breeders overseas with, several being sold to the United States, Canada, continental Europe and Russia.

Rhoson Pluen (f. 1992), by Rhoson Taranki out of the Downland Rosewood daughter Pennwood Santa Monica, was one of the last mares to be sold when the Rhoson ponies were being reduced, and she found like-minded breeders when purchased by Les and Lorraine Partridge for their Rosedale Stud at Cwmbran in South Wales. Named after Rose Farm, where Les was raised into a dairying enterprise, their stud would be famed for its quality Section B with Downland breeding, but prior

Rhoson Gauguin, shown with success before his export to the Buttfield family in Western Australia.

to this they were breeding palominos and Mountain ponies. Their decision to breed Section Bs coincided with their interest in the British Riding Pony at a time when Section Bs had sufficient quality to show in both riding pony and breed classes.

Originally Les bought the cream Section A mare Cynan Sunflower which, put to Downland Chevalier, produced a colt named Rosedale Muskateer (f. 1969), the first homebred colt used at the stud. His son Muskatel was out of Mrs Partridge's homebred Chevalier mare Castellybwch Misty Morn, whose dam, Criban (R) Misprint, had been running wild until consigned to the sales at Talybont-on-Usk; born in 1972, Muskatel would be their next stud stallion until the following year an outstanding colt was born to Downland Flair (another product of Downland Chevalier), whose dam was the Section A Cnewr Fanny. Flair's colt by Downland Mohawk, named Rosedale Mohican (f. 1973), would put the stud on the map in the show ring – he won extensively, with his best result coming at the Royal Welsh, where he took the male championship in 1977 under Vivian Eckley and again in 1982 under Eckley's daughter, Jocelyn Price, who took him a step further when he went overall champion. He was a small, pretty stallion with great quality, something he passed on to his stock. He twice appears in the pedigree of the stud's more recent homebred colt, Rosedale Oberon, a black colt born in 1998 which enjoyed success when shown – he was champion at the Bath and West in 2000, champion at the Royal Welsh Winter Fair in 2001 and reserve champion at the West Midlands Stallion show in 2002. In admiring how the Partridges had brought her bloodlines together to produce Mohican (Mohawk on a Chevalier mare), Mrs Cuff remarked: 'You have put the jigsaw together quicker than anyone else without using any more'.

Oberon's story begins with the Partridges' foundation mare, Downland Wild Honey (f. 1956), bred by Llewellyn Richards by Bwlch Valentino out of the dun Section A mare Criban Dun Rose. Partridge Senior bought her daughter, Downland Honey Bee (f. 1964) from Mrs Webster (Varndell Stud) for his son and future daughter-in-law as a wedding present in 1972. Put to Downland Dauphin, Wild Honey had previously produced, for Mrs Cuff, Downland Wonderland (f. 1963), which the Partridges later bought from Mrs Grant-Parkes when she was giving up her Section Bs; in turn, Wonderland to Tanlan Julius Caesar produced the filly Rosedale Fairyland in 1974. (They sent five mares to Mrs Gowing's stud in Norfolk that year, including four Section Bs.) Fairyland began a dynasty of 'Fairy' names in the stud, including her daughter Fairy Tale (f. 1981 by Rosedale Mohican); put back to her sire, Fairy Tale produced Fairy Ring (f. 1985), the dam of Oberon (f. 1998), enhancing the Downland bloodlines since he was by Pennwood El-Dorado. Oberon proved his worth as a sire. To their Fayre Oaks purchase Caraway Royal Anthem (bred in Scotland by Karen Slight, and a Royal Welsh winner), he produced Royal Applause – top price for a Section B at the High Flyer Sale at Fayre Oaks in 2006 at 4,400 guineas – and then a sister, Rosedale Royal Approach, which sold for 1,800 guineas at the same venue in 2009.

The stud also benefited from its own purchase at Fayre Oaks when the 13-year-old black mare by Downland Mohawk, Downland Melissa, was secured from the Talponciau Stud for 2,000 guineas. Typical of her breeding, she excelled in her new surroundings at Rosedale, breeding winner after winner; Rosedale Marghuerita won at the Royal Welsh, while her siblings Madeline and Mozart were placed second and third respectively. Downland Minette (f. 1968) was another to come from Fayre Oaks. She was first purchased by Miss Ferguson from Mrs Cuff and sold on for ridden show pony classes, where she did well for the Daffurn family, for whom she qualified for the ridden 12.2 hands class at the Horse of the Year Show. She went on to breed a host of small-height ridden ponies by Rosedale Mohican, including the lead rein Rosedale Moquette (ridden by the Partridges' daughter Sarah) and Marionette, which won the First Ridden class at the Horse of the Year Show in 1997 for the Prosser family.

Among the stallions which stood at Rosedale was Sianwood Goldrush, a chestnut son of Downland Krugerrand and Cennen Galena (f. 1973), a grey mare by Downland Chevalier out of Coed Coch Nora (Coed Coch Gildas x Coed Coch Gala, both by Coed Coch Berwynfa); Nora was purchased for 300 guineas by Emrys Bowen at the 1969 Fayre Oaks Sale. Bowen, from Llandeilo in West Wales, was a well-known figure in many walks of life for his voluntary work: he was a Council member of the WPCS, Chairman of the Editing and

Emrys Bowen points out the attributes of Sianwood Bayleaf to HRH The Prince of Wales, his co-judge for the overall championship at the Royal Welsh Winter Fair in 2001.

Finance committees, its Veterinary Officer and President in 1988; he was also Chief Veterinary Officer for the Royal Welsh Agricultural Society as well as honorary veterinary officer for a number of other societies. His pony breeding enterprise was supported by his wife Mary, a nurse also from Llandeilo, who settled there after qualifying at Liverpool University and a brief period of practice in Leek, Staffordshire. He was popular among the Welsh pony and cob fraternity, who often sought his wisdom and advice, and a great friend of Mrs Cuff since the time of her stay at Llanddeusant, some ten miles away.

Downland Chevalier was greatly admired by Bowen and formed the basis of his breeding policy, either through direct crossing on Coed Coch Nora or through his sons. Typical of the successful Chevalier cross on Berwynfa was Galena, a big winner when shown in hand; she was a winner of the prestigious Rogers Aviation British In Hand Championship at the Bucks County Show in 1980 and champion at the Royal Show the following year. Nora provided an excellent dam line for the Downland stallions used at Cennen, which included Downland Kestrel by Romance; Downland Yeoman, Krugerrand and Chivalry, all by Chevalier, and Rotherwood Tomahawk by Mohawk. Kestrel was shown in hand and then under saddle by Jane Hill (Bannut Stud), who rode him to many championship victories during 1976 and 1977 before his sale at Fayre Oaks in 1977 for 1,800 guineas to Australian breeder Carl Powell. Cennen Pioneer (f. 1976), by Downland Yeoman out of a mare by Gredington Mynedydd, was another successful stallion to go 'Down Under'.

Abercrychan Spectator.

Downland Yeoman also crossed well on another of the foundation mares at Cennen, Cusop Sprightly Lass, a daughter of Cusop Hoity Toity, which was purchased at the Fayre Oaks sale in 1975 for 450 guineas – most successful among her fillies was Cennen Serenade, born 1977. Interestingly, it was when crossed back to Downland breeding that she performed best as a matron, producing winner after winner, all much in demand by other breeders. Song, by Downland Rembrandt, was top-priced Section B foal at Fayre Oaks in 1991 and a winner for Vic Green when shown by David Puttock; Sonata (f. 1983 by Rotherwood State Occasion) was a winner during every year of her youngstock days, including reserve youngstock champion at the Royal Welsh in 1985; Sonatina (f. 1992) was sold to Mr and Mrs Morton in Canada and was female champion at the Toronto Winter Fair within four days of arriving as a foal; and Seraphine (f. 1998 by Sianwood Arcade) was champion at the Royal Show when shown as a brood mare by the Carrwood Stud. Cennen Serenade and two of her fillies, Song and Symphony, along with Coed Coch Nora and her filly, Arbennig, have formed the basis of the Celton Stud, another supporting Downland lines.

The Cennen Stud was one of several situated in the Towy Valley, named after the river that flows through

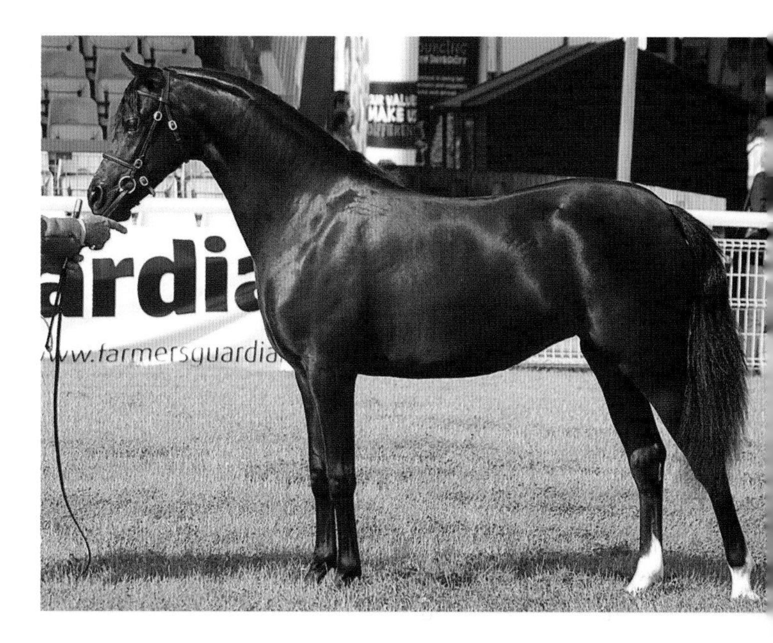

Sianwood Antonia (f. 2008), Youngstock Champion Northleach.

West Wales to the Irish Sea. The valley itself gives its name to a stud owned by 1999 WPCS President Will Jones who, along with his wife Eira, bred some of the best riding ponies in Britain. Their Section Bs were less well known, although their resident stallion, Quern Fjian, appears on the top line of many good pedigrees. They lived close to Llandovery, as did John James and his family, who had started their Sianwood Stud during the 1980s. John had a childhood interest in ponies and produced in-hand ponies such as Cusop Steward for Mr and Mrs Eckley before assisting with local pony breeders, Mr and Mrs Baker-Jones, for whom he produced riding ponies for the show ring. He and his wife Ann both rode and encouraged their children, who competed successfully on homebred ponies – breeding the Welsh Section B was an obvious choice.

Cantref Gorse (f. 1963), an FS mare by Bolgoed Automation, was their first brood mare, although she soon gave way to two special ponies, Cennen Galena (f. 1973) and Abercrychan Antonia (f. 1987). Galena was purchased from her breeder at a time when he was drastically reducing his numbers and she proved an instant success, straight away breeding three stallions for her new owners. The previously-mentioned Sianwood Goldrush (f. 1989) was kept on, but his full brother, the grey Sianwood Silvermine (f. 1990), was first used

Sianwood Silvermine, Ridden Champion WPCS Centenary Show 2002.

Rosedale Marionette winning at the Horse of the Year Show in 1997.

then gelded – whereupon he became a big winner in First Ridden classes. In 1993, Galena produced a son by Downland Arcady named Sianwood Arcade, who would take over from Goldrush as resident stallion and proved a huge success when used on the stud's other favourite, Abercrychan Antonia.

Born in 1987, Antonia was bred only a few miles away by Miss Margaret Thomas, whose FS mare Antonia was by the celebrated Reeves Golden Lustre, the Ceulan-bred colt owned by Mrs Gilbert. She had previously bred an extremely renowned mare by Downland Chevalier called Abercrychan Antonella, a big winner in hand and a very successful brood mare for Mrs Vivian Appell at her Harmony Stud. Antonella's son by the British Riding Pony, Oakley Bubbling Spring, Harmony Bubbling

Champagne, was champion at the Horse of the Year Show in 1983. Antonia was differently bred – her sire was Pennwood El-Dorado and her dam, Abercrychan Novella by Abercrychan Spectator out of Antonia. Unlike Antonella, she was put to Welsh stallions but, like her, she excelled as a matron, especially when crossed with Goldrush and then Arcade. Her daughter by Goldrush, Sianwod Bay Leaf, won twice at the Royal Welsh while her full sister, Antoinette, was retained for breeding and has already bred a champion to Arcade called Sianwood Caradog, champion at Monmouth as a yearling. Arcade on Abercrychan Antonia has produced champion after champion including the firstborn, Sh-Boom, which sold for 4,000 guineas as a yearling at the Fayre Oaks High Flyer Sale, Amelia – a winner at the Royal Welsh in 2009 and 2010, when she was also a winner under saddle – and the black filly born in 2008, Antonia, champion youngster at Northleach.

It was of considerable advantage to breeders in that area that they provided a good selection of stallions for other breeders, unaware how 'green' the concept would become in the 21st century –albeit that the stallions invariably had a leaning towards Downland bloodlines. Many were breeders who combined the breeding of ponies with other livestock. One such was Clive Morse, who very successfully bred Charollais sheep like his near-neighbour Elwyn Davies, who bred them along with quality Section Bs at his Sunbridge Stud. He owned the chestnut stallion Abercrychan Spectator, which he used on his homebred mares by Downland Manchino (f. 1974), a dark brown son of Downland Romance and Downland Misty Morning. Based on the foundation mares Anglesey Aurora, Brockwell Louise, Hever Cadi, Twylands Firefly and Downland Tigermoth, Davies was well known for his quality ponies – including the future stallions Sunbridge Tiger Tamer (f. 1979) by Manchino out of Tigermoth (sold for top price of 1,900 guineas at Fayre Oaks in 1980) and Sunbridge Alicante (1978) by Spectator out of Aurora, sold the previous year to Mrs Taylor for stud duties at her Mark Oak Stud. Alicante progressed to a useful show career in the hands of Sandy Anderson, who went on to form his own Thistledown Stud.

The Downland stallions would feature further afield, such as in the Tetworth ponies bred by the Honourable Maureen Rose-Price, daughter of the Irish peer and World War I naval captain, Lord Dunboyne of County Meath, and wife of Lt Col Robert Caradog Rose-Price DSO OBE, formerly of the Welsh Guards. Mrs Rose-Price's sister, the Honourable Mrs Monck, bred Mountain ponies with the Yaverland prefix on the Isle of Wight. Col and Mrs Rose-Price lived at Tetworth Hall, Ascot, a stylish 17th-century house that was latterly owned by the Radisson Hotel chain magnate, Jasminder Singh. Mrs Rose-Price's stud of Welsh Section Bs lasted less than 20 years with the last two foals, Tetworth Catelpa and Tetworth China Doll, registered in 1983. She started with three well-bred mares – Cusop Glamour by the Criban Victor son, Cusop Architect, Beckfield Anastasia by Kirby Cane Stormer, a grandson of Tanybwlch Berwyn and Downland Sandalwood by Downland Dauphin out of Reeves Golden Samphire. From Bonus she kept Tetworth Gavotte (f. 1970) by Lydstep Barn Dance; from Anastasia she kept two mares, Czarina (f. 1966) by Downland Chevalier and Polonaise (f. 1969) by Barn Dance; Czarina provided her with another two, Swan Lake (f. 1971) by Barn Dance and China Doll (f.

Tetworth Nijinski led by his breeder Mrs Rose-Price prior to his sale to the Laithehill Stud.

1978) by Downland Mandarin; and from Sandalwood she kept for breeding her daughter by Solway Master Bronze, Tetworth Terracotta (f. 1968).

Mrs Rose-Price liked to show her mares and youngstock and she did so with much success, although she had very strong opposition at the time. Downland Sandalwood beat many of them at Northleach in 1966 when her foal, Tetworth Tropical Romance by Downland Romance, also won. Tropical Romance was sold at the Fayre Oaks Sale in 1967 to Robert Hensby for his Laithehill Stud in Yorkshire. Sandalwood later became a principal mare at Mrs Eileen Burrough's Hale Stud in Sussex, where her FS 2 mare, Elphicks Firebird (Solway Master Bronze x Reeves Fairy Lustre) bred a host of good colts including The Hale Friday, Felix and Friday. One of her most famous stallions was The Hale Florin (f. 1983), a son of Hale Friday and Tetworth Harlequinade (Tetworth Nijinsky x Tetworth Terracotta). Other mares that bred well for her were Belvoir Sweet Fern (by Kirby Cane Plunder) and Varndell Vanity (by Downland Mandarin), which she successfully crossed with Friday and Felix. Vanity's son by Carolinas Purple Emperor, The Hale Valentine (f. 1987), was typical of her ponies which had great success under saddle.

The third of Mrs Rose-Price's foundation mares was Beckfield Anastasia, roan in colour like her half-brother Beckfield Ben Hur by Trefesgob Benedict – as one of the top working hunter ponies in the country, Ben Hur won the famous Bombay Hunt Cup at the 1970 Ponies of Britain Show. The roan colour ran through to Czarina, whose first son by Barn Dance was Tetworth Nijinsky (1969), a cream like his sire and grand dam Lydstep Blondie. Nijinsky was almost unbeaten in yearling classes when shown and was used at home for several seasons before joining Tetworth Tropical Romance at Laithehill, where the cream colour persisted and proved highly profitable at a time when dilute colours became very fashionable.

Chapter XIV

Baledon, Skellorn, Rotherwood, Varndell, Colbeach, Cottrell, Mintfield, Lemonshill, Carrwood, Thornberry

Lemonshill Falcon, second in the Section B class at the Horse of the Year Show in 2012.

Baledon Squire, Champion Royal Welsh 1978.

Other breeders took advantage of the Coed Coch bloodlines as a basis for their studs, which would excel in a new age when showing in hand had great intrinsic value and type within Section B was beginning to settle into an acceptable pattern. One such breeder/exhibitor was Ann Bale-Williams, who was keen on ponies from an early age, riding her ponies at shows in the Cheshire area where she lived with her parents.

Her first venture into the Welsh breeds came through Section A, when her first major prize winner was purchased from Lord Kenyon. In 1965 Ann's parents returned to Gredington, where they purchased the Section A, Coed Coch Dunos, as her 21st birthday present and later purchased the yearling Section B filly Gredington Blodyn (f. 1964), by Coed Coch Berwynfa out of Gredington Saffrwm (Criban Victor out of the Royal Welsh champion Gredington Milfyd). Blodyn was an instant success in breeding classes as a youngster and as a mare; she twice won the Royal Welsh and was female and reserve overall champion in 1971. As a brood mare, she will go down in the history of the breed as being one of the most successful of all time.

As well as being a good producer for the show ring, Ann Bale-Williams had a good eye for a pony and chose well when she selected Downland Chevalier for her prize-winning mare, whose offspring by him were very successful – three of them, Baledon Squire (f. 1969), Flower Girl (f. 1971) and Flower Girl Too (f. 1976), all won at the Royal Welsh. Blodyn's filly, 357 (f. 1970 by Solway Master Bronze) was also a good winner and a prominent brood mare in the stud, and was successful in her own right through her offspring by Coed Coch Berwynfa, Rose Queen (f. 1975) and the stallion Baledon Czar (f. 1977). Blodyn bred excellent females and males: two of her colts made a special name for themselves, starting with the most famous, Baledon Squire, a bay

Gredington Blodyn by Coed Coch Berwynfa, foundation mare at Baledon Stud.

colt by Chevalier. His outstanding show career spanned six seasons, during which time he won 36 first prizes from 52 outings as well as five WPCS medals and 22 championships. He was also a successful sire at the stud and among his winning progeny was Baledon Pioneer (f. 1975), a colt out of Clan Prue which won the Melbourne Royal in 1976 after his export to Australia; his son Baledon Nobleman (f. 1974) also won when exported to the United States. Blodyn's colt, by the 1971 Royal Welsh champion Downland Mohawk, Baledon Chief (f. 1975), was also a good winner for the stud and stood second at the Royal Welsh as a two-year-old.

Miss Bale-Williams had shown that the cross of Downland Chevalier on a Coed Coch Berwynfa mare was one of the major triumphs of breeding recorded within Section B, and proved that it was not confined to one mare. As circumstances allowed, she did it once more, this time with Coed Coch Priciau, Lt Col Williams-Wynn's Royal Welsh champion which had been given to her following the Colonel's death in 1977. Sadly Priciau died months after her bay colt by Chevalier was born, but what a colt he turned out to be. By any standards, Baledon Commanchero, born in 1980, had a good career when shown in hand, twice winning and reserve male champion at the Royal Welsh. However, it will be for his exemplary performance under saddle that he will be best remembered. Eight times a qualifier for the Olympia Finals and four times the ridden champion at the Royal Welsh, he was the supreme overall performance champion at the 9th International Welsh Show held at Peterborough in 1992. His talents were not confined to the flat classes, as he was also a great success over jumps – he was awarded the coveted Brodrick Memorial Trophy in 1998 in recognition of his contribution to the Welsh breed.

The Baledon Section Bs were always in high demand and this was reflected in the prices they achieved at auction. At the 1990 Fayre Oaks Sale, an outstanding consignment was brought forward for open auction, achieving the three top prices for the whole sale. The 12-year-old black mare Baledon Czarina (f. 1978), a sister to Czar and granddaughter of Gredington Blodyn, made 2,600 guineas, the highest price paid for a Section B at auction up to that time; she went to Mrs Weston for her Seaholm Stud. The eight-year-old Blodyn daughter Outdoor Girl (f. 1982) made 2,300 guineas, while her half-sister, Bronze Poppy (f. 1981) made 2,100 guineas to the Toya Stud in Devon. Given that the next highest price at the sale was for a Section A mare which made 1,400 guineas, these prices were outstanding.

Baledon Rose Queen (f. 1975) became a foundation mare at the Skellorn Stud in Cheshire, where the Wainwrights had already established at Lockgate Farm, near Macclesfield, a successful stud of Mountain ponies. They started breeding Section Bs when Judy Wainwright took a 12 hands Section A mare, Brynpica Butterscotch, to Baledon Squire; the result was a smart bay colt, Skellorn Music Boy (f. 1977), which would become the resident stallion at Skellorn. This mating was repeated to produce Skellorn Bittermint, who became the foundation of the Eeabrook Stud. Gredington Heddiw (Ardgrange Debonair x Gredington Milfyd), a mare spotted on a visit to Gredington and considered the 'ideal' type for Music Boy, producing in 1980 his first foal, Skellorn Charmaine, a big winner in hand and then influential at the Linksbury.

Skellorn Diadem, Royal Show Champion 1995.

As Music Boy was pretty, small and full of Welsh character, he was best suited to bigger, stronger mares such as Minsterley Crystal (f. 1975 Wingrove Sion x Crawel Carianne), which was saved from slaughter at Beeston Market. As luck would have it, she was carrying Skellorn Diadem (f. 1981), a cream filly by Weston Sugar Bar (f. 1974 Weston Gigli x Weston Pearl), the dam of a host of top-class ponies such as the full sisters by Skellorn Consort (f. 1988), Precious Doll (f. 1997) and Barbie Doll (f. 1999) as well as

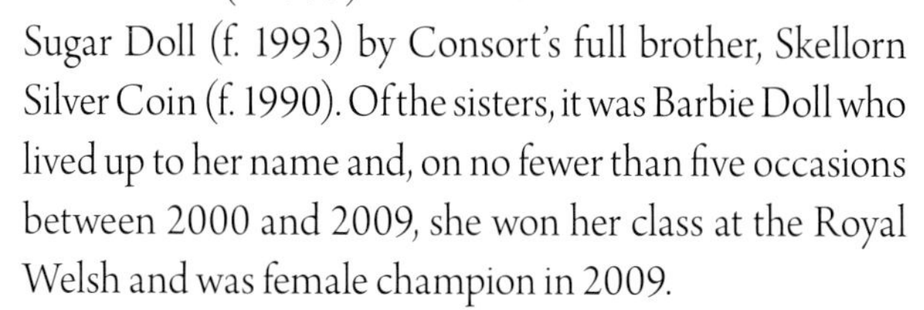

Sugar Doll (f. 1993) by Consort's full brother, Skellorn Silver Coin (f. 1990). Of the sisters, it was Barbie Doll who lived up to her name and, on no fewer than five occasions between 2000 and 2009, she won her class at the Royal Welsh and was female champion in 2009.

The stud's first Royal Welsh win had come in 1986 with Skellorn Rose Princess (f. 1985 Skellorn Music Boy x Baledon Rose Queen) when she took the youngstock and reserve female championships; she won again, this time under saddle, in 1993, when she also qualified for Olympia. The previous year she had won both the barren mare and ridden Section B classes at the International Welsh at Peterborough, when ridden by Kerry Wainwright, who took over the stud on the death of her mother, Judy. Riding her homebred Section Bs had become a tradition for Ms Wainwright, who also qualified for the Olympia Ridden Native Finals both her stallions Skellorn Consort and Silver Coin (both by Downland Gold Sovereign out of Skellorn Anita by Music Boy); she was an obvious choice to ride Baledon Commanchero, whose record under saddle was remarkable by any standards.

At the same time, Mrs Elizabeth Mansfield's highly successful Rotherwood Stud of Welsh ponies was also being modelled on the Berwynfa/Chevalier cross. Mrs Mansfield (now Mansfield-Parnell following her marriage to Tim Parnell, a dairy farmer as well as former racing driver and manager of the BRM Formula 1 team) has run a Section B stud in parallel with a top-class stud of British Riding ponies. Rotherwood has been a name synonymous with the best of breeding and the highest show ring accolades for more than 40 years, during which time her Rotherwood Section Bs have taken no fewer than four overall Royal Welsh championships.

Neither Mrs Mansfield-Parnell's father, a prominent lawyer, nor her mother had any interest in horses, although they never discouraged their daughter, who was pony mad and desperate to ride. The family home had a large garden with room enough for a stable and a pony, which led to a horse and over time fox–hunting, which she enjoyed. It was at this point that her heart lay between two dashing men who followed two sports connected only by speed – one was Tim Parnell and the other Rodney Mansfield, a promising young National Hunt jockey. She settled for the latter but, ironically, following his untimely death, she married Tim Parnell and now shares his abiding love for Formula 1 racing, in which he maintains an active role.

Rodney Mansfield's family were farmers in Ashby de-la-Zouch with a strong tradition of hunting and racing, as well as breeding show hunters – but not ponies. The young Mansfields settled at Nook Farm, a property already owned by Rodney's family situated on the outskirts of Ashby. In 1970 the old red brick stable yard was complemented by a newly-built house, and a decision taken to breed ponies, which would not only provide an interest for the new bride but also bring in a small income to the farm. The breeding programme was initially based on Mountain ponies but soon shifted to Welsh Section Bs, and Mrs Mansfield-Parnell bought the best her limited funds could buy. Her keen eye for a good pony was supported by Bill Newborough, an old-fashioned horseman who was her travelling companion, mentor and friend. With his ability to turn horses out to the highest standard, he was well placed to help and advise on her new interest in the show ring, where the Rotherwood ponies would excel. Once a few rosettes had been gained, Mrs Mansfield's success soon grabbed the attention of her husband, who joined her in an interest that was quickly developing into a lucrative business at Nook Farm.

Travelling the length and breadth of the country with a car and trailer, Newborough and Mrs Mansfield sought out bargains and gradually developed a taste for quality ponies with excellent conformation, movement and breeding. A visit to the sales at Hay-on-Wye brought the first stallion to Rotherwood – Bryn Nipper (f. 1965), a small-height Welsh Section B bought from the Towy Valley Stud; Nipper eventually was sold to the Netherlands, where he became a successful breeding sire. It was time to seek out mares for the stud and Gredington was an obvious choice as they were among the leaders in Section B at the time.

1965 marked the real starting point for the Rotherwood Stud, as it was during the summer of that year that Gredington Tiwlip (f. 1962) came to Nook Farm as a three-year-old in foal to Criban Victor; she bred 19 foals at Rotherwood, where she remained for the rest of her life. By Coed Coch Berwynfa and out of the Royal Welsh champion Gredington Milfyd, Tiwlip (with the similar concentration of blood as Blodyn) was bred in the purple and was a perfect foundation for a stud which, over time, would be associated with only the very best. She possessed an ability found in the best of mares

Left: Rotherwood Honeysuckle shown by Rodney Mansfield, Royal Welsh Champion 1978.

Below: Rotherwood Penny Royale, Royal Welsh Champion 1983.

Rotherwood Golden Rose, Champion at the Great Yorkshire Show 2009.

to breed very much to her own type with a keen family likeness – like the proverbial peas in a pod. Her first two filly foals, Rotherwood Crocus (f. 1966) and Columbine (f. 1967), were by Chirk Caradoc, but it was her third filly, Rotherwood Honeysuckle (born in 1968), which set the pattern of Welsh ponies associated with Rotherwood Section Bs. She was by Downland Chevalier, without doubt the most successful choice of stallion for Tiwlip; at a time when Chevalier was not available to outside mares, five others by him followed, including Rotherwood Commander (f. 1971), Lilac Time (f. 1973), Lavender (f. 1976), Love-in-the-Mist (f. 1978) and Cavalier (f. 1980). Given that Tiwlip was three-quarters sister to Miss Bale-Williams's Gredington Blodyn, it was no surprise that the cross worked so successfully.

Like Blodyn and her offspring, Tiwlip was a champion in the show ring herself but, more significantly, her progeny did even better – with the Royal Welsh once more proving a worthy benchmark. Honeysuckle took the Rotherwood prefix right to the top in the Welsh Pony rings: she was twice overall champion there in 1970 as a two-year-old (when she beat Bowdell Quiver for the overall title) and then eight years later as a brood mare, when she bettered Baledon Squire under Mrs Towers-Clark. It was at the 1978 Show that she had a foal at foot named Rotherwood Penny Royale, a filly that would emulate the great success of her dam at major shows throughout Britain; she took the Royal Welsh title in 1983, two years after Honeysuckle's full sister, Lilac Time, was judged champion by Harry Chambers. Honeysuckle and Penny Royale were the top show mares of that particular era, although it was Honeysuckle which was the personal favourite of Mrs Mansfield-Parnell. Between them, they would provide their breeder with an abundance of females with which to expand her stud.

Two of Tiwlip's Chevalier offspring were sold to Australia. The first, born in 1971, was Rotherwood Commander, a striking stallion which was successful in the show ring and as a sire. He crossed well with Coed Coch Bugeiles, which had now joined the stud, producing Rotherwood May Morning in 1974, a mare which was retained for breeding. Commander was the choice of Eric Dudley for his Kirreway Stud, where he had gathered a few quality mares from Mrs Cuff; by any standards he was a huge success there. The second to go to Australia was Rotherwood Love-in-the-Mist (f. 1978).

Mrs Cuff generously allowed her Royal Welsh champion, Downland Mohawk, to leave Wales to stand at stud with Mrs Mansfield-Parnell and to him Tiwlip produced a striking bay colt, Rotherwood Tomahawk (f.

Cottrell Stud card featuring Rotherwood State Occasion.

1981), which was the top price of 2,000 guineas in 1982 when purchased by Emrys Bowen for his Cennen Stud in West Wales.

Honeysuckle's 1978 daughter Penny Royale produced a number of outstanding colts, including Royal Chief (f. 1983 by Downland Mohawk), Prince Regent (f. 1988 by Brockwell Prince Charming, Casino Royale (f. 1987 by Ardgrange Dihafel) and Spy Catcher (f. 1994 by Orielton Aristocrat). Casino Royale was purchased as a foal by Greg Higgins from Australia and was shown for him before his export; Casino Royale was champion at Lampeter in 1988 as a yearling and returned later that year to Wales for the Royal Welsh, where he was youngstock champion. Meanwhile Rotherwood Lilac Time (f. 1973 Downland Chevalier x Gredington Tiwlip) produced two outstanding colts, Student Prince in 1979 by Keston Royal Occasion and Crown Prince (f. 1988) by Rotherwood Prince Regent, which was sold to the United States.

There were several top-class stallions used at Rotherwood over the years but none to compare with Keston Royal Occasion, who left an indelible mark on the Section B during his time there. Mrs Mansfield-Parnell was offered him by his breeder, Mrs Janet Meyer, at a time when she was reducing her pony breeding activities at Keston. (Prior to her move there, she had registered her Welsh ponies with her Clyphada prefix, the name associated with Royal Occasion's dam, Periwinkle, which was by Solway Master Bronze out of Kirby Cane Plume.) For several years previously, Mr and Mrs Meyer had gathered some beautifully-bred mares at their stud as well as a son of Downland Chevalier and Coed Coch Gold Mair (f. 1959) called Downland Mandarin. They showed their ponies extensively around the major shows with success, including Periwinkle and her son Keston Royal Occasion, which was a riding pony youngstock champion at the Ponies of Britain Stallion Show in 1973 as well as champion at the Bath and West. He was known to everyone as 'Cowboy' as a result of his dam foaling just as his breeders were about to leave home for the Royal Première of the film, *Midnight Cowboy*; there would be no glittering occasion for them that evening!

Miss Elspeth Ferguson presents Mrs Mansfield-Parnell with the championship trophy following the victory of Keston Royal Occasion at Northleach in 1978.

In 1976, the same year that Commander flew out to Australia, Mrs Mansfield-Parnell received a phone call asking if there was a place for Cowboy at Rotherwood; she jumped at the chance as she thought him to be an ideal outcross for Tiwlip and her growing family. Tiwlip produced two good colts by him, Rotherwood Sovereign (f. 1977) and Monarch (f. 1979), but it would be her daughter, Honeysuckle, that bred the best offspring by Cowboy. They were her 1978 filly, the Royal Welsh champion Rotherwood Penny Royale, and her 1979 colt, Rotherwood State Occasion, reserve supreme at the Royal

Show as a yearling and reserve champion to his sire at the West Midlands Stallion Show in 1981 – the only time he was shown that year and after which he was sold to Mrs Johns-Powell (Cottrell Stud). Rotherwood State Occasion was nine times winner of the WPCS Sire Rating competition – following a run of success established by his grand sire, Downland Mandarin, and his sire, Keston Royal Occasion.

Keston Royal Occasion had a remarkable temperament both in and out the stable, which enhanced his excellent show career both in hand and under saddle while at Rotherwood. This included the supreme at Newark, champion at The Royal, champion at Northleach and reserve for the Lloyds Bank In Hand Qualifier for the Horse of the Year Show at Devon County; he twice qualified for the National Pony Society Ridden Native Final at Olympia, where he stood reserve champion in 1979 ridden by Gillian Sant. Some 22 years later, Rotherwood Wild Snowdrop by Royal Occasion's grandson, Rotherwood Prince Regent, took the best of breed at Olympia under David Blair, who was also the last person to judge Keston Royal Occasion in Britain.

Cowboy also bred well for other breeders, a fact reflected in his four-year run of success in the Ponies of Britain Progeny Award Scheme –he led the WPCS's In Hand Sire Rating Scheme in 1980 and again from 1982 to 1984. At Rotherwood he produced from another of Mrs Mansfield-Parnell's foundation mares, Chirk Delightful, the chestnut colt Rotherwood Royalist (f. 1981), male champion at the Royal Welsh in 1984, and to Weston Lark she produced Rotherwood Goldfinch, resident stallion at the Aytounhill Stud in Scotland. Lark produced a further Royal Welsh victory for Rotherwood when her Ardgrange Dihafel filly, Lorikeet (born 1987), won the mare class and female championship in 1996 under Mrs Mountain.

After waiting three years to fulfil a dream to take Keston Royal Occasion home to his Mirinda Stud in Victoria, Australia, Peter Barry from Melbourne eventually persuaded Mrs Meyer and Mrs Mansfield-Parnell to part with him, albeit aided and abetted by an undisclosed but reportedly record sum of money. He was every bit as successful in his new country, where his use on riding pony mares was a triumph as it had been in Britain. The Australian breeders loved him and he died there in 2001. In Britain he left behind a son, Brockwell Prince Charming out of Brockwell Puss – which was used extensively at Rotherwood before being sold to the Millcroft Stud – and a daughter out of Honeysuckle, Rotherwood Royal Honey, which was put to Orielton Aristocrat in 1991 to produce another Royal Welsh winner, Rotherwood Aristotle, which stood at stud at home before joining Paula Cullen's Paulank Stud in Ireland. Penny Royale, meanwhile, was put to the popular cream stallion Cottrell Artiste, producing a pretty little colt which would go on to do well in ridden classes, but not before he sired Rotherwood Golden Rose (f. 2004), whose dam, Rotherwood Honey Rose (f. 1985), was a daughter of Downland Mohawk and Rotherwood Honeysuckle. Rose has been the latest success story from Rotherwood, as she has been shown successfully as a brood mare – taking, from only a few outings, the Section B championship in 2007 at The Royal Show and in 2009 at the Great Yorkshire.

Although Downland Mandarin has been somewhat overshadowed by his son, Keston Royal Occasion, he

Colbeach Lady Levanne, Champion at the Great Yorkshire Show 1984.

did exert quite an impression wherever he went. When Mrs Meyer decided to sell him, her friend and neighbour, Olive Webster, immediately jumped into her car and drove to collect him just in case she had a change of heart. Her Varndell Stud was founded on two mares, Varndell Vanessa (Brockwell Cobweb x Downland Honeybee) and Cusop Rhythm (Solway Master Bronze x Cusop Rhyme) – both mares bred well to Mandarin, especially Rhythm and her daughter by Chirk Caradoc, Varndell Rosette. Of Rhythm's offspring, two colts by Mandarin are of special note – Varndell Reveille (f. 1964) went to Australia, where he had a major impact on Section Bs, and Varndell Right Royal (f. 1977) which, after a good show career in hand with Mrs Webster, settled at the Boston Stud near Wetherby in Yorkshire.

Mandarin was leased to the Colbeach Stud in Lincolnshire for the 1981 season and was subsequently sold to Mrs Renita White, who had founded the stud in 1965 with the help of her husband, Norwood White – he remained there for the rest of his life. Mrs White came from a hunting family who farmed a large part of Lincolnshire; her father, Joseph Ward, lived at Moulton Park in Spalding but took a hunting lodge at Belton House, Grantham, and hunted with the Belvoir Hunt for 68 seasons. While his daughter did not quite share his love of fox-hunting, she did share his love of horses and was attracted to ponies when her own daughter, Virginia, took up showing as a child. As a pony breeder, Mrs White developed a taste for quality riding ponies, which she bred most successfully. However she also had a love of Welsh ponies – the quality Welsh Section Bs with Downland breeding had special appeal, so she started a small stud based on these bloodlines at her home at Colbeach, which quickly became too small for her needs so the family moved to a larger property in Leicestershire.

The nucleus of her stud was four well-bred mares – Surreyhills Duchess Anne (Downland Chevalier x Belvoir Honeysuckle), Towyvalley Quail (Downland Chevalier x Bowdell Quill), Lechlade Polyanthus (Solway Master Bronze x Downland Dresden) and Isley Walton Diorite (Kirby Cane Golden Rod x mare by Solway Master Bronze). Towyvalley Quail was bought for 550 guineas and proved a useful purchase when her son, Colbeach Bonaparte, sold for 1,300 guineas at the 1975 Fayre Oaks sale, when the average was a mere £166. Duchess Anne was purchased at Fayre Oaks in 1970 and produced two outstanding fillies, which would later cross well with Mandarin. The first was Colbeach Lady Levanne (f. 1974) by Coed Coch Endor (double cross Coed Coch Berwynfa), who was later exported to Denmark, and the second Lady Lobelia (f. 1980) by the Chirk Crogan son, Seaholm Fabian. Both 'Ladies' were extensively shown at the major shows, like all the best ponies from Colbeach, taking championships at The Royal, Great Yorkshire, National Pony Society and many county shows.

Endor, originally purchased from Charles Castle's Watermans Stud, was replaced by the young Seaholm Fabian, whose dam, Downland Fandango, was by Downland Romance. Next came the Rotherwood Commander colt, Overstone Country Boy, which was used during the stud seasons 1978 to 1981, after which Downland Mandarin held court. Mrs White had already enjoyed success with her mares when they were put to his son, Keston Royal Occasion, producing the good winner, Colbeach Royal Secret, in 1981.

The spin-off from the successful Berwynfa/ Chevalier/Keston Royal Occasion cross was felt most strongly at the Cottrell Stud, where Rotherwood State Occasion reigned supreme until the death of Mrs Johns-Powell, whereupon he returned to his birthplace in 1999. His success at Cottrell was immediate; in 1985 he headed the WPCS Sire ratings for Section B, an achievement he repeated another eight times with a run of success only interrupted by Abercrychan Spectator in 1990 and ended by State Occasion's son, Cottrell Artiste, in 1999.

Cottrell, an estate owned by the Johns-Powell family since 1941, is situated at St Nicholas a few miles west of Cardiff and is well known as a country club and golf course – but it was not always so. During World War II the large 16th century mansion house was used to accommodate soldiers and the Fire Service. However it became derelict and was demolished in 1970 – the Johns-Powell family lived at the estate manager's house at Sheepcourt Farm, Bonvilston, not far from Cardiff Airport. It was here that Pat Johns-Powell, a keen horsewoman herself, encouraged her daughter, Susan, who rode top-class riding ponies and show hacks. Her interest in Welsh ponies, like so many breeders, started with a few Section A mares, but the grass at Cottrell was much too rich for them so she decided to concentrate on Welsh Section Bs. She became a member of the WPCS in 1969 and was elected its President in 1998, but sadly was not able to fulfil her commitment due to her untimely death later that year.

Mrs Johns-Powell not only brought together the best bloodlines available in the Stud Book at the time but also proved what a good eye she possessed through the success of her own ponies. Time has shown how easy it has been for breeders to spend large sums of money for top-class ponies which never produce others of merit. Mrs Johns-Powell paid those large sums and her purchases produced champion after champion – surely the mark of an accomplished breeder. We have already seen how she bought the best of mares from the country's top breeders: Revel Glimpse and her daughter Weston Glimpse from Jack Edwards, as well as Weston Rita (Weston Paul x Weston Romany), second top price at the 1972 Fayre Oaks Sale, Weston Odette (full sister to Weston Choice) and Weston Moll Flanders by Weston Gigli. From Mrs Cuff she purchased Downland Almond, Downland Merry Month (Downland Mohawk x Downland Love-in-the-Mist) and Downland Jasmine (Downland

Krugerrand x Downland Jamila), Carolinas China Rose (Solway Master Bronze x Downland Dresden) and the 1981 Royal Welsh champion, Rotherwood Lilac Time.

It could be argued that, with such a wonderful array of mares, she could scarcely go wrong and her choice of State Occasion would make sure that she didn't. She admired him so much when she saw him in 1979 shown with his dam that she tried to buy him then; it took her another two years to seal the deal. Prior to this Mrs Johns-Powell had used Solway Master Bronze, which she found crossed well – Revel Glimpse producing, in 1976, Cottrell Lisa, which went on to breed well to State Occasion; her fillies Cottrell Lapwing (f. 1988) and Cottrell Liberty (f. 1989) by him won extensively, including the Royal Show and Royal Welsh, where Liberty was reserve youngstock champion in 1991. Having produced three top-class ponies to Almond by Keston Royal Occasion, which included the ridden winner Cottrell Sandpiper, she then put her to Solway Master Bronze in 1982 with the resultant colt, Cottrell Sandalwood, winning Royal Welsh as a yearling and going on to be a major prize winner in West Wales for the Fairywood Stud.

Weston Glimpse must be considered to be one of the most successful mares registered in Section B of the Stud Book, as not only did she reign over the Royal Welsh rings herself but so too did her offspring by Rotherwood State Occasion. In 1992, her bay filly Cottrell Charm caught the eye of Mrs Robina Mills to stand overall Champion and in 2000, her full sister, Cottrell Royal Glance, a grey like her dam, was female champion and overall reserve for judge Len Bigley.

Downland Almond (f. 1975 Downland Mohawk x Lechlade Angelica) added to the laurels gathered by the Cottrell ponies as her State Occasion progeny won year after year at the Royal Welsh. Cottrell Artiste won as a two-year-old for his new owner, Mrs Gilchrist-Fisher, and Cottrell Amethyst stood reserve female champion in 1988. The highest family achiever was Cottrell Aurora, a chestnut filly born in 1993, which was overall champion at the Royal Welsh in 1995; despite being a firm favourite with Mrs Johns-Powell, somehow Mr and Mrs Grant managed to persuade her to allow Aurora to join their Bamborough Stud in New South Wales, Australia. Almond's colts by State Occasion also fared well, with Cottrell Ambassador taking the Supreme at Glanusk while his younger brother, Cottrell Aur, topped the Section B foal prices at Fayre Oaks in 1992, when

Cottrell Lapwing (f. 1988) with her colt foal enjoying a new life in Scotland.

he sold to Meirion Davies for 1,000 guineas. Aur went on to stand reserve youngstock champion at the Royal Welsh the following year, having joined Cottrell Fabergé at the Davies' Rhoson Stud, which had enjoyed much success with this son of State Occasion and Carolinas China Rose. Fabergé was overall champion yearling at the Ponies Association UK Show in 1993 and champion youngstock at the Royal Welsh the following year after he qualified for the In Hand Finals in the Royal Show at Wembley. His brother, Cottrell Royal Consort, was youngstock champion at the Royal Welsh in 1999.

A problem of finding a mate for the State Occasion daughters, which had been retained in the stud, eventually arose so the hunt was on for a stallion. The first used was Pennwood El-Dorado; next Mrs Johns-Powell settled on a very attractive young colt named Cwrtycadno Cadfridog, which she had seen shown with success by Hugh and Jane Edwards (Sarnau Stud). Previously supreme native champion at the 1992 West Midlands Stallion Show aged three, Mrs Powell brought Cadfridog out in 1996 to take the Royal Welsh overall Section B title and also the Templeton Horse of the Year Show qualifying title. He bred well at Cottrell, including the 2012 Horse of the Year Show ridden Section B winner Cottrell Leonardo (dam Cottrell Liberty), then at the Weydown Stud following his retirement to Hampshire, where he made his mark for Mr and Mrs Hounsham.

It would be accurate to say that no stud in the history of the Welsh Section B has produced more winners at the Royal Welsh than Mrs Johns-Powell with her Cottrell ponies. After her death, Mrs Johns-Powell's daughter-in-law, Caroline, kept the stud going for a few years, but it was eventually dispersed. Nevertheless, the bloodlines have

Above: Cottrell Artiste.

Below: Mintfield Song Thrush, Royal Welsh Champion 2004.

continued under the guidance of Edwin Prosser, Mrs Johns-Powell's great friend and showman, who led most of the Cottrell ponies to their greatest victories. Prosser's grandfather had been a shepherd on the Brecon Beacons, where horses and ponies became part of his childhood and an interest he wanted to share with his daughter when he purchased a Section A from the Bryn Stud in South Wales. A few more Section As followed, and the show ring beckoned – his best victory with Mountain ponies came in 1976, when he took the championship at Glanusk with Foxhunter Clove.

His first Section B was Cottrell Marksman, a son of Solway Master Bronze and Revel Glimpse, which he showed with success; eventually the Mintfield Stud (named after his grandfather's cottage) grew and stables were built on his father's allotment at Cwmavon, Port Talbot. Other Cottrell ponies to follow included the colt Cottrell Royal Consort (twice reserve male champion at the Royal Welsh), and the mares Cottrell Anwyl (Pennwood El-Dorado x Cottrell Amethyst) and Cottrell Sunflower (Rotherwood State Occasion x Weston Moll Flanders). Pennwood Indiana and Nefydd Antonella were other mares to join the Mintfield Stud.

During the heady days of Cottrell success, Prosser was largely engaged in showing ponies for Mrs Powell at major shows, although he always kept a few of his own on the back burner to be shown locally. Over time, his own breeding came to the fore and ironically it was Cottrell Sunflower's crossing with Cwrtycadno Cadfridog that gave him his favourite pony and greatest showing success. Mintfield Song Thrush, foaled in 1996, was judged light horse supreme at the Royal Welsh in 2004 and qualified for the in hand final at the Horse of the Year Show. Using tried and tested bloodlines from Weston, Downland, Rotherwood and Cottrell, he models the type of ponies he aims to breed on his favourite mare, which has bred well for him. Unlike Mrs Johns-Powell, who was a very quiet, self-effacing lady, Prosser has taken an active role in the WPCS, being a Council member from 2004 to 2010 and President from 2013 to 2014, and overseeing the move of the Society's office from Aberystwyth to Bronaeron, taking an active role in the renovation of the new building.

Cottrell Artiste was one of Prosser's charges when he showed him as a yearling in 1987, after which he was sold to Mrs Gilchrist-Fisher from Lemon's Hill Farm, Hemyock, Devon, who took the farm's name for her Section B stud (she had previously lived near Basingstoke, where she used the Baughurst prefix). Mrs Gilchrist-Fisher's interest stemmed from her childhood in Ireland, although she had first learned to ride while her father, Captain Robert Ross Stewart, had been on loan to the Australian Navy and took command of the destroyers Hobart and Australia. In 1941 he returned to Ireland to command the Londonderry Escort Force and was then sent to Scapa Flow. Ireland was the home of her mother, who came from a well-known hunting family; her brother, Lawrence Hastings, stood at stud at Friarstown the renowned National Hunt sire, Cottage (sire of Grand National winner Sheila's Cottage) and also bred the celebrated show hunters, Mighty Atom, Mighty Grand and Mighty Fine, Nat Galway-Greer's Dublin champions.

The first Welsh ponies came by way of those purchased for her daughter Rosanne, who first had the Section A, Mountain Puffin, and then the Section B, Droveside Rumpus. Through her local Southern Counties Welsh Pony and Cob Association, Mrs

Gilchrist-Fisher became great friends with several prominent breeders, including David and Lorna Reynolds (Springbourne Stud) – whose daughter, Cerys, shared a pony club interest with Rosanne at a time when jumping and hunting was preferred to the show ring. Inspired by Rumpus, Mrs Gilchrist-Fisher decided to embark on a breeding programme of Section Bs and in 1976 spotted a good filly foal by Downland Tanglewood (by Downland Chevalier) at the Uplands Stud, which she was allowed to purchase only on condition that she also bought her dam, Shimpling Enchanté, bred by Mrs Gowing from Norfolk. Enchanté was by Mrs Gowing's resident stallion Tanlan Julius Caesar, a son of Downland Chevalier, while her dam, Godolphin Marguerite, was in turn by Downland Dauphin. With the best of Downland blood flowing through her veins, it was little wonder that she crossed well with more of the same and in 1979 she produced an outstanding colt by Downland Mandarin, Baughurst William of Orange, which attracted the attention of Australian buyers and subsequently made his name for the Owendale Stud in Victoria.

Her filly, Uplands Étoile, became a foundation mare at the Lemonshill Stud and made a great start as a brood mare when her yearling colt by Brockwell Prince Charming, Lemonshill Little Emperor, became Mrs Gilchrist-Fisher's first Royal Welsh winner in 1987. Étoile was put to a number of good stallions, but none worked better than Cottrell Artiste, which by this time was resident with her in Devon. His first crop of foals was outstanding with her colt by him, Lemonshill Limelight (f. 1989), living up to his name when he stole the show as a yearling. From his early outing at the Newark and Notts, where he was champion Section B, and then

Above: Baughurst Aderyn, yearling filly.

Below: Uplands Étoile, winner as a yearling at Northleach in 1977.

Judy Creber In Hand qualifier for the Horse of the Year Show, to the Royal Welsh, where he was also champion, he was one of the most successful colts to appear on the show scene for years. He also marked the beginning of a remarkable and lasting partnership for the stud with Mark Tamplin from Rudry, Glamorgan, whose parents bred the Griashall ponies. Tamplin was recommended to Mrs Gilchrist-Fisher by Mrs Bachman (Carolinas Stud) and Limelight would be the first of many champions produced by him bearing the Lemonshill prefix. Limelight continued to be shown with success and was then sold to Gill Simpson, who showed him in ridden classes where he won at the National Pony Society Show and also qualified for Olympia; later he was sold again, to the Duntarvie Stud in Scotland, where his showing career continued – including winning the coveted Glyn Greenwood championship at Ponies UK. When Cottrell Artiste was leased to Mrs Simpson for her Wortley Stud in Derbyshire, he produced two outstanding winners, Wortley Fisher King (f. 1996), Royal Welsh male champion in 1998, and Wortley Celebration (f. 1998), champion Show Hunter Pony of the Year at the Horse of the Year Show in 2010.

Limelight's younger brother, Hylight, also won at the Royal Welsh and was youngstock champion as a yearling in 1991, although his best victory came at the International Welsh Show held at Peterborough in 1992, when he was judged overall youngstock champion; Hylight was selected by the Johnson family for their Stoak Stud based at Llandyn Hall, the former home of the Edwards and their Weston Stud. Having notched up three Royal Welsh winners, Étoile demonstrated her worth when she produced yet another, this time a filly to Rotherwood State Occasion – Lemonshill May Queen enjoyed a great yearling season, taking the championship at Cheshire followed by a win at the Royal and reserve youngstock champion at the Royal Welsh.

The choice of Cottrell Artiste could not have worked better on Mrs Gilchrist-Fisher's other foundation mares, Coed Coch Aden and her elegant daughter by Downland Mohawk, Baughurst Aderyn. Aden (f. 1972) was selected by Mrs Gilchrist-Fisher on the back of the success demonstrated by Mrs Mansfield-Parnell at Rotherwood with Coed Coch Bugeiles by Coed Coch Berwynfa. Aden was bred by Lt Col Williams-Wynn and bought by Miss Ferguson for John Hurst as a base for breeding riding ponies at his Runnings Park Stud near Worcester. By Berwynfa and out of Coed Coch Deryn, a daughter of Criban Victor and Sensigl (by Tanybwlch Berwyn), she was bought from an advertisement in 'Horse and Hound'. The foal she was carrying was a filly by Downland Cavalcade named Baughurst Constellation (f. 1983), which to Cottrell Artiste produced a colt, Lemonshill Great Occasion. He was chosen by the Tamplins for their show mare Laithehill Pavlova, which duly produced a cream colt in 1997, Griashall Kiwi; he was twice male champion at the Royal Welsh, first in 1999 when he took the overall award and then in 2010, when he stood overall reserve under Daydre Chambers.

Before her success with Artiste, Aden bred a useful black colt, Lemonshill Royal Flight, (f. 1988 by Rotherwood State Occasion), now at stud in Sweden, and then a host of fillies including Lemonshill Lapwing, the top-priced Section B yearling at Fayre Oaks in 1993 when sold for 2,000 guineas to the Countess of Dysart. That same year she produced an outstanding filly, Lemonshill

Lemonshill Alarch, Champion Royal Welsh 1997.

Angelica by Artiste, which was a big winner of youngstock classes including the Royal Welsh, Royal of England, Essex and Cornwall – to name but four. Angelica was retained as a brood mare and again she proved her worth when her yearling son by Soudley Taliesin, Lemonshill Top Note (f. 2002), was youngstock, male and reserve overall champion at the Royal Welsh in 2003. Meanwhile Aden's daughter, Baughurst Aderyn (f. 1984), proved to be an ideal mate for Artiste, with a host of lovely fillies to her credit including arguably the best, Lemonshill Alarch (f. 1994), Songbird (f. 2000) – which has been retained – and Firecrest (f. 1992), yet another Royal Welsh winner, which sadly died after a short career as a brood mare. Responding to the challenge of finding a suitable stallion for her Artiste mares, Mrs Gilchrist-Fisher chose several stallions for Firecrest, although her best was her last – by Douthwaite Signwriter, a grey colt born in 2005, Lemonshill Falcon. A champion in hand, he also impressed under saddle in 2012, when he was second in the Ridden Section B Final at the Horse of the Year Show having made the front line on each of his two previous appearances.

Lemonshill Alarch was first shown for her breeder but was sold to Patrick and Karen Cheetham from Cheshire, who enquired to Lemonshill for a filly to join their Cheshire County winner, Carrwood Casuarina. Karen already had an interest in pony breeding through her parents, Tom and Celia Harrison, at their Carr Farm on the Derbyshire/Cheshire border, where work horses and hunters were augmented by ponies for Karen to ride and jump. Their first mare was the Welsh pony mare, Morwyn Nicotiana, which they showed successfully and bred to both Mountain Pony and Section B stallions including Downland Gold Sovereign, the sire of Casuarina. Following Karen's marriage to Patrick in 1986 and work commitments in Australia, it was on their return to England in 1995 that they decided to resurrect their stud and Lemonshill Alarch was purchased. She proved to be an excellent choice, as she provided her new owners with a lifetime result when Mrs Colbatch-Clark made her female and overall reserve champion at the Royal Welsh in 1997 – the Cheethams' first visit to the Show as exhibitors.

Alarch was beaten that day by Carwed Charmer, a stallion which would become a major player within the Section B both in terms of his own show ring performances and those of his offspring. He had a lasting effect on the Carrwood Stud through his son Orpheus (f. 2000) out of Isley Walton Athena, one of four mares selected as a foundation for the stud. Bred by Mrs Shields from Castle Donington, Derbyshire, Athena's sire was Rotherwood Aristotle and her dam, Colbeach Madonna (Downland Mandarin x Isley Walton Diorite by Kirby Cane Golden Rod). She bred a lot of colts, several of which went on to successful careers under saddle – such as Carrwood Xanthus (f. 2003), winner at Ponies UK 2011, and Carrwood Pegasus (f. 2007), the first winner of the Tagg La Liga Award for ponies placed in qualifiers for the Horse of the Year Show in 2012.

Unshown, Orpheus (f. 2000) was sold to the Hilin Stud, and won at Glanusk as a two-year-old. It was, however, a case of sold but not forgotten as his progeny out of Alarch proved to be outstanding and his daughter Adeniog (f. 2007) has been retained in the stud and takes over from her dam as a show mare. Her full brother, Redwing (f. 2008) was a winner at the Bath and West before his sale to Sweden, and another brother, Carrwood Phoenix, born in 2006, has been the only pony sent out to be shown. Leading professional Colin Tibbey (Lacy Stud) was responsible for his many wins, which included the Royal Welsh in 2008 – that year Alarch had three winners at the Royal Show: Phoenix, Adeniog and Bronzewing. The last–mentioned, foaled in 2004 by Moelview Charmer Boy, has also been Royal Show male champion as well as youngstock champion at Lampeter as a yearling; he is retained at Carrwood for stud duties.

Bronzewing's daughter, Carrwood Summertime (f. 2010) – another big winner – is retained in the stud to replace her dam, Cennen Seraphine, a granddaughter of Downland Arcady and Rotherwood State Occasion. Seraphine was purchased at Fayre Oaks as the third foundation mare for the stud and proved a great show mare, with the Royal Show championship to her credit among others. Her son by Carrwood Phoenix, Carrwood Pharaoh (f. 2011), joins Bronzewing for stud duties, thus allowing an outcross of sorts for the other two young mares retained in the stud, Songlark and Kittiwake, daughters of Alarch by Orpheus.

The success of Carrwood, based on sound dam lines, demonstrates the ability of a small stud to achieve as much as, if not more than, some of the larger ones. This could equally be said for Mrs Johns-Powell's near neighbour, Fiona Leadbitter, who used the Cottrell stallions extensively once she embarked upon breeding Section Bs in 1982, when the first foal carrying her Thornberry prefix was born. Following her family's move from London to South Wales, Miss Leadbitter's first Welsh Section Bs were the Cottrell-bred geldings Plover, Snuffles and Viking, purchased from their breeder with whom she and her parents became great friends. By this time she was smitten with the breed, so a Section B mare was sought and an advert in 'Horse and Hound' led them to Devon, where they purchased a grey two-year-old filly, Cwmwyre Samantha (f. 1978) by Ardgrange Dihafel. Samantha won on her first show ring outing at Lampeter, where she was also judged female champion, and within

two years she also stood reserve champion at The Royal. As a brood mare she proved exceptional, accounting for championships at major shows such as The Royal, Northleach, Ponies UK, Three Counties and Shropshire and West Midlands – but most prestigious was her overall championship at the Royal Welsh in 1987. By the age of thirteen she had won the brood mare class there on no fewer than four occasions.

Cwmwyre Samantha's success was not limited to the show ring, for she was also an outstanding breeder, unlike some show mares. It may have been a case of beginner's luck, but her first foal (born in 1982) was Thornberry Royal Gem, a filly by Keston Royal Occasion, one of his last before his export to Australia. Following an outstanding youngstock career for her owners, she was sold to the Millcroft Stud and soon became its leading

Carrwood Bronzewing.

Cwmwyre Samantha, foundation mare at Thornberry Stud.

show mare by twice winning championships at The Royal and Northleach as well as female champion at the Royal Welsh in 1993, having won the barren mare class there as a four-year-old. Samantha also bred the Royal Welsh winner Thornberry Demelza (f. 1986 by Abercrychan Spectator); she was youngstock, female (beating her dam) and overall champion in 1989, judged by Janet Evans (neé Edwards, of Weston Stud). Male champion that year was the yearling colt Linksbury Celebration (f. 1988) by Cusop Steward and a grandson of Keston Royal Occasion on his dam's side. Bred by the Landon family from West Wales, he was another example of top-class Section Bs emerging from a small stud, a pattern of breeding which was emerging towards the end of the 20th century.

Thornberry Royal Gem, Northleach Champion, shown by John Carter.

Samantha seemed to breed well to any stallion to which she was put, including Downland Mohawk to which she produced the Royal Show winner, Thornberry Smoke Signal. However it was to Rotherwood State Occasion that she would consistently produce top-class foals starting with a colt, Thornberry Gamekeeper, born in 1985. He was almost unbeatable as a yearling, accounting for championships at The Royal and Northleach as well as reserve male champion at the Royal Welsh as part of his dam's winning progeny group. Gamekeeper's full brother Thornberry Royal Diplomat joined Royal Gem at the Millcroft as a future stud stallion; he was shown successfully and qualified for the Creber In hand Final at the Horse of the Year

Show in 1990. A full sister, Thornberry Debutante (f. 1989) was retained as a mare for the stud as well as three others out of Cwmwyre Samantha – Morwenna by Mollegaards Spartacus, Love Song by Aston Love Knot and Samantha by Eyarth Rio.

Several outcrosses were brought into the stud to expand the mare base and of these Rotherwood Lorikeet (f. 1987 Ardgrange Dihafel x Weston Lark) would take the stud to the top once more at the Royal Welsh, when she was female champion in 1996. She bred well to Thornberry Gamekeeper, producing Thornberry Ptarmigan and Thornberry Serin, both champions under saddle; the latter was retained as a brood mare. Work and family commitments have restricted the stud's showing activity in recent years, although 2012 was a landmark for Thornberry at the Royal Welsh with a second place for the three-year-old filly, Thornberry Arbenig (Ptarmigan x Samantha), who brings together the best of the stud's bloodlines.

The success of the Thornberry ponies, considerable by any standards, has flown in the face of a desire to breed up-to-height ponies, since the majority carrying the prefix have been well under 13 hands. Height does not appear to have been a priority, while Welsh character has – probably carried into the gene pool by Section A blood found a few generations back. It identifies a discrepancy between what judges accepted in Section B classes during the latter part of the 20th century and what judges have been generally reluctant to accept as the breed moved into the 21st. The quest for height in the ridden Section B has placed demands on the breed, whose infusion of height through Arab and Thoroughbred sires through the Foundation Stock Register in the Stud Book has diluted as the years have rolled by. Nevertheless the Section B appears to have stabilised in terms of type, with the height settling on or around the 13 hands mark.

Chapter XV

Laithehill, Paddock, Thistledown, Moelgarnedd, Millcroft

Paddock Camargue at the Horse of the Year Show in 1990.

Belvoir Zechin in Australia.

Mention has already been made of the Laithehill Stud in West Yorkshire owned by Robert Hensby, one of the most enthusiastic exhibitors of Section Bs over a 40-year period. He described his entry into breeding Welsh ponies as 'purely accidental', although he had an interest in ponies as a child when he competed at local shows in jumping classes and mounted games. From school he joined his father's agricultural contracting business at Denby Hall, Flockton near Wakefield, where he bought his first Welsh pony as a teenager in 1963 with the intention of breaking and selling her on. Purchased from well-known Arab breeder Glyn Greenwood, he was able to register her in the Foundation Register of the Welsh Stud Book and named her Laithehill Swallow, the first pony to carry this prefix. Swallow's second place at the Otley Show in 1964 was enough to fire the enthusiasm of Hensby, who immediately became committed to showing and breeding ponies for the rest of his life. The pages of the WPCS's annual *Journal* opened his horizons and he was drawn to stallions such as Solway Master Bronze, Coed Coch Berwynfa, Brockwell Cobweb and Lydstep Barn Dance – more importantly, he was drawn to the Welsh Section B. The Stud Book shows us that he continued with Section As for almost a decade but the last was registered with the Laithehill prefix by 1973.

The die was cast as early as 1964 when he visited the Gredington Sale at Bangor-on-Dee with £200 in his pocket; he bought his first Section B, a filly named Greenlinks Belladonna (by Solway Master Bronze), the first Welsh pony registered by Anne Berryman from Slough in Buckinghamshire. Mrs Berryman became well known in and around the Home Counties, where her Greenlinks ponies were shown with success and received universal approval for their performance under saddle. She had a very successful stallion in Hever Olympic (Keston Royal Occasion x Hever Guinevere by Solway Master Bronze), who bred well with her mares including Royal Occasion's dam, Clyphada Periwinkle, and Lechlade Angelica, which was dam of the stud's stallion, Greenlinks Olympic Cowboy (f. 1991). Belladonna was the first to be shown by Hensby in Section B classes and the first to breed a Section B foal for him when put to Mrs Lord's son of Coed Coch Berwynfa, Hollytree Prys.

Without doubt his most fortuitous purchase was that of Tetworth Tropical Romance (f. 1966 by Downland Romance x Sandalwood, a daughter of Reeves Golden Samphire), for she immediately put her stamp on the stud by producing three fillies which became its nucleus. The first was Laithehill Bronze Cupid (f. 1969) by Solway Master Bronze, followed by Cameo (f. 1970) by the Section A Belvoir Talisman, and Harvest Dance (f.

1971) by Lydstep Barn Dance. With three homebred mares requiring a stallion, Hensby and his great friend and mentor, Mrs May Parkin, who bred a few Section Bs nearby at her Gunthwaite Stud, decided to visit the Ponies of Britain Stallion Show at Ascot in 1969, where they much admired the stallion Belvoir Zoroaster. A visit to the Duchess of Rutland's Belvoir Stud led to the joint purchase of a cream colt foal, Belvoir Zechin, a full brother to Zoroaster. Zechin enjoyed the most successful season when shown, with wins at Ponies of Britain, The Royal, Royal Welsh and Great Yorkshire. Hensby bought out Mrs Parkin's share of the partnership and stood Zechin at stud; his first foal was Tropical Romance's fourth, a colt named Laithehill Zwanziger (f. 1972), which was sold to Mrs Martens from the Netherlands, who was well known for her bus trips to England and Wales with Dutch enthusiasts.

For the next five years Zechin bred some first-class stock for both Hensby and other Section B breeders, who took full advantage of his Downland bloodlines. Belladonna's filly by Hollytree Prys, Laithehill Madonna, had bred two nice foals to Belvoir Talisman, but it was her 1972 colt, Laithehil Zircon, by Zechin that most impressed. Zircon was sold to Mrs Parkin, who showed him most successfully before selling him to Australia, where he continued his successful run. Ironically it was a surprise phone call from Down Under in 1976 that placed Hensby in a dilemma as the caller, Mrs Berry, offered a high price if he would sell Belvoir Zechin. It proved too much of a temptation and Zechin soon headed for Australia, but fortunately he left some lovely mares behind to continue his lines – however this did leave Hensby with ten mares to cover and no stallion. Luck was on his side when he received, out of the blue, a letter form Mrs Rose-Price asking him if he could be interested in buying her young stallion, Tetworth Nijinski, which he already knew – there was no hesitation and the cream stallion joined the Laithehill mares in Yorkshire in 1977.

In the meantime, Hensby had bought in a few mares, including Kirby Cane Plume, which was on offer when the Droskyn Stud was disbanded; put to Mrs Parkin's well-known roan stallion Radmont Tarquin, she produced Plush in 1978 and Pavlova in 1981 by Tetworth Nijinski. Both mares added a valuable outcross to Hensby's breeding. Despite her small height, Plush was a top show mare, winning at The Royal and four times at the Great Yorkshire Show. By comparison, Pavlova was full height and bred some top show winners such as Harlequin and Pirouette, the latter sold to Mark Tamplin for whom he subsequently bred Griashall Kiwi, Valentino and Swanlake. Harlequin and Pavlova were by Abercrychan Spectator, the next stallion to come to the stud through another phone call – this time from Elwyn Davies, who had decided to disperse his Sunbridge Stud. Knowing that Hensby had admired his other stallion, Pennwood El-Dorado, which had been sold, he offered him Spectator. Hensby was particularly impressed by his small head, big front and spectacular movement so he jumped at the chance – it was yet another wise move, as he became a major influence in the stud for the years ahead.

Two other stallions would have a major impact on the stud. The first was Chamberlayne Don Juan, the Lechlade Scarlet Pimpernel son, destined for Australia but at the time standing at stud in the North of England with Dorothy Addison. He crossed very well with Laithehill Harvest Dance, producing in 1981 a filly called Cordelia, one of Hensby's favourites and an ideal model for a Section B in the eyes of her breeder. First prizes at The Royal and Royal Welsh at five years of age reflected her success when shown in hand, and she was also a winner when ridden by Hensby's daughter, Victoria, who had inherited her father's interest in Welsh ponies. Cordelia's daughter Ophelia (by Abercrychan Spectator), foaled in 1991, twice won the young mare class at the Royal Welsh but sadly died aged nine, leaving an orphan foal, Laithehill Oberon, whose sire, Heaton Romeo, made 10,000 guineas for his breeder, Julie Perrins, at the 2005 High Flyer Sale at Fayre Oaks when purchased by a Mr Booth. Romeo won the Royal Welsh in 1996 as a yearling; his dam, Katrina of Kirkhamgate was by Radmont Tarquin out of a mare by Belvoir Zechin. By this time Romeo's sire, Eyarth Rio had become a good show winner and a successful sire in his own right.

Laithehill Cordelia.

Oberon very quickly established himself as one of the top Section B stallions in the country and joined the breed's elite by winning the WPCS Sire Rating Award for four consecutive years from 2008 to 2011. He has also made a name for himself in the sale ring, as two of his offspring have achieved notable prices – Laithehill Mustard Seed out of Cordelia's other Spectator daughter, Titania, sold for 3,500 guineas at Fayre Oaks in 2005 to Clive Morse, who also purchased that year the yearling filly Laithehill Valeta, by Heaton Romeo, for 3,000 guineas. Valeta's full sister, Laithehill Mazurka, cream like her sire, became a prominent show mare for Hensby, but it was her filly by Oberon, Laithehill Concerto (f. 2009), which exceeded all expectations by selling to Mrs Pollett from Bristol for 6,000 guineas at the High Flyer sale at Fayre Oaks in 2010; a big winner for Hensby, she has continued in the same vein for her new owner. In 2012, another Oberon offspring topped the sales when his palomino son Laithehill Wild Wind, a double cross to Eyarth Rio, sold for 3,400 guineas to Gretchen Aitken for her Family Partners stud farm of Section Bs in Oregon, United States.

Fellow Yorkshire breeders Bernard and Maureen Butterworth had gathered several well-bred mares from different bloodlines for their Paddock Stud situated near Hebden Bridge, the popular former mill town close to the Lancashire border. Having bred seven

Royal Welsh Gold Medal winners, six for Section B and one for Section C, their success as breeders has been exceptional. They were both raised on dairy farms high on the Pennines, where a work horse was kept as well as a pony, purchased from the local market at Halifax. Mrs Butterworth's father kept a Welsh cob for work on the farm while his brother, George Redman, was prominent in the Hackney Society, where he competed along with famous names such as Haydon, Black and Hanson. (He was a well-known judge, a Council member of the Hackney Society and its President in 1961.) Following their marriage, the Butterworths never travelled far from their roots and a dairy business was sold in 1962, by which time they had bought a Victorian wool mill in Paddock Old Town overlooking the River Calder and Hebden Bridge. Previously owned by Abraham Gibson, a well-known local benefactor, the five-storey mill was powered by generators driven by water from the moors to the north and benefited from the softest water in the Bradford area for washing the wool. Butterworth took advantage of this and for 30 years he and his wife produced as many as 25,000 treated fleeces annually for the domestic market.

The Butterworths entered the world of the Welsh breeds inspired by their neighbour and good friend Glyn Greenwood, who kept Mountain ponies and Arab horses – so it was perhaps a case of familiarity rather than choice that initially led them to the Welsh. At this time one of their regional representatives in Wales (they had a large poultry enterprise prior to the fleeces) told them about the wonderful Section C ponies he had seen, belonging to Lady Chetwynd. A visit to Wales was followed by the arrival of two Section C fillies bred at the Cwnlle Stud in West Wales by Lady Chetwynd's celebrated stallion, Lyn Cwmcoed – one for their son Fraser and one for Mrs Butterworth. Too young to be ridden, the fillies were put in foal resulting in a beautiful filly born to Cwnlle Lily in 1973 named Paddock Dawn. As Fraser grew, so did the ponies and the best of breeding available in Welsh Cobs was purchased from the leading studs – a pattern they would follow when they moved into Section B.

During a visit to the Merioneth Show while holidaying in Wales, they were captivated by a beautiful Section B filly, Weston Choice, which had swept the boards that day. Instantly they knew that the future of their Paddock Stud lay in this section of the Stud Book and in no time the Cobs were gone and the search for quality, well-bred Section Bs was on – they were no less thorough with this than they had been with the Section Cs and Ds. They were again drawn to Wales, first to Downland in the south and then to Coed Coch in the north, as they had already seen the benefit of the cross of Downland Chevalier on mares by Coed Coch Berwynfa. Mrs Cuff refused to sell their first choice of Jamilla or, during a subsequent visit, their selection of a chestnut filly foal by Chevalier, Downland Duet. Fortunately, Mrs Cuff relented and Duet did go to Paddock Farm, where she became a great show ring success when shown during her first year.

The Butterworths also gathered a small band of well-bred mares – Kiltinane Butterfly, Brockwell Japonica, Downland Siskin and Weston Madeleine, a full sister to the filly they had so much admired, Weston Choice. The sale of Madeleine at Fayre Oaks with her filly foal by Downland Cognac, Myth, may have proved premature, as she went to David Williams from Bala as a foundation for his Moelgarnedd Stud. It was to Mrs Cuff that they

Above: The Butterworths' prolific home bred winner, Paddock Gemini, shown by the author at the Royal Welsh in 1983.

once more turned to secure a stallion and they chose Downland Toreador (f. 1971) by Chevalier. Doubling up the Chevalier blood, they put Toreador to Duet to produce her first foal in 1978, the rich chestnut coloured colt Paddock Gemini. Not shown until he was five, he immediately took the Butterworths to the very top in the show ring, qualifying for the In Hand Finals at the Horse of the Year Show in 1984 and 1985, and supreme pony champion at The Royal Show in 1984 under saddle he qualified for Olympia. The same cross produced for them a very pretty little filly named after her dam – she was the only female from the original group of mares that would be kept on for breeding.

By this time the success of the Baledon and Rotherwood Section Bs had provided a successful model for breeding Section Bs and the Butterworths now turned to Lt Col Williams-Wynn and his Coed Coch ponies carrying Berwynfa blood. Lunch at the Colonel's lovely home at Plasnewydd was followed by a walk round the fields and once more the Butterworths chose one of the best when they saw for the first time Coed Coch Gala. She was not for sale at the time, but fate played its hand five years later with the death of Williams-Wynn and the opportunity to buy Gala at the stud's dispersal sale in 1978. The story has already been told (Chapter 11) of how they bought her full sister Penwn, the previous lot to Gala, just in case they were outbid on her sister, which they preferred – but that would not be the case and both mares headed to Yorkshire. History shows that it was a wise move to buy Penwn, as she provided them with a remarkable female line on which to model their stud, while Gala gave them an outstanding stallion in Paddock Camargue (f. 1984) by Downland Chevalier,

Below: Coed Coch Gala at Paddock Stud.

Paddock Sahara, Champion Royal Welsh 2005 for Sandy Anderson and shown by Craig Elenor.

which would be both a show-stopper and a perfect mate for the Penwn females. Camargue was one of the best males to emerge from the Chevalier/Berwynfa cross – he won championships at major shows from his yearling days, although 1990 proved to be one of his best, with ten championships. He was judged pony supreme at The Royal and champion at the Great Yorkshire, where he also qualified for the Judy Creber In Hand final at the Horse of the Year Show. The following year he was champion at Glanusk, the Bath and West and Northleach, and male and overall reserve champion at the Royal Welsh. Camargue was shown by Norman Fairbank, who has helped with the stud since 1989.

The three-year-old Berwynfa colt Coed Coch Lygon was also purchased at the 1978 sale as part of their breeding plan but although he worked in parts he found no long term future in the stud. His best get was a daughter named Paddock Garland out of his grandmother, Gala

(f. 1982) – she won the Royal Welsh as a novice mare in 1989 and her son by Camargue, Paddock White Lightning, would be a winner in the show ring before standing at stud with Lady Dysart. Camargue would bring all the necessary Coed Coch blood to the gene pool at Paddock, particularly through the Penwn line. Mrs Cuff remained loyal to the Butterworths and allowed them to send Penwn to Chevalier, producing a filly, Princess Charming, in 1980 and the colt Orion in 1983. Princess crossed well with Camargue, genetically her equal. His first foal, Paddock Picture, born in 1989, was sold to Mrs Elenor for her son Craig to show, which he did with great success – Craig later became one of Britain's most successful professional showmen. As a yearling Picture beat Camargue in the Section B championship at the National Pony Society Show under Mrs Jocelyn Price and the following year won the Breeders' Award for the best young pony at The Royal. As a brood mare she won the overall championship at the Royal Welsh in 1998, followed up with reserve female awards in 2002 and 2005. Picture's full brother, Riverdance (f. 1995), was retained in the stud and principally used on Camargue daughters.

Coed Coch Penwn's 1988 daughter by Gemini, Paddock Penwn, bred exceptionally to Camargue – their daughter Pretty Polly won extensively, including adult female champion Section B at the WPCS Centenary Show held on the Royal Welsh Showground in 2002. To Riverdance, Pretty Polly bred three top-class foals – Peaches (f. 1989), a champion in the Netherlands before returning to join the Linksbury Stud, Sahara (f. 1999), overall champion at the Royal Welsh in 2005 and Petra, winner of the novice mare class at the Royal Welsh in 2009. Paddock Penwn's other Camargue daughter,

Paddock Northern Lustre, Champion Royal Welsh 2007.

Peioni, bred well to a selection of stallions, producing Paddock Hyperion by Riverdance, Parasol by Eyarth Mercury, Alesha by Moelview Prince Charming and Jamila by Eyarth Rio. All four have been winners at top shows, although pride of place must go to Paddock Alesha, supreme champion at Glanusk in 2010, the same year she was overall champion at the Royal Welsh under Daydre Chambers. In 2011 she claimed the Brightwell's

Section B final at the Welsh National Show and was also reserve adult champion at the International Welsh Show at Builth Wells and again a winner at the Royal Welsh itself, but this time as a barren mare.

Camargue showed his ability to outcross when Paddock Silver Lustre bred winner after winner by him. By Watermans Mandolin, a chestnut by Downland Mandarin, Silver Lustre's dam was the well-known Reeves Fairy Lustre, which the Butterworths had purchased along with her daughter, Droskyn Fairie Queen (by Rotherwood Commander), when Nigel Sykes dispersed his Droskyn Stud. Lustre's first foal by Keston Royal Occasion, Paddock Fairy Lustre, was shown successfully and then sold to the Millcroft Stud, where she made a significant impact. Silver Lustre made her presence felt at home when put to Camargue – in 1995 they produced a filly from the last crop of foals by him before his export to Sweden. This was Paddock Northern Lustre, Edwin Prosser's Section B champion at the Royal Welsh in 2007 for Sandy Anderson (Thistledown Stud).

Camargue's influence at Paddock can only be matched by his influence at Mats and Cecilia Olsson's Kulltorp Stud in Sweden, where half of the resident mares and the young stallion, Kulltorp Juventus, are by him. Inspired by the Coed Coch ponies which he had seen on trips to Britain, Mats Olsson's search for new Section B blood for his stud initially took him to Downland in 1994, whereupon he decided that what was required was a son of Chevalier out of a Berwynfa mare – Camargue obviously fitted the bill. He later returned to Mrs Cuff, where he purchased the entire crop of foals in 2000 including a colt, Downland Wild Fowler by Downland Arcady out of Downland Moonwalk, which is used to cross on the Camargue mares.

In 2000 Kulltorp Yes, a son of Paddock Camargue out of Springbourne Brocade (granddaughter of Menai Shooting Star and Coed Coch Berin), was leased to the Thistledown Stud in Leicestershire, specifically to cross with the Coed Coch Gala daughters Paddock Gala's Gem and Paddock Garland, the Royal Welsh winner. The resultant fillies, Thistledown Summer Gala and Thistledown Gabby (f. 2001) were put to the stud's resident stallion, Paddock Sahara. With only Menai Shooting Star providing an outcross four generations back, it can be said that the Downland/Coed Coch magic continued to work when Summer Rain (f. 2008) won at Northleach and Goodness Gracious (f. 2007) came second at the Royal Welsh in 2009. Arguably Paddock Garland's most successful daughter has been Thistledown Garbot (f. 1998 by Paddock Gemini), placed in the mare classes at the Royal Welsh and the dam of a host of good winners including Thistledown Galileo, Gigi and The Governor.

This success reflected more than just a degree of faith placed in the Butterworths' breeding programme as other Paddock mares at Thistledown included Paddock Northern Lustre by Camargue and Plush by Paddock Gemini out of the Coed Coch Penwn daughter, Princess Charming. Paddock Plush put to Delvers Tarragon by Bunbury Thyme (son of Lydstep Barman) produced Thistledown Pretty Woman, which was reserve champion at Northleach in 2008. Paddock Sahara's championship at the Royal Welsh in 2005 followed up by Northern Lustre's championship in 2007 did much to enhance the reputation of Sandy Anderson's Thistledown Stud of Welsh Section Bs at a time when his Mountain ponies were riding the crest of a wave. However the success at

Thistledown Alhaarth, Royal Welsh 2007.

the Royal Welsh in 2005 was enhanced by a unique achievement when Jane Williams (Sarnau Stud) selected Sahara's daughter Thistledown Cormelia (f. 2004 out of Mynach Cornflower) for the female championship, the first time that the male and female Gold Medals at the Royal Welsh have gone to a father and daughter.

Sandy Anderson hails from Loch Watten in Caithness, some 15 miles as the crow flies from John O'Groats, Britain's most northerly village. His family were involved with Clydesdale horses and Shorthorn cattle as well as the local Post Office, but the prospect of playing football professionally followed by university took him first to Central Scotland and then to London, where he secured his first job. In his working life he proved to be both very good at his job and ambitious as he followed promotion in the transport industry. From a leading post with the American firm TIP Trailers, he was head-hunted to lead the privatisation of British Rail under the Conservative government in the early 1990s. With the impending election of a Labour government and the likely U-turn in policy, the taste for investment ran cold for entrepreneurs and Anderson, along with his colleagues, pursued a successful management buy-out. Luck was on their side when a major investor came along almost immediately and huge profits were enjoyed by all concerned including Anderson, who became a millionaire over a period of a few weeks and many sleepless nights. He has since built up a large investment portfolio which includes ownership of Grandstand Media, the organisation which took over the running of the Horse of the Year Show.

At the time he was living with his family near Tring in Hertfordshire, where he and his wife, Georgina, kept a pony for their girls which just happened to be Welsh. Having attended a few shows, Anderson had a taste for breeding and in 1989 he registered his first foal, Thistledown Northern Star, a Section B by Burstye Flavius out of Lydstep Joyful, a daughter of Abercrychan Spectator. Section A and B ponies were both kept initially, although a small number of Welsh Cs were introduced in 2000. As the number of ponies increased the need for a larger property became obvious and a small Thoroughbred stud farm on the outskirts of the village of Costock in Leicestershire was purchased in 1994. In 20 years since the move, the stud has become the largest in the country, if not Europe, with a presence in the show ring for both Section A and B which has seen no equal. Anderson's drive in business is matched with that for breeding and showing Welsh ponies, so it is little wonder that he sought the best bloodlines to develop his interest in Section Bs. As well as producing ponies from home, particularly mares, he also engages the use of professional producers for his ponies and Craig Elenor from Thirsk in Yorkshire is his preferred producer for Section Bs.

It would be fair to say that Anderson's early taste in Section B was for a more quality type of pony, to which end he purchased two mares from the Rhoson Stud, Rhoson Ffansi by Rhoson Taranaki and Rhoson Aida by Downland Rembrandt. A stallion he used early on was Gigman Jacana, bred by successful South Wales breeder Sue Easton by Rotherwood State Occasion out of a mare by Downland Mohawk. The outcome of the mating with Rhoson Ffansi was Thistledown Flair (f. 1997), typical of her breeding and appealing to like-minded judges. Flair was second at the Royal Welsh and the Welsh Centenary Show in 2002, champion at the Three Counties in 2005 and a winner at Northleach in 2006. Meanwhile Aida was crossed with a relative newcomer to Thistledown, Mompesson Wild Party (f. 2003), a son of the Royal Welsh champion Carwed Charmer, to produce an outstanding colt in Thistledown Alhaarth (f. 2006), which was unbeaten as a yearling in 2007 when a winner at the Royal Welsh, The Royal and the Cheshire and reserve champion at Cheshire County. Wild Party was also responsible for the Cheshire reserve youngstock champion, Thistledown Sea Shanty (f. 2010) out of

Thistledown Arctic Fox made 9,000 guineas to go under saddle at the High Flyer Sale, Fayre Oaks, in 2011.

Loveden Sea Pearl, and the chestnut colt, Thistledown Eye of the Tiger (f. 2009), second at the Royal Welsh in 2011, whose dam Thornberry Eirianwen has twice won there – in 2006 when two years old and in 2009, when she took the reserve female championship. Wild Party was also sire of Thistledown Glamour Chick (f. 2011 out of Thistledown Gabby), which in her only two outings in 2012 won both the Royal Welsh and National Welsh.

Various stallions have been introduced to the stud over the years, some with more success than others. With such a wealth of Paddock breeding, it has been essential that crossing with them gets results and Anderson did just that with another outcross, Mynach Buccaneer (by Moelview Mohawk), which produced the 2006 Royal Welsh winner Thistledown Night Shift (f. 2005) out of the Gala's Gem daughter Thistledown Gala Night, as well as the Three Counties winner, Thistledown Sheer Lustre (f. 1999), a daughter of Paddock Northern Lustre. Her half-sister by Rotherwood Sate Occasion, Thistledown Golden Lustre, was successful as a brood mare, with the Northleach championship among her credits; her son by Sahara, Thistledown Wicked, was chosen by Robert Grant for his Bamborough Stud in Australia in 2012. Buccaneer also sired the 2011 Royal Welsh Winter Fair foal winner, Thistledown Golden Zumba out of Thistledown Garbot.

Showing mares has become something of a passion for Anderson and one of his most successful next to Paddock Northern Lustre has been Cottrell Lara, by Cwrtycadno Cadfridog out of Cottrell Liberty. Throughout 2005–2006 she swept the boards by winning at the major shows, with her best results coming at the South of England and

Cottrell Lara, Champion at Lincoln County in 2008.

Kent County Shows in 2005 and 2006 respectively, when she not only won the Section B championship but also qualified for the Cuddy In Hand Final at the Horse of the Year Show. Anderson took full advantage of Mrs Johns-Powell's bloodlines when the Cottrell Stud was dispersed by adding the Cottrell mares, Neptune, Nell Gwynne and Anya to his brood mare list which included Cottrell Lara and Cottrell Rose of China. The latter bred the Royal International Mini Champion, Thistledown Rose of Sharon, while Anya bred two outstanding colts. Thistledown Arctic Fox (f. 2008 by Paddock Sahara) was top price at the High Flyer Sale at Fayre Oaks in 2011, where he made 9,000 guineas to go under saddle. His winning half-brother by the Lampeter champion Stoak Tuscany, called Thistledown Arctic Wind, sold for 4,000 guineas to Mats Olsson in 2010. In an interesting development for Thistledown, new blood was brought into Thistledown from Mark Bullen's Imperial Stud in 2012 – this included the mare Imperial Jessandra, along with her colt foal Imperial Geronimo by Imperial Pavarotti, a son of Eyarth Tigra. As a yearling Geronimo won at the Royal Welsh, Great Yorkshire and Cheshire Shows where he was also overall youngstock Welsh champion.

The impact Paddock breeding had in this stud was comparable to what it had on others, albeit through only one female and not several as we have just observed. They could scarcely have been further apart – Moelgarnedd, situated near Bala in picturesque North Wales, and Millcroft, situated close to the holiday town of Dawlish in Devon. The former belonged to David Williams, whose interest in Section B was truly kindled following a visit to the Weston Stud, where Jack Edwards had the task of inspecting a filly for the FS Register. All went well and the two men hit it off straight away, Williams very much in awe of Edwards whose knowledge and skill as a breeder, coupled with friendly advice, impressed from the outset. He was also most impressed with the ponies, particularly the Section Bs which he admired for their Welsh character, a quality that has been associated with the Moelgarnedd ponies since the first ponies were registered with the prefix in 1981. Williams describes himself as a hobby breeder, a pleasure he combines with a hectic work schedule in the food industry.

It is not surprising that he concentrated on Weston breeding when selecting foundation mares for his stud, and he attributes much of his success to Edwards and his family. Weston Graceful by Weston Gigli and Llandyn Briallen by Chirk Crogan were his first purchases, and he was lucky to purchase from the Butterworths at the Fayre Oaks Sale the 11-year-old mare Weston Madeleine for the bargain price of 500 guineas. According to Williams, she had a 'rather insignificant' late filly foal at foot by Downland Cognac, by the name of Myth. Both Madeleine and Myth would play a major part in the future success of the Moelgarnedd Stud. Myth developed over the winter and embarked upon a successful showing career the following year, when she stood reserve in youngstock championship at the Royal Welsh to the yearling colt, Weston Consort. Myth may have enjoyed a successful career in the show ring, although this would be eclipsed by her success as a matron – her first foal, Moelgarnedd Miri (f. 1986) by the 1972 Royal Welsh champion Brockwell Chuckle followed in her dam's footsteps, both in the show ring and as a brood mare.

Williams experienced a high level of success over the ensuing years based on a very strong mare line, something that has become noticeable in the history of the Section B and a factor common to most livestock breeding. It could be argued that his success stemmed not from Paddock Myth but from her grandmother, Weston Princess Mandy, the dam of Williams's favourite, Weston Choice. His next step was to double up the bloodline with Miri by covering her with Brookhouse Chuckler by Weston Neptune out of a daughter of Princess Mandy, Weston Diana, which Chuckler's breeders, Mr and Mrs Hughes, had purchased for 1,100 guineas at the 1978 Weston Sale. The outcome was everything that Williams could have wished for, as he not only had an excellent show filly in their filly, Moelgarnedd Madonna, but in time, like her dam, she would become an exceptional breeder, too. Madonna was not the only show winner from Miri – others included the colts Mandolin, Montana and Cadfridog, all exported as future stallions. However, exceptional among her offspring has been Moelgarnedd Myrddin (f. 2000), many times a champion as a youngster when shown in hand in Britain and junior champion at

Moelgarnedd Stadros, Youngstock and Male Champion at the Royal Welsh in 2013.

the International Welsh Show at Aachen in 2003. He was produced, like all the Moelgarnedd ponies, by Wales's leading in hand producer, Colin Tibbey, a showman very much in the mould of the famous men from Coed Coch and Gredington and every bit as effective in his job. Myrddin has since gone on to a sparkling career under saddle in the hands of Eleri Marshallsay – he qualified for both the Horse of the Year Show and the Olympia ridden final (where he was best of breed) and was supreme ridden mountain and moorland at the British Show Pony Society Heritage Championships in 2005, the same year that he was judged ridden champion at the International Welsh Show in Belgium.

His breeding combines the best of both Williams's mares as his sire, Moelgarnedd Dewin, is Madonna's son by the well-known Royal Welsh champion, Carwed Charmer. Dewin (f. 2007) had an exceptional show record as a yearling, winning eight major shows including the Royal Welsh and standing reserve champion at the Great Yorkshire; he was subsequently sold to Sweden.

Moelgarnedd Myrddin, Brynseion Ridden Champion 2007.

Dewin's sister Dymuniad won the Royal Welsh as a foal and his full brothers Moelgarnedd Calypso (f. 2001) and Derwyn (f. 2002) have been principal winners in the show ring. Calypso made the top price of 5,300 guineas at the 2001 Fayre Oaks Sale. In the ownership of David and Elizabeth Lynes from Anglesey, Derwyn was youngstock, reserve male and reserve overall champion at the Royal Welsh in 2005 – the same year that his championship at the Scottish Horse Show led him to the Cuddy In Hand Final at the Horse of the Year Show, where he was runner-up in the pony section. In a showdown between the two brothers at the National Welsh Show, Calypso got the nod over Derwyn in the Rhydspence National Section B Championship.

The most recent of Moelgarnedd colts to hit the top has been the bay colt Moelgarnedd Stadros, born in 2010 by Hilin Carnedd out of the imported mare, Weston Park Musk Rose. Wishing to introduce an outcross for his mares while retaining the Weston Princess Mandy bloodline, it was during a holiday to Australia that Williams once more turned to the Edwards family. He spotted a beautiful bay filly by the imported stallion, Eyarth Sama, and decided that she fitted the bill in all aspects. The move has proved successful and Stadros looks set to take over as senior sire at Moelgarnedd; in 2012, among other accolades achieved, he was judged champion at the National Welsh Show at Malvern by Mrs Mansfield-Parnell, a week after he had been supreme native pony at the NPS Championship Show while in 2013 he was judged youngstock and male champion at the Royal Welsh Show. This demonstrates that all of Williams's success has not relied on one mare alone, but certainly much of it traces to a special one, Weston Princess Mandy through Paddock Myth. She was the pony responsible in the case of Moelgarnedd, just as Paddock Fairy Lustre was at the Millcroft Stud in Devon, where her influence has been felt along with two other highly significant mares, Briery Starlet and Thornwood Penny Royal.

The Millcroft Stud was started in the early 1960s by Betty Knowles and her daughter, Frances, who was mad about ponies from an early age and took a keen interest in breeding them. The Knowles family was involved in grain milling at the popular Devon coastal resort of Dawlish; Knowles Mill, one of two 19th-century water-powered mills in Dawlish, is now in the care of the National Trust. Betty and Frances Knowles started with riding ponies, but their interest changed to the Welsh breeds and quickly settled on the Section B, although three of their foundation mares came from Section A – Criban May Rose, Criban Symphony and Twyford Quickstep. Weston Blue Petal by Chirk Crogan and Cusop Tiptoes by Coed Coch Pawl launched them into Section B, as well as the stallion, Sinton Whirligig (f. 1965) by Sinton Giration out of Coed Coch Pefr. Whirligig was produced for the show ring from Elspeth Ferguson's Rosevean Stud and shown by the familiar figure of Robert Gilbert, her showman and head groom. He won extensively at major shows such as the National Pony Society Show, where he was reserve champion in 1972; among the best of his offspring was Millcroft Ragtime (f. 1970) out of Bwlch Redwing.

Whirligig's days in the stud were numbered when Frances Carter fell in love with a colt foal named Aston Love Knot, by Downland Romance out of a mare by Coed Coch Planed. Small at only 12.2hh, grey and very pretty, his Mountain pony breeding proved an ideal match for the gathering herd of mares at Millcroft Farm. He was such a success that Whirligig's services were no longer required, resulting in his sale to Western Australia in 1975. Love Knot remained a firm favourite throughout his life at the stud and died there aged 30.

As the stud began to establish itself, more mares were sought, and it is interesting to note that, like Cusop Tiptoes, the preference was for FS1 mares

Millcroft Suzuya.

within the Welsh Stud Book. They included Reeves Silver Lace (Ceulan Gondolier x Ceulan Silver Lustre), Belvoir Mignonette (Belvoir Gervais), Bwlch Redwing (Gredington Mynedydd), Criban Ninon (FS by Bwlch Valentino), Elphicks Half Moon (Solway Master Bronze), Abercrychan Donna Bella (Downland Chevalier), Lydstep Lady's Slipper (Downland Chevalier) and, most importantly, Briery Starlet (Brockwell Cobweb x Criban Ester). Lady's Slipper provided the mother and daughter team with a great deal of publicity as she was a major winner at shows, her best result coming at the Royal Welsh in 1977 when she also qualified for the Horse of the Year Show. Briery Starlet was successful in the show ring, but it was her breeding success that set her apart from the others in the stud.

Starlet carried the Briery prefix of prominent Arab breeders Major and Mrs Hedley from Windermere, who had purchased her FS dam, Criban Ester by Bwlch Valentino, for 420 guineas from Mrs Binnie at Fayre Oaks in 1961. Their intention was to cross her with their Arab stallions to produce children's riding ponies, so they had no use for her filly foal by Brockwell Cobweb, which they sold back to Mrs Binnie – but not before naming her Briery Starlet. It was as a four-year-old that Starlet moved to Millcroft, where she remained for the rest of her life, producing 17 foals. The early ones, all by their resident stallion, Aston Love Knot, proved so good that the full sisters, Millcroft Serina (f. 1971), Simona (f. 1973) and Suzuya (f. 1975), were all kept within the stud as foundation mares. Other daughters, such as Silhouette (1981) by Downland Caribou and Soubrette (f. 1983) out of Suzuya by Ardgrange Difreg, were also retained for breeding. The grey mares from Millcroft were familiar figures in the show ring. Suzuya was probably the best of them and twice won at the Royal Welsh among many major victories, especially in the South West where the Carters were great supporters of the shows and did much to promote the Section B. Her first win at the Royal Welsh came in 1980, followed by another in 1985 when she was also female champion and reserve overall to Downland Gold Leaf. That year she was champion at both the Devon County and the Royal Cornwall, where she stood reserve in the Lloyds Bank qualifier for the Horse of the Year Show.

Frances Carter died at a relatively young age in 1993, but not before she had sought out fresh bloodlines for her much-loved Millcroft ponies. On the advice of Dr Wynne Davies she persuaded the Waxwing Stud to part with a young mare shown by them for the Butterworths, from whom she had been purchased. This was Paddock Fairy Lustre, who had already proved her worth by coming second at the Royal Welsh as a two-year-old. She was shown successfully in brood mare classes by Frances' husband John, who had developed an interest in the ponies from his wife as well as an expertise in showing them. In recent years he has been joined by their son Anthony, who shows various forms of livestock as well as the ponies. Fairy Lustre immediately crossed well with Love Knot and in 1985 she produced a colt to him, Millcroft Copper Lustre, which claimed the youngstock and male championship at the Royal Welsh in 1987.

Ardgrange Difreg (f. 1975), a double cross of Chirk Deborah with the Downland stallions Romance and Chevalier as grand sires, had done the stud well, but a youngster, Thornberry Royal Diplomat (f. 1987), came to Millcroft with the best of credentials with Rotherwood State Occasion as his sire and Cwmwyre Samantha his

Left: Millcroft Dahlia, Champion and Reserve Cuddy Qualifier, Royal Welsh 2008.

Opposite: Millcroft Iska Roc.

dam. Samantha's daughter, Thornberry Royal Gem (f. 1983 by Keston Royal Occasion), also joined the stud, bringing it much glory in the show ring, the climax of which was female and reserve overall champion to the stallion Linksbury Celebration at the Royal Welsh in 1993.

Meanwhile in 1990 Royal Diplomat qualified for the Judy Creber Finals at the Horse of the Year Show when champion at Devon County, and won at the Royal Welsh. His son out of Paddock Fairy Lustre, Millcroft Royal Lustre (f. 1991), also won at this show in 1998. His stock have been outstanding, particularly out of Millcroft Penny Lane, a daughter out of Thornwood Penny Royal, another purchase from Waxwing by the Carters which would produce the next generation of Millcroft winners. By Varndell Right Royal out of Hever Grania, Penny Royal was bred in Yorkshire by Dorothy Addison, a small breeder who has made her own mark, not only with this mare but also with the sire of Penny Lane, Thornwood Royalist by Chamberlayne Don Juan (by Lechlade Scarlet Pimpernel) out of Downland Dabchick by Chevalier. Royalist was lightly shown, with success, although he proved his real worth for several studs over the years and appears in the pedigrees of many a successful pony. Many breeders consider him to have been one of the most underestimated stallions during his lifetime. Both Thornwood Penny Royal and Millcroft Penny Lane were successfully shown in mare classes, the latter female champion at the Royal Welsh in 1994 and champion at Northleach in 1997. Her offspring by Millcroft Royal Lustre have been exceptional including Geisha (first at the Royal Welsh in 2005 and 2008), Dahlia (overall champion and supreme champion Welsh for the Tom and Sprightly Trophy at the 2008 Royal Welsh) and Dow Jones (Working Hunter Pony champion at the Horse of the Year show in 2012).

Although the most recent success of the stud has come through Thornwood Penny Royal and Paddock Fairy Lustre, the Starlet line remains in the stud some 50 years on through her daughter Suzuya, which produced to Ardgrange Difreg, Millcroft Dipity and Millcroft Soubrette (Royal Welsh barren mare winner in 1995), and Millcroft Sparking Occasion by Rotherwood State Occasion. To Thornwood Royalist, Dipity produced in 1998 the stallion Millcroft Ghost – yearling winner at the Royal Welsh in 1999 – which was subsequently gelded, while Sparkling Occasion's granddaughter by Ghost, Millcroft Lotus (f. 2003), won the novice brood mare class at the Royal Welsh in 2010. Ghost has enjoyed an enviable record under saddle, having qualified for the

ridden classes at the Horse of the Year Show no fewer than five times. His most outstanding offspring has been the stallion Millcroft Iska Roc, one of the most accomplished and successful Section B stallions to have competed since the days of Baledon Commanchero. Born in 2005, Iska Roc was first and reserve male champion at the Royal Welsh as a three-year-old, only to return the next year when he was champion of the ridden section and qualified for the Horse of the Year Show. His record at the Horse of the Year Show has been remarkable – over a four-year period he has won both the breed ridden class and the native working hunter pony class over jumps. In 2010 he was best of breed at Olympia, while in 2012 he won the Ponies UK Bulgaria Working Hunter Supreme Pony Championship as well as the British Young Pony Event Championship at Burghley. All the while he has been ridden and produced by Sam Roberts, one of the country's leading young professionals.

The stud's emphasis on the performance qualities of its ponies has been largely promoted by Maggie Carter, who married John after the death of his first wife; Maggie has been very much the driving force behind the success of Iska Roc. His grandmother, Tricula Rosie Wren, another daughter of Thornwood Royalist, was bred by Joyce Coltart from Ayrshire, who also bred her dam, Tricula Jenny Wren (Twylands Firecracker x Monkham Snow Bunting), a champion show hunter pony and champion ridden Welsh pony at the Royal Welsh when qualifying for Olympia in 1982. Tricula Rosie Wren won the novice brood mare class at the Royal Welsh in 1995, while Rosebowl, her foal at foot by Millcroft Royal Lustre, won the filly foal class. Rosebowl's full sister, Rosette (f. 1996), took the championship at Glanusk as a yearling, while Rosebowl was youngstock and reserve overall champion at Glanusk in 1998. Seven years later she was champion at Northleach as brood mare with Iska Roc as a foal at foot. Other than Iska Roc, Rosebowl also bred Millcroft Cappuccino, second at the Royal Welsh for the Cadlan Valley Stud in West Wales, where she is a resident. 1994 also saw the arrival of another Royal Welsh winner for the stud when an outcross mare, Downland Tahiti, produced a Royal Lustre colt, Millcroft Riviera, which has been kept entire.

It is interesting how the impetus for show ring success at the Millcroft Stud has not diminished over a 50- year period and it remains a major contender in the breeding as well as the performance rings. However, the Carters stressed the emphasis they place on 'breeding Welsh ponies for their riding qualities', which was not the case for all breeders whose goals remain in breed classes alone. With a growing market for the ridden pony and high prices paid for those with ability in dressage, showjumping and eventing, perhaps market forces will influence the direction of the Section B in the years ahead.

Chapter XVI

Hilin, Eyarth, Heniarth, Telynau, Moelview, Waxwing, Mynach

Carwed Charmer ridden by Mari Evans, Supreme Ridden Champion at the Welsh International Show 2000.

The Evans 'Hilin' children. From left to right,:Arwyn, Mari, Meriel and Elin.

The story of the Welsh Pony, Section B, owes its origins to a demand for ponies suitable for children to ride during the early part of the 20th century. We recall Miss Brodrick showing Tanybwlch Prancio at the London shows, and Mrs Inge, Mrs Binnie and Mrs Crisp breeding ponies for their own children to hunt and enjoy. Throughout the book we have seen how some breeders have consistently promoted their ponies through their success in ridden classes. In a poignant comment found in the obituary for their treasured stallion Carwed Charmer (which had won the ridden championship at the International Welsh Show in the Netherlands the year before his death), Dewi Evans and his family wrote: 'I suppose that one regret is that our plans to ride him this year more extensively after retiring him in hand last year were tragically curtailed.'

Carwed Charmer (f. 1991) had in many ways catapulted the Hilin Stud belonging to the Evans family into the spotlight – initially through his performances at the Royal Welsh Show but latterly through the large number of winning progeny produced during his lifetime and after his premature death in 2002, the same year that he recorded his third male championship at the Royal Welsh. His record at the show was remarkable by any standards and showed that a limited showing career in no way diminished the opportunity for success – he was lightly shown as a youngster and mainly competed at shows in Wales such as Glanusk, Clwyd, Meirioneth and Aberystwyth, where he recorded championship victories. Having won at the Royal Welsh as a two-year-old, he went on to be overall champion in 1997 and in 2000, the year that he was also judged reserve supreme Welsh. If the WPCS In Hand Sire Ratings are anything to go by, Carwed Charmer rates alongside the likes of Downland Chevalier and Rotherwood State Occasion. It was in 1999 that he first made the top three – then for six years (2000 to 2005 inclusive) he headed the list. Two of his sons have also featured – Eyarth Tayma, which won in 2007 and 2008, and Carrwood Orpheus, which came second in 2002, 2008 and 2009. All three stallions have stood at stud at Hilin as well as the Tayma son, Hilin Carnedd, which came third in the Sire Rating in 2011.

The Hilin Stud is situated in the picturesque lakeside setting of Cwrt-y-Llyn ('house by the lake') at Pentrefoelas, some six miles east of Betws-y-Coed at the northern edge of the Snowdonia National Park in North Wales. The farm lane meets the main road at Cernogie Mawr, a former staging post for coaches travelling the main artery from London to Holyhead – the lake provided water for the horses and fish for the travellers who stopped there, including Queen Victoria. The neighbouring property of Bryn-Heilyn was farmed by Cyril Lewis (Rhyd-y-Felin), who was a great friend of John Evans, Dewi's father. The story is told how one day the two friends were looking round the mares and, on coming to one which had just foaled, Lewis lifted the foal's tail and commented, 'Another damn colt!' Little did he realise the importance of that colt – it was the highly influential Rhyd-y-Felin Selwyn.

A rocking horse proved inadequate to satisfy the interests of John's son, so a pony was purchased and this was the start of an interest that has grown and now passed on to Evans' own family. Eirian Evans (née Williams) was brought up on a farm on the Coed Coch Estate, where her father first bred Mountain ponies (Gwyndy Ucha Stud) and then Section Bs. Eirian's first pony, Cilrhedyn Moonlight, was purchased in the 1958 Coed Coch Sale and when put to Coed Coch Berwynfa bred Gwyndy Ucha Bronwen, later sold to Sir Harry Llewellyn. Cwrt-y-Llyn has been home to ponies for the Evans' four children, who competed in both showing and jumping classes, as well as to well-bred riding ponies and horses, often including a Thoroughbred and Welsh cob stallion which stood at stud. As young adults they continue to enjoy showing the ponies and their son Arwyn, now living with his wife at Bryn-Heilyn, as well as being chief showman breaks horses as a sideline to his job on the farm.

Evans' mixed-stock farm has seen increasing numbers of Section Bs since the first was registered with the Hilin prefix in 1968, when his parents gave him a mare and stallion along with life membership of the WPCS. The mare, purchased as a yearling at the Menai Bridge sale for 45 guineas, was Llwydiarth Colomen (f. 1961), by Coed Coch Siarlys; she was first ridden by Evans' sister, and then bred four foals, all of which were colts. The stallion used was the Section B, Crwydryn Berwynedd, a grandson of Coed Coch Berwynfa with a dam line tracing back to Ceulan Revolt; John Evans had bought him at Fayre Oaks. As soon as she changed ownership to his son, Colomen bred five consecutive fillies, which quickly provided Evans with a nucleus of mares with which to start his stud; in 1973 her filly, Wennol (f. 1968), produced another, Wenora, which would become the lasting tribute to Colomen as the stud started to grow. Stallions would become a feature of the stud over the years and the careful selection of individual traits as well as bloodlines has ensured its success in both the show and sale rings. Wenora was by the first of them, Eyarth Dubonnet (f. 1970), a son of Ardgrange Debonair and Tanlan Swynol, the foundation mare at the Eyarth Stud.

Merddyn Fancy and Shawbury Primrose, by Criban Victor, were put to Dubonnet and it would be the latter whose offspring, Hilin Tulip (f. 1973), would carry her bloodlines on to the next generation along with another daughter, Hilin Violet (f. 1975) by Ardgrange Dihafel. In the meantime, Hever Omega (f. 1976) by Keston Royal Occasion joined the stud for stallion duties; purchased cheaply as a foal, he came from the original Coed Coch Perfagl line bred by Lady Astor and brought with him Solway Master Bronze through his dam, Hever Katriona. Both Omega and his sire had a major impact on the stud when crossed with the well-bred Tetworth Nijinski daughters, Tetworth Vodka (out of Tetworth Czarina) and Cabaret (out of Cusop Glamour). Typical of Evans, both mares were spotted at the Fayre Oaks, the former bought for 150 guineas and the latter privately from Maureen Rose-Price when she failed to make her reserve. Omega on Vodka produced Czarina (f. 1979) and Virginia (f. 1980), while Cabaret to Keston Royal Occasion produced Hilin Royal Cabaret (f. 1979), the dam of the 'C' line at Hilin which was so successful. Royal Cabaret's 1987 daughter, Hilin Ceri by Longfields Longshoreman, was successful in the show ring for the Evans family, winning around the North Wales circuit as well as twice coming second at the Royal Welsh, in 1990 and 1991. Eyarth Tayma sired two good winners out of her, Hilin Tahiti (f. 2003) and Hilin Tesni (f. 2000). Tayma also sired the 2012 Royal Welsh male and overall reserve champion, Hilin Tattoo (f. 2003 out of Glansevin Garland), which a few weeks later also took the Brightwells Section B final at the National Welsh Show at Malvern under Australian judge John While. Cabaret's daughter, Caryl by Hever Omega, produced Hilin Teresa (by Eyarth Tomahawk), which in turn to Eyarth Tayma produced the Lampeter winner Hilin Horatio (f. 2004), which made 4,300 guineas to Mrs Bates at the 2010 High Flyer Sale at Fayre Oaks.

Hilin Virginia made a breakthrough for the stud when in 1983 she produced an exceptional chestnut colt by Evans' latest acquisition, Twycross Personality, purchased at the 1980 Fayre Oaks Sale for 300 guineas. Personality was a son of Tetworth Nijinski, like his grandmother, and out of Drayton Sunflower (by Solway Master Bronze) bred by the Viscountess Melville, whose Section Bs were very successful under saddle. Virginia's colt, Hilin Caradus, showed a lot of the Welsh character associated with Master Bronze, something that he too would pass on to his progeny. When shown as a three-year-old Caradus was male champion at Anglesey, overall champion at Merioneth and first at Cheshire. Reflecting on securing Welsh type within the breeding at Hilin, although Longshoreman was by Downland Chevalier, his dam, Bywiog Sailor Beware was by the small typey Section B stallion Kirby Cane Shuttlecock (sire of the Royal Welsh champions Revel Glimpse and Reeves Fairy Lustre) and his grandsire, one of Lady Wentworth's breeding by Coed Coch Glyndwr.

Most famous of the Caradus offspring was his previously mentioned Royal Welsh winning son, Carwed Charmer (f. 1991), bred at Carwed Fynydd near Denbigh by Evans' cousin, Mrs M D Williams, who leased him Charmer's dam, Carwed Petra (f. 1971), which had been a good winner around the highly competitive North Wales circuit. She was by Georgian Red Pepper, whose mare line goes back to Springbourne Ladybird, a granddaughter of the Arab, Shazda, but whose sire was Solway Master Bronze. Thus Solway Master Bronze came to Charmer from both top and bottom of his pedigree. Despite his premature death, six of his daughters were kept in the stud and he was also popular with other Section B breeders keen to take advantage of his good breeding and showing record. The Eyarth mares seemed to cross well with him and Eyarth Tayma (f. 1997) out of the Eyarth Celebration mare, Arabella, was selected by Evans as an outcross for some of his mares, as was Carrwood Orpheus (f. 2000), a son of Isley Walton Athena. With Charmer's early demise, his sons were a godsend for the Evans family, who fortunately had secured his bloodlines for the stud through these two colts.

Caradus certainly had put his stamp on his stock, none more so than one of the stud's top show fillies, Hilin Seren y Bore (f. 1997), whose dam Valentine's Carys was bred by Downland enthusiast Jim McNaught,

Hilin Seren Y Bore winning at Glanusk with her breeder Dewi Evans and judge, Ann Bigley.

and was another Fayre Oaks bargain. Valentine Carys was by Downland Gold Sovereign x Downland Celeste. Seren y Bore proved to be a real money spinner for the stud, with four of her foals reaching high prices at Fayre Oaks – in 2006, Serena by Carrwood Orpheus made 3,200 guineas, in 2007, Hilin Da Vinci by Glansevin Grafitti (by Carwed Charmer) made 1,700 guineas, in 2008, Hilin Renoir by Carrwood Orpheus made 1,300 guineas and in 2009 his full brother, Hilin Cappiello, brought 1,400 guineas.

The foal crop at Hilin was an essential part of the farm income, so Evans was always on the lookout for mares that might suit his stallions. He had a very good eye and a nose for a bargain. Two mares which did Charmer extremely well were Downland Joyous Occasion (Keston Royal Occasion x Downland Joyous), which had been gifted to Evans by Stanley Griffith, and Georgian Ginette (Gunthwaite Briar x Georgian Geraldine). Ginette produced a colt to Charmer, Hilin Etifedd (f. 2003), which has been retained in the stud and successfully started a career under saddle; he sired the top-priced foal at the 2006 Fayre Oaks when Hilin Mona Lisa sold for 3,200 guineas and in 2009 her full sister Marbella sold for 1,600 guineas. The fillies were out of Hilin Branwen, a daughter of Boston Bentick (Varndell Right Royal x Boston Bodecea by Downland Mohawk), also dam of the stallions Hilin Glyndwr (f. 2000) and Hilin Peredur (f. 2002). The Downland breeding seemed to work again when Joyous produced in 1995 a filly, Hilin Hudoles. Doubling up the Charmer bloodline by putting Tayma on Hudoles worked a treat when she produced two colts, Hilin Carnedd (f. 2004) and Hilin Cato (f. 2009). Carnedd was male champion at the Royal Welsh in 2007 as well as champion at Cheshire and youngstock champion at Glanusk. Cato was youngstock, reserve male and reserve overall champion at the Royal Welsh in 2008 and was later sold to France after winning the 2009 International Welsh Show held in the Netherlands.

Carnedd's value as a sire was reflected in his entry to the WPCS Sire Rating competition in 2011, when he came third. The stud had a clean sweep in the competition in 2008 when Eyarth Tayma took first place, Carrwood Orpheus second and their sire, Carwed Charmer, third. Tayma first appeared in the top three in 2006 and led the chart in both 2007 and 2008; his full brothers, Eyarth Sama and Tigra have been exceptional sires in Australia. In many ways the Section B breeders of North Wales have been spoiled from the earliest days of Mrs Inge then Miss Brodrick, who not only helped create the breed but did much to support and encourage it by making available top-class sires for others to use. North Wales has remained a stronghold, and in some ways has made the breeding of Section Bs somewhat easier, as studs have provided a wide range of stallions from which others can benefit. Hilin has been such a stud, and one family to have made full use of its stallions has been the Williamsons from the hill farm of Clegyr Mawr near Melin-y-Wig in Clwyd, whose Eyarth Stud quickly became established after its inception in the early 1970s.

For many enthusiasts Jess Williamson has been the driving force behind the stud's success, although she would be quick to remind them that it was a family affair headed in equal measure by her father and mother, who each brought different qualities to the breeding of Welsh ponies at Clegyr Mawr. The Williamson family was one of five which came to the Vale of Clwyd from Cumbria around 1870. Jess's father was a renowned stockman, having won at the age of 19 the 'New Judging Competition' at the famous Livestock Show at Smithfield as far back as 1932; he obviously had an exceptional eye for farm livestock, a skill he brought to the selection of ponies for the stud. He first managed two farms in the area before purchasing Clegyr Mawr in 1961. Jess's mother was the daughter of a Cheshire stockbroker with a love for ponies; she met her husband while she was working as a land girl in the area, bringing with her a love for ponies which had developed since she was a child. Together with her husband's skill at selecting them, it is little wonder that their daughters, Mo and Jess, would share their joint experience and become involved in breeding, Mo preferring Welsh cobs and Jess Section B. Jess married a neighbouring farmer, Selwyn Parry, who shares his wife's interest but takes a back seat while their son Joe has enthusiastically joined their farming enterprise, which includes the breeding and showing of the ponies.

It was Mrs Williamson who sparked the interest in breeding Welsh ponies long before the Eyarth Stud was registered, when she took a little mare to Miss Brodrick's Tanybwlch Berwyn, a familiar story for many enthusiasts in that part of North Wales. The Williamsons and their daughter were in awe of the Weston ponies, Choice in particular, which was a big winner at the shows they attended. Somewhat inspired by her, the stud took its first steps when Mr Williamson purchased the two-year-old Tanlan Swynol for 170 guineas at the Bangor-on-Dee May sale in 1968. Bred by Mrs 'Winnie' Evans at her family's Tanlan Stud on Anglesey, Swynol (by Mrs Borthwick's Cusop Sheriff colt, Trefesgob Timothy) was twice second when shown as a foal. She was followed by Crossways Champagne, Polaris Primula, Coed Coch

Weston Twiggy, foundation mare at Eyarth Stud.

Bettrys, Bacton Cefnella and Marchwiel Melissa; the first Eyarth ponies registered in 1970 were Eyarth Bettula by Brockwell Chuckle out of Bettrys and Dubonnet out of Swynol.

Having made a start, two further mares would determine the destiny of the stud – first Leighon Glamour and then Weston Twiggy. It was on a trip to Fayre Oaks in 1970 that the filly foal Leighon Glamour was spotted and purchased for 150 guineas. She was by Downland Dandini out of Leighon Butterfly, which carried much Mountain pony blood as her sire was Sinton Gyration and dam Gredington Obringa by Coed Coch Planed. Unwittingly the Williamsons were embracing the 'Weston' policy of embedding Section Bs with Section A blood. Glamour was shown with success and bred well to the Rotherwood stallions – firstly Commander, whose son out of her, Eyarth Musketeer (f. 1977), was sire of Geufronuchaf Miss Royal, dam of many of the best Telynau ponies. (Her last foal, Eyarth Gemma (f. 1990) by Rotherwood Carnival was second as a brood mare at the Royal Welsh in 2000 for Clive Morse.) Returning to Leicestershire in 1977, this time to Keston Royal Occasion, she produced Eyarth Celebration, a colt foal which was unbeaten in foal classes. His win at the Royal Welsh in 1979 under Mrs Egerton set him on a remarkable journey which would take him to the very top as a show stallion and one of the most influential stallions to appear in the Welsh Stud Book – next to Rotherwood State Occasion, he became the most influential of Keston Royal Occasion's offspring.

In 1979 Celebration was in good company when he stood reserve youngstock champion to Weston Rosebud and reserve male champion to Cusop Banknote at the Royal Welsh – his win was a milestone in the history of the Eyarth Stud as it set a marker for things to come. One aspect of the stud which has set it apart from all others has been its policy of using a range of stallions, seldom fixing on one and restricting homebred stallions to one, Celebration. By using a range of bloodlines, it has made it easier for purchasers to breed them on and, importantly from a commercial point of view, return for future purchases without compromising their breeding plans. 1979 would also be significant in the stud's history because the three-year-old fillies Weston Crystal and Weston Twiggy were purchased at the final Weston Sale of that year. Both mares bred well for their new owners, although it was Twiggy which would prove exceptional. She had been reluctantly offered on the final sale, as Edwards and his family struggled to reduce their Australia-bound stud to a manageable number. It was Edwards who tipped the Williamsons the wink by telling them to ignore the colour (cremello), while assuring them that she would not have been for sale had it not been for a restriction on numbers for the flight – well made and extremely well bred. By Weston Chilo out of Llynsun Blue Mist, she was a double cross of Gorsty Firefly and half-sister to the successful Weston ponies, Rosebud and Picture. It is well known that the family regretted not taking Twiggy with them to Australia, but little did they know that the stallions they would later import for their Weston Park Stud would be her direct descendants.

Twiggy's cross with Eyarth Celebration became one of the happiest marriages recorded in the Stud Book, although Twiggy herself seldom held centre stage at the stud where eyes were firmly fixed on her many beautiful daughters. Of her sixteen foals, all but six were by Celebration and, of these, four were by Carwed Charmer. Of the progeny by Charmer, Eyarth Athene (f. 1995) stands out as the best and certainly most admired, so much so that Jess Williamson gave the palomino filly to her son to make sure that she would never succumb to the substantial offers made for her from around the world. Only lightly shown as a youngster, she was youngstock champion at the 1998 Royal Welsh. Her offspring have been successful, too – Eyarth Apollo was Section B ridden winner at the Horse of the Year Show in 2006 and Eyarth Mercury was champion at Lampeter and stood at stud at both Paddock and Telynau before his sale for 3,100 guineas at Fayre Oaks in 2009.

When the grey filly Eyarth Arabella, by Celebration out of Twiggy, was foaled in 1982, the Williamsons might have been forgiven for thinking it was a lucky break for them as breeders but, as the years went on and more and more lovely fillies from the partnership appeared, they knew that they had done something special. Along with the full sisters Zsa Zsa (f. 1983), Isabella (f. 1990) and Ophelia (f. 1998), Twiggy had provided her owners with the foundation on which they would build their future success. In the meantime a colt, Eyarth Harlequin, was foaled to Twiggy and Celebration in 1984; he didn't quite make his reserve at the 1985 Fayre Oaks sale, but was spotted by the author and David Blair, who had established the Waxwing Stud in Scotland – Harlequin would head to Fife instead of back home to Clwyd. The following year he embarked on a showing career, winning at Lampeter, Royal Highland and East of England shows, only to be sold to Dutch breeder, Peter Steeghs, for whom he was an outstanding stallion. Such was his success as a sire that in 1998 the Netherlands WPCS awarded him the elite 'Preferente' status, given to mares or stallions

Eyarth Harlequin as a two-year-old.

whose offspring have proved to be exceptional in the show ring or in performance classes. In an interesting move, Harlequin returned to Wales in 2000 when he was purchased by the Rhoson and Heniarth Studs as an outcross for their mares.

The success of Harlequin took Best and Blair on a trip to the Eyarth Stud in 1986, where they saw Celebration for the first time – they knew his sire very well but this was the first experience of his son, and they were more than impressed. After a great deal of persuasion, the Williamsons agreed to allow Celebration to go on lease to Waxwing for a season on condition that he was shown at a few shows – this he did with unprecedented success, taking the show ring by storm and thrusting the Eyarth Stud into the spotlight in 1987. In a handful of appearances, he was reserve supreme and overall mountain and moorland champion at the West Midlands Stallion Show, supreme at Glanusk, champion at the Royal Highland and National Pony Society Shows, reserve champion at the Great Yorkshire and winner of the prestigious Glyn Greenwood championship at the Ponies UK Show. Sadly his success was never enjoyed by Mr Williamson, as he passed away before his favourite Section B, Celebration, had worn his first ribbon as a stallion.

The Twiggy fillies by Celebration crossed very well with the stallions at Hilin and Arabella produced Eyarth Cordelia (f. 1989) by Hilin Caradus; when crossed back to Celebration she produced an outstanding filly, Eyarth Windflower (f. 1997), six times a winner at the Royal Welsh, youngstock reserve champion in 2000, overall champion in 2003 and female champion in 2006. Eyarth Pansy, her full sister foaled 2003, has been retained in the stud. Carwed Charmer became an obvious choice for Arabella and a succession of colts was produced for the benefit of other breeders – they included Eyarth Sama (f. 1996), Tayma (f. 1997), Tigra (f. 1999) and Savero (f. 2000). Their full sister Eyarth Nata has been successfully shown locally and took the Section B championship at the National Foal Show in 1998 – she has been retained along with her filly by Celebration, Eyarth Kachina. Charmer was also selected for Arabella's full sister by Celebration, Eyarth Isabella, and she too produced a

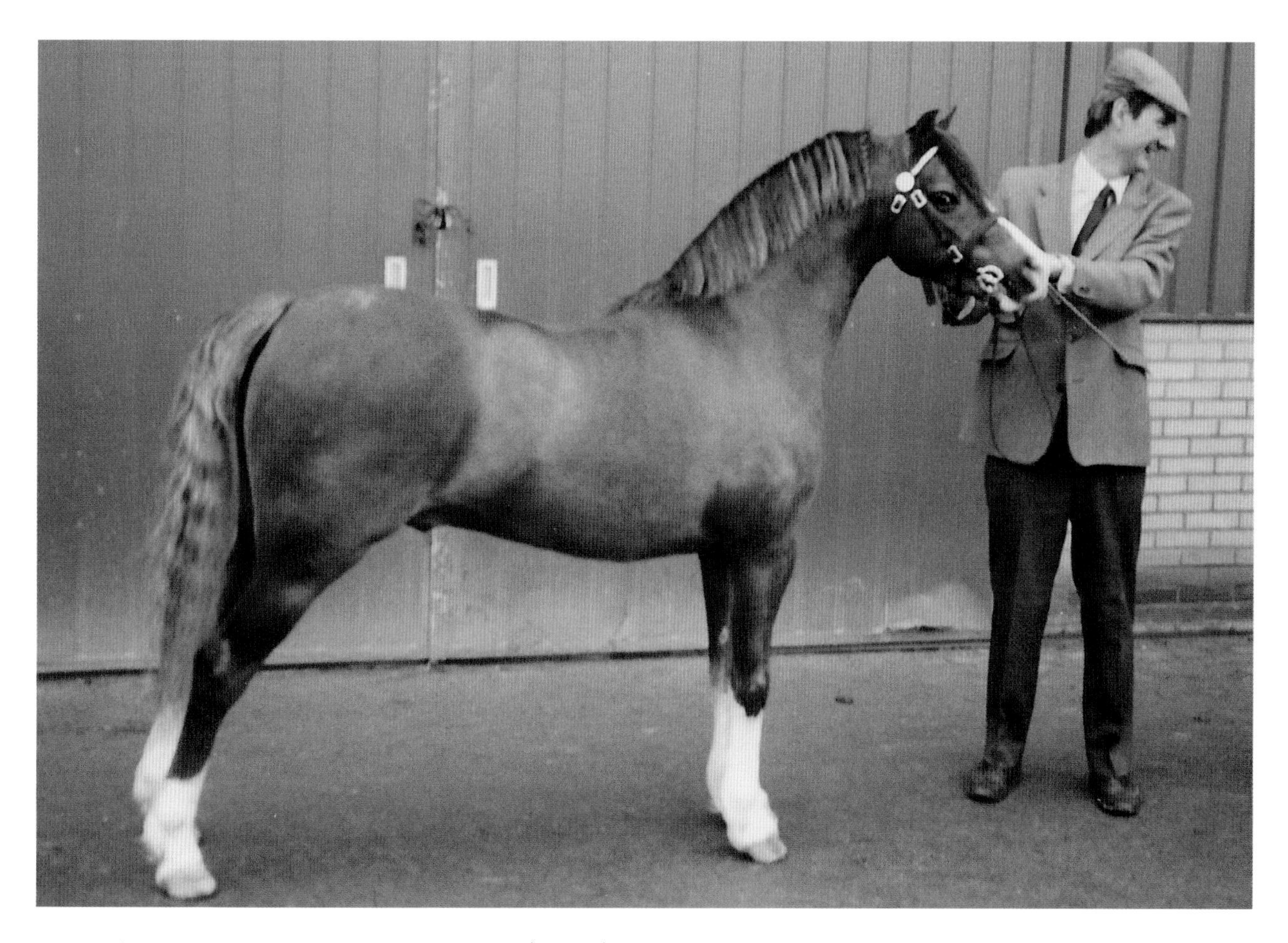

Eyarth Rio, shown by David Jones, Telynau Stud, while on lease to Holland.

series of colts sold on for stud work – Ramah (f. 1997), Hansel (f. 1998), Simeon (f. 2001) and Shammah (f. 2002). Her bay daughter by Charmer, Eyarth Davita, was retained in the stud.

Eyarth Zsa Zsa (Eyarth Celebration x Weston Twiggy) was nicknamed the 'Little Sea Horse' when she was foaled in 1983 due to her elegant long neck and beautifully tapered head. She was shown extensively with success, winning the novice mare class at the Royal Welsh in 1990 along with her winning colt, Eyarth Tomahawk (by Pennwood Milan), which later became the resident stallion with the Moelview Stud. Interestingly the stallion choices for her bucked the trend set for her siblings, although once more Hilin provided the majority of them. Again she was a prolific colt breeder and her first was Eyarth Rio (f. 1987) by Hafodyresgob Buzby (by Nefydd

Eyarth Beau Geste.

Autumn's Chuckle). Rio was rejected by his dam and had to be hand-reared; he was even travelled to shows with the Williamsons when they showed their other ponies, so that he could be fed. Precocious and highly attractive, he was spotted by David Jones and Geraint Thomas for their Telynau Stud, for which Rio was a prolific winner in the show ring as well as a prolific breeder of top-class Section Bs. Between 1997 and 2000, Zsa Zsa produced three colts by Boston Bentick – Eyarth Beau Geste (f. 1997), Thowra (f. 1999) and Mowgli (f. 2000), which took up stud duties in Scotland. Both Beau Geste and Thowra were winners at the Royal Welsh and Mowgli second, but it is the dun, Beau Geste, which has made a name for himself as a breeding sire at the Waxwing Stud. Zsa Zsa's foals have always been in great demand, so two of her fillies were retained – Titania (f. 2003) by Carwed Charmer and Lili Marlene (f. 2004) by Carrwood Orpheus, which has followed her dam's success in the show ring and like her, 21 years later, won the novice mare class at the Royal Welsh in 2011.

As we have seen, the relationship between breeding policies at Eyarth and Hilin was extremely supportive and many others benefitted from them. The show ring success of Eyarth Windflower has brought her many admirers, although it has to be said that the production skill of her owners, Richard Miller and Meirion Davies of the Heniarth Stud, cannot be underestimated and has become their hallmark over the years in both Sections A and B. Davies' roots were established in Section B through the ponies he bred with his parents at the

Heniarth Wood-Wind.

Rhoson Stud in West Wales, not far from the Heniarth base at Ferryside in Carmarthenshire. Welsh-speaking like his parents, Davies has pursued a career in television as Head of Children's Programmes with the Welsh television channel S4C; following this he was Head of Content when he was charged with developing the Royal Welsh Show coverage, before joining the independent sector in 2012. Miller's creative talents have developed as a contributor to various publications, although it is as a professional stud manager and show ring producer that he is best known. His early interest in the Welsh breeds was expressed through Welsh cobs both at home and abroad, firstly with Miss Wheatcroft with her Sydenham cobs in Gloucestershire, and later with (Mrs) Hope Ingersoll at Grazing Fields in Massachusetts, United States.

Having specialised initially in Section As, their move into Section B was prompted by the visit to Eyarth where they spotted the potential in Windflower – something they exploited to its fullest, both in the show ring and in the paddock. Their choice of Mrs Mansfield-Parnell's Section B stallion Lemonshill Top Note proved a good one and subsequent visits to Top Note have proved equally successful. The first foal born, Heniarth Wood-Wind (f. 2007), began his career in style by winning the foal class at The Royal. As a yearling he was champion at Lampeter and reserve champion at Glanusk. At Cheshire in 2010 he was overall champion Section B and runner-up for the Cuddy Final at the Horse of the Year Show. His progeny have shown promise, outstanding among them being the French-bred colt, Adagio de l'Aurore

Six-time Royal Welsh winner (overall champion in 2002 and 2003) Eyarth Wind Flower, shown by Richard Miller for the Heniarth Stud winning the In Hand 'Cuddy' Supreme at the last Royal Show of England which was held in 2009.

(dam by Kirby Cane Statecraft); produced by Heniarth in 2011, he took the yearling supreme at Lampeter, reserve champion at the Bath and West and youngstock supreme at Northleach. In a move to bring back to Wales bloodlines exported over the years, Heniarth imported from Australian breeder Wendy Trimble, Tooravale Houston (f. 2012), a colt by Weston Manhattan (Weston Cottonwood x Weston Mary Ann) and out of Imperial Honora (Eyarth Tigra x Derralea Honour by Weston Chippendale).

As previously mentioned, two studs situated in North Wales selected Eyarth stallions –Rio for Telynau and Tomahawk for Moelview. When Rio joined the Telynau Stud in 1988, it was already well established by David Jones and Geraint Thomas at Geraint's family home of Gwylfa, not far from Caernarfon, looking out to the Menai Straits to Anglesey. David had been brought up with Welsh mountain ponies at his grandparents' home in Montgomeryshire, where they bred unregistered ponies during the 1940s and 1950s; he also had an interest in sport and music. Jones later took up education and taught in North Wales, where he also bred and showed a few Section A ponies. Thomas was interested in and worked with horses, like his father. He was a talented rider and rode competitively in pony classes through to horses before his family had to concentrate their efforts into farming. It was not long before ponies arrived back at Gwylfa, although they were Sections Cs – an early passion with the breed which brought much show success.

Telynau Royal Charter qualifying for the Templeton In Hand Final at the Horse of the Year Show 1998.

Jones and Thomas jointly formed the Telynau Stud (Welsh for 'two harps') in 1988. On a visit to Eyarth to look at a filly in February 1989, they passed on her, but the cheeky two-year-old chestnut colt, Rio, immediately caught their attention. He had returned home unsold from Fayre Oaks the previous year, when they had noticed him but had no thoughts of buying a colt – this time they would not miss the opportunity and a deal was struck. Unconcerned with height, above all they wanted visitors to recognise that the Section Bs bred by them were unmistakably Welsh in character. Rio's big front, beautiful head and extravagant action were qualities that particularly appealed to his new owners, who quickly set about forming a band of 'old-fashioned' mares around him – which they did in some number. The foundation mares whose blood has remained within herd include Tanlan Gwenora (Farasi White Knight), Cottrell Sapphire (Solway Master Bronze), Geufronuchaf Miss Royal (Eyarth Musketeer), Starlyte Fairy Footsteps (Downland Mandarin), Penucha Tsient (Eyarth Celebration) and Douthwaite Swan Lake (Tetworth Nijinski).

The records show that Geufronuchaf Miss Royal's mating with Eyarth Rio has been a marriage made in heaven – seven full sisters from the union have remained in the stud. Foaled in 1980, she was bred on nearby Anglesey by Gwilym Griffiths, who bred a few Welsh ponies at Geufron Uchaf, which gave his ponies their prefix. It is often said that the 'breeding will always out', and in Miss Royal's case it most certainly did – and with the best of back breeding no one should be surprised that she became a notable brood mare in the annals of the Welsh Stud Book. Her sire was Griffiths' resident stallion, Eyarth Musketeer, a son of Rotherwood Commander and Leighon Glamour and half-brother to Eyarth Celebration; on her dam side, Belan Morning Mist was by the Weston Stud's great breeder Gorsty Firefly out of a mare by Downland Chevalier. The addition of Rio's impressive pedigree to the genetic pool was always expected to produce something good, although no one could have thought it would be replicated with such consistency. Of her Rio colts, Telynau Royal Charter (f. 1993) and Royal Anthem (f. 1999) would form the basis of studs on opposite sides of the United States, Anthem to Virginia and Royal Charter to California. Indeed it was Royal Charter who raised the profile of Telynau as a Section B stud, although by then Eyarth Rio had already embarked on a successful show career – having been a winner at Lampeter in 1990, champion at the Royal Highland in 1991 and third at the Royal Welsh in both years. Telynau Filigree, foaled the same year as Royal Charter, had already put down a marker for her sire by winning foal classes at The Royal and Shropshire and West Midlands shows in 1993. Royal Charter made his mark the following year, however, when he started the season by winning at the West Midlands Stallion Show before going on to win at both Glanusk and the Royal Welsh.

The following year Royal Charter embarked upon another great season, recording reserve championships at Lampeter, Bath and West and Cheshire. Meanwhile his sire's reputation as a show stallion was gathering momentum and he took both the breed championship at one of the early shows, Newark and Notts, as well as the qualifying ticket for Templeton In Hand Final at the Horse of the Year Show in 1995. For the next five years Rio competed with the best and finally, in 2000, having taken the championship at the Great Yorkshire Show, once more qualified for the Horse of the Year Show. He didn't always have it his own way, though, and in 1996 he had to give way to his son at Lampeter, where Royal Charter was judged champion. This was the start of a very satisfying show season for the colt, whose youngstock career was complete with wins at Glanusk, The Royal and the Royal Welsh, where he also took reserves for both the youngstock and male awards. As a mature stallion, he was judged champion Section B at The Royal in 1998 and, like his sire before him, qualified for the in hand finals at the Horse of the Year Show; the next year he took many tricolours including those on offer at Northleach, Great Yorkshire and Cheshire. Telynau Royal Charter was used for stud duties before his departure to the United States and proved successful when crossed with the stud's Cottrell Sapphire (Solway Master Bronze x Revel Glimpse), providing a future mare for the stud, Telynau Sceptre (f. 2000).

While basking in the glory of their winning colt, Jones and Thomas had more to celebrate when Geufronuchaf Miss Royal produced three fillies in quick succession – Telynau Miss Royal (f. 1995), Royal Gala (f. 1996) and Royal Diadem (f. 1997). All three were retained as brood mares, along with four other full sisters by Eyarth Rio – Royal Celebration (f. 2002), Royal Heiress (f. 2003), Royal Opera (f. 2004) and Royal Heritage (f. 2006). Miss Royal made an early impact by winning at both Lampeter and Glanusk as a yearling, a feat repeated by her younger sister, Royal Diadem, in 1999, which was bettered in the Glanusk championship by another sister, Telynau Royal

Thornwood Royalist at Stanley Grange Stud.

The fact that Eyarth Rio stood top of the WPCS In Hand Sire rating championship on three occasions (1997, 1998 and 1999) measures his worth as a sire. The task at hand for his owners was to find stallions which would follow on his success, especially when crossed with his many daughters out of Geufronuchaf Miss Royal. In 1993, Rio was leased to Peter Steeghs in the Netherlands, who had enjoyed huge success with Eyarth Harlequin. Rio's influence is still felt through his many offspring and one in particular stands out – Breeton Dai, a major winner himself and sire of many. His son, Stougjeshoeve Escudo, made history when he was Robert Hensby's choice as youngstock champion at the 2012 Royal Welsh, the first overseas-bred pony to take this award. Meanwhile in Britain, the Rio son Soudley Taliesin (f. 1996 out of Rotherwood Primrose) has proved his worth as a sire of top-class ridden ponies, particularly for the Annandale Stud, whose show hunter ponies Annandale Darcy (f. 2002), Motivator (f. 2005) and Maria (f. 2007) have all qualified for the Horse of the Year Show and another daughter by Taliesin, Crystal Vision (f. 2003 x Millcroft Cascade), bred by Mrs Parker, was judged Supreme Pony of the Year in 2011.

The stallions which proved successful as crossing sires for Rio stock at Telynau included Laithehill Allegro (f. 1989), a grey by Abercrychan Spectator out of Weston Anita by Weston Charmer. He was successful when

Gala, which stood reserve there and second at the Royal Welsh. Both grew into beautiful mares and were shown successfully in brood mare classes. In 2002 Gala gained second at the Royal Welsh, followed by a win in the junior mare class at the WPCS Centenary Show – for which Jones was Honorary Secretary. The following year Gala was winner of the Rhydspence Section B final at the National Welsh Show at Shrewsbury. Royal Diadem's career outshone all other mares in the stud, beginning in 2003 when she was champion at The Royal and reserve at Cheshire, where she went one better the following year to stand champion Section B. The 2008 Royal Welsh arguably witnessed her greatest victory when she stood reserve female to the overall champion, Millcroft Dahlia, judged by Christine Jones (Bunbury Stud).

Above: Eyarth Zsa Zsa.

Opposite: Moelview Chieftain.

shown, and crossed well with Gala and Diadem, producing Miss Regalia (f. 2000) and Royal Diamante (f. 2002) respectively. Eyarth Mercury (Eyarth Celebration x Eyarth Athene) was leased from the Butterworths – he returned to Yorkshire in 2008 to the Great Yorkshire Show, where he stood champion. Again he crossed well with the aforementioned mares, producing Glory Be to Gala and Royal Delight to Diadem. The palomino, Telynau Royal Delight, has already shown promise in the show ring when as a yearling she was judged second at Glanusk in 2008 during the show's 50th anniversary. More recently Jones and Thomas have brought in a total outcross for their mares with the introduction of the aged stallion, Mynach Flower Power, a full brother to the Royal Welsh winner and Royal Show champion Mynach Mayflower by Thornwood Royalist. He had formerly stood at the Wian Stud on Anglesey, having been purchased at Fayre Oaks as a yearling in 1992. A daughter by him out of Royal Heritage won the Royal Welsh Show foal class in 2012.

Unquestionably Eyarth Rio put the Telynau Stud on the map in the same way that Eyarth Tomahawk had done for another in North Wales, the Moelview Stud based in the picturesque town of Dolgellau, situated on the southern edge of the Snowdonia National Park and former county town of Merionethshire, now Gwynedd. Robert Jones was a local butcher with an interest in Arabs and part-breds, and his son Richard kept a few Shetland ponies. In order to sustain his son's interest, Jones decided to purchase a few Welsh Section Bs as a hobby which his family could enjoy. Well known and popular within the local community, Jones was well placed to find the ponies they required from local studs such as Hilin and Eyarth, which influenced the type of Section B ponies they would breed. Initially Hilin Caradus and Hever Omega were selected for their mares and later Carwed Charmer, but it was the purchase of the winning Royal Welsh colt foal Eyarth Tomahawk (f. 1990 Pennwood Milan x Eyarth Zsa Zsa) that would set them on a path to success, both in the show ring and as breeders. Jones had admired him as a foal when shown with Zsa Zsa and tried to buy

him, but he was on offer at the time to the Netherlands; undaunted, he waited for the Fayre Oaks Sale, where he bought him for 700 guineas. A favourite of both father and son, Tomahawk was the mainstay of their show for 18 years, winning championships at all the North Wales shows – he won the Royal Welsh as a foal and again as a three-year-old, the same year that he headed Carwed Charmer for the championship at Anglesey.

There was a good local show circuit, which suited their early needs, but in many ways it was the American judge and prominent Welsh breeder, Mrs Hetty Abeles (née Mackay-Smith), who brought the stud to a wider audience, nationally and internationally, when she judged their young Section Bs at the WPCS's 9th International Show at Peterborough in August 1992. Tomahawk was second to the eventual youngstock champion, Lemonshill Hylight, and their filly Nannau Charm won the three-year-old filly class. Tomahawk's credentials have already been observed, but Charm and her Nannau prefix were newcomers to exhibitors outside the North Wales circuit. Her breeder, Mrs R E Dallimore, whose husband was a prominent vet in the Dolgellau area, was a hobby breeder like Jones – they also bred her dam, Rebecca (Eyarth Celebration x Minsterley Carol), which, put to Hafodyresgob Buzby, foaled Nannau Charm in 1989. She resisted offers by Jones to purchase her from the time she was a foal and it was as a three-year-old that he eventually managed to buy her for his son, but only on condition that the Shetland ponies were sold.

Nannau Charm was much admired that day in August – her breeding value was established the very next year when she bred an outstanding colt by Eyarth Tomahawk named Moelview Mohawk, a good winner when shown as a foal and yearling, but unsold when offered at Fayre Oaks in 1994. Mohawk caught the attention of the Waxwing Stud, who subsequently visited Dolgellau and purchased him for their mares in Scotland. The following year he was lightly shown, winning at both the Royal Highland and Royal Welsh. Mohawk was an early indicator of how successful this cross would be – in 1995 it produced Moelview Chardonnay, a filly retained as mare and dam of Rioja (f. 2003 by Eyarth Rio), a colt purchased for use in the Telynau Stud. The following year, Carwed Charmer was tried out on Charm with the result a free-moving colt called Moelview Charmer Boy, which went on to breed top-class winners for his owner Jane Blackburn – particularly Westaire Saffron Lace, reserve female champion and gold medal winner at the Royal Welsh in 2011. While on lease to the Paddock Stud, Charmer Boy also bred the 2009 Royal Welsh champion, Paddock Alesha.

Eyarth Tomahawk crossed well with Hilin Cerian (by Hilin Caradus) to produce in 1999 the filly Moelview Crystal, which has become a successful breeder for Mandy Jones, also from Dolgellau. Crystal's colts by Eyarth Tayma, Melau Montana (f. 2008) and Melau Morocco (f. 2005), enjoyed good show records at the Royal Welsh - particularly Montana, with wins to his credit in 2009, 2010 and 2011. Nannau Charm also provided Robert Jones with a colt, Moelview Cherokee (f. 2001), which has already repaid his owners with a mare, Moelview Lunar, third at the Royal Welsh in 2012. She is out of Moelview Moonshine (f. 1996 Eyarth Tomahawk x Merryment Over the Moon), which also bred a useful colt, Moonstruck (f. 2011) by Moelview Prince Charming. Prince Charming is the latest of

Jones' stallions to appear under saddle, the other being Chieftain, who captured the male championship at The Royal as a two-year-old in 2006 and later in his career qualified for the ridden Section B class at the Horse of the Year Show, where he was placed in 2011 and 2012.

Moelview Prince Charming (by Moelview Mohawk) was sold in utero when his dam, Llangeitho Princess Royal, was sold for 3,000 guineas at the High Flyer sale at Fayre Oaks in 2006. She was offered by the Waxwing Stud, which had already bred a full sister to Prince Charming, Waxwing Royal Princess, top-priced Section B foal at the Fayre Oaks Sale in 2000 at 2,300 guineas; shown in South Wales with success by Mr and Mrs Jones (Rockbury Stud), she justified their selection by winning the two-year-old filly class at the Royal Welsh two years later. The most famous of the mare's offspring has to be Waxwing Prince Hal, winner at the Royal International Horse Show under saddle in 2010 and first at the Horse of the Year Show the following year. Their dam, Llangeitho Princess Royal, was no stranger to the sale ring – she had been top-priced foal at 1,400 guineas in 1996 and top-priced female at 2,000 guineas in 1998. She was bred at Tregaron in Ceredigion by Harold Morris who, along with his daughter Heather Isherwood, bred only a few ponies originally sourced at Fayre Oaks. Cawdor Brenda (f. 1963 by Llanarth Nightshade) was purchased from Lady Seton and Wingrove Foxtrot (f. 1964 by Downland Chevalier) was bought from his breeder, Iorworth Williams, from Carmarthen. In 1972 they produced a filly, Llangeitho Duchess, which would provide them with a host of winning ponies including the stallion Llangeitho Twerp (f. 1976), unbeaten in ten outings as a yearling and then unshown until 1993 when, on lease to Telynau, he was both a winner and champion when lightly shown. Twerp was by Sarnau Showman while his brother, Llangeitho Sovereign (f. 1987), was by Abercrychan Spectator; Sovereign on his half-sister Llangeitho Bouquet (Keston Royal Occasion x Llangeitho Duchess) produced Llangeitho Princess Royal in 1996.

The Waxwing Stud in Scotland took its name from HMS *Waxwing*, the Fleet Air Arm base near Dunfermline which serviced the busy dockyard of Rosyth during World War II. The stud was established in 1972 by schoolteachers Tom Best and David Blair at the site of the old HMS *Waxwing* base before moving it to Saline in rural West Fife in 1993. The stud's success has stretched over 40 years across four sections of the Welsh Stud Book – it has the distinction of having bred in hand winners of Section A, B, C and Part-bred at the Royal Welsh. In addition, it has bred ridden winners in Section A, B and C at the Royal International Horse Show and the Horse of the Year Show, where the stud's Section A, Waxwing Thumbs Up, was judged supreme pony in 2007. Their Section C, Waxwing Rheel, having won there in 2011, repeated his success in 2012, completing a record nine consecutive appearances at this prestigious show.

The foundation mare at Waxwing was Sparkler of High Tor (Weston Glow Worm x Springbourne Jewel), purchased as a two-year-old from Dr Wynne Davies in 1972. She bred several part-breds which qualified for the ridden finals at the Horse of the Year Show before settling on Section Bs, initially to the Rotherwood stallions. To Keston Royal Occasion she produced Waxwing Royal Mint (f. 1979), which was Royal Highland champion in 1982 under A L Williams; next followed the full sisters by Rotherwood State Occasion, Waxwing Poppy Day and

Foals looking down towards the yard at Waxwing.

Wood Sorrell, both top-priced fillies when sold at Fayre Oaks. Sparkler bred well to whatever she was put, and her crossing with Eyarth Celebration (on lease to Waxwing) was the grey, Waxwing Celandine (f. 1988), which was Highland Show champion in 1996.

Having produced ponies for the show ring for other breeders as well as themselves, several top-class stallions were made available to the stud including Paddock Gemini, Eyarth Harlequin, Eyarth Celebration, Downland Yeoman, Downland Warrior, Thornwood Royalist and Hever Quiver. It was while Thornwood Royalist (by Chamberlain Don Juan) was in residence that Waxwing became the home of Miss Margaret Gethin's Mynach

Mynach Moon Flower, Champion Royal Highland 1989, shown by co-owner of the Waxwing Stud, David Blair.

ponies, which had relocated from Devon to Inverness following the death of her partner, Major John Bateman. Miss Gethin originated from the Home Counties, the daughter of Lt Col F D S Gethin, RA; she was a driver in FANY (First Aid Nursing Yeomanry, also known as the Princess Royal's Volunteer Corps) during World War II. After the war she met Batemen on an equestrian course and found she shared his interest in fox-hunting, racing and polo. They settled first in Sussex, where they ran a livery yard and showed hunters, hacks and cobs. Their neighbours were John and Alison Mountain at the Twyford Stud, who encouraged them to breed a few ponies to maximise their grazing, and Section Bs were chosen. Coming from a riding background, there was no surprise that the Foundation Stock breeding appealed to them, and the Duchess of Rutland's Belvoir ponies fitted the bill quite nicely. Three fillies, all registered as FS1, were purchased – Belvoir Turquoise (f. 1963 by Wickenden Osprey), Belvoir Tiger Lily (f. 1963 by Belvoir Gervas) and Belvoir Samantha (f. 1960 by Belvoir Gervas).

Brockwell Berwyn, Solway Master Bronze and his son, Elphicks Master Romeo, lived locally and were used before the palomino Section A Twyford Scamp (f. 1967), by Coed Coch Asa out of the beautiful cream mare Revel South Wind, was purchased following Miss Gethin's move to Ashburton, Devon in 1967. Not surprisingly, Solway Master Bronze suited Tiger Lily, whose dam, Belvoir Tangerine, was by the Arab, Azym. Three of their daughters were retained – Madonna Lily (f. 1966), Miss Lily (f. 1968) and Day Lily (f. 1979). Miss Lily was a prolific winner in both ridden and brood mare classes and she won the Royal Welsh in 1974.

RWAS · CAFC
1904

SIOE FRENHINOL CYMRU
ROYAL WELSH SHOW

Opposite: Mynach Miss Prim Reserve Champion Royal Welsh 2006 and 2007 shown by the author for Miss Gethin.

Above: Mynach Oriana, three times winner at the Royal Welsh, and here taking the championship at the Great Yorkshire in 2007.

The greatest contribution to the Mynach stud came from Belvoir Samantha, whose daughter by Brockwell Berwyn was the first filly to be bred at the stud. This was Mynach Moonshine, whose daughter by Lechlade Scarlet Pimpernel, Mynach Moon Flower (f. 1976), would see Miss Gethin's Mynach prefix triumph at major shows well into the 21st century. Moon Flower was a remarkable mare that bred well to any stallion to which she was put. She was champion at the Royal Highland in 1989, when the first of her exceptional offspring by Thornwood Royalist, Mynach Mayflower, was the winner of the foal class. Mayflower won extensively, including the Royal Welsh in 1990, the same year that Moon Flower produced her full sister, Mynach Flower Song, which has been one of the few to have won at the Royal Welsh in all three of her youngstock years under

The author seen here with Waxwing Penny Gold when she won the yearling filly class at the Royal Welsh in 2008.

different judges. Two full brothers by Royalist were also produced, Mynach Flower Power, which was sold to the Wian Stud in Anglesey as a yearling, and Moonwalk, a big winner under saddle for Margaret Holt from Lancashire, who qualified him for the ridden Section B class at the Horse of the Year Show.

The combination of the Solway Master Bronze bloodline with that of Downland Chevalier in Mynach Mayflower worked perfectly with the next stallions to arrive at Waxwing, which by this time had become the Mynach Stud's permanent home. In 1996, the year when she was champion at The Royal and a winner once more at the Royal Welsh (produced by Miller and Davies), Mayflower had a chestnut filly, Mynach Mimosa, by Moelview Mohawk; Mimosa was an early winner when she won at the National Pony Society Show at Malvern. More importantly, she crossed well with Eyarth Beau Geste to produce two Royal Welsh winners, Mystical and Miss Prim. Mynach Mystical was the winning colt foal at the Royal Welsh before his export to North America, where he was later overall champion at the Toronto Winter Fair in 2011 and 2012. Miss Prim (f. 2005) was youngstock and overall reserve champion at the Royal Welsh in both 2006 and 2007.

Mares like Mayflower are often referred to as 'golden' mares, for they are very special and fairly elusive but are recognisable for their unique breeding capacity. Her Mohawk offspring included the Mynach ponies Pinafore (f. 1998), champion Section B at the National Pony Society in 1999; Hornblower (f.,1999), champion Section B at The Royal as a yearling and many times Horse of the Year Show qualifier under saddle; Buccaneer (f. 2000), stud stallion at the Thistledown Stud and another many times qualifier for the Horse of the Year Show under saddle; Oriana (f. 2001), three times winner of the Royal Welsh and four times champion at the Royal Highland; and Aurora (f. 2002), twice reserve champion at the Great Yorkshire and dam of Royal Highland champion Mynach Sun Shade by Eyarth Beau Geste. Mayflower herself bred well to Beau Geste, producing the Royal Welsh winners, Mynach Oceana (f. 2003), reserve youngstock champion in 2004, and Mynach Arcadia (f. 2004), a winner in 2007.

Eyarth Beau Geste has proved more versatile than merely crossing well with one mare line, although his book is annually limited to the mares at Waxwing. His strike rate of winners at the Royal Welsh has been quite remarkable for a stallion with such a small number of offspring – other than the four Mynach youngsters, he has produced the yearling winners Wedderlie Mardi Gras (f. 2001), Waxwing Penny Gold (f. 2008) and Waxwing Glint (f. 2011). Mardis Gras was bred in the Scottish Borders by Mrs Jennie Campbell, a well-known breeder of Aberdeen Angus cattle and a great supporter of the Welsh Section B since the days of breeding them for her daughters to ride and hunt. She had already bred Wedderlie Martina (f. 1993) – a half-sister to Mardi Gras by Eyarth Rio – which was reserve female champion at the Royal Welsh when shown, like her brother, by the Waxwing Stud. Following youngstock championships at both the Royal Welsh and the Centenary Show in 2002, Mardi Gras was sold to Gretchen Aitken and her family in Oregon, for whom he has been an outstanding success in both the show ring and as a sire; he was twice winner of the Welsh Pony and Cob Society of America Sire of the Year, in 2010 and 2011. Coincidentally it was the American judge and accomplished breeder, Mrs Gail Thompson (whose Gayfield Stud rates among the best in the United States), who chose Waxwing Penny Gold to head an entry of 58 – the largest for Section Bs recorded in the history of the Royal Welsh. Penny Gold is one of a several champions out of Millcroft Pleasure, whose son by Beau Geste, Waxwing Penny Farthing, was a winner at the Great Yorkshire and champion at the Royal Highland in 2009.

Beau Geste's most recent Royal Welsh winner was Waxwing Glint, a daughter of Cwrtycadno Glain (f. 2002), one of two full sisters loaned to the stud by their owner/breeder, Mrs Megan Lewis. A Silver Medal winner at shows in and around her home near Llandeilo, Glain has suited the Eyarth stallion with a daughter by him, Waxwing Glide, in Sweden, and a son, Waxwing Glimmer, a palomino which was supreme champion at the National Welsh Foal Show in 2011 and champion Section B at the Royal Welsh Agricultural Society's Winter Fair in 2012.

It would appear that breeders have continued to travel along a successful path and that some are successfully combining new and established bloodlines. As we come towards the end of the story for the present time, we shall see how a few more have helped to advance the progress of the Welsh Section B into the 21st century.

A championship card for the 2013 Royal Welsh Show which celebrated its 50th year at its permanent site at Llanelwedd in Mid Wales

Chapter XVII

Cwrtycadno, Sarnau, Bronheulog, Fayre, Glansevin, Ernford, Carolinas, Llanarth, Horsegate, Cadlanvalley

Cwrtycadno Perlen, Overall Welsh Section B Champion at the Royal Welsh in 2013.

Talhaearn Eirlys y Pasg, Royal Welsh Female Champion 1994.

In 2008 Cwrtycadno Glain and her full sister, Cwrtycadno Perlen, were made available to the Waxwing Stud from their breeder, Megan Lewis, a remarkable woman who completed an amazing journey on horseback from Beijing to London between the 2008 and 2012 Olympic Games. Brought up in the Far East where her father was a head teacher, she studied at the University of London's School of Oriental and African Studies and spent ten years teaching in London before returning to West Wales, where her father's family farmed. Although her interest in horses also included competitive endurance riding and breeding a few Arabs, Ms Lewis initially turned her attention to Welsh cobs because she felt that Section Bs appeared to be more riding pony than Welsh in character.

A chance visit to the less popular Friday sale at Fayre Oaks in 1987 changed that view when she found a seven-year-old mare full of Welsh character. This was Talhaearn Eirlys y Pasq (f. 1979), which she purchased for 520 guineas from the Evans family of Abergele in North Wales following the death of their father. The grey was by a local stallion, Nefydd Autumn's Chuckle, which carried both Downland Dauphin and Romance on either side of his pedigree, and out of Keston Fantasia (by Solway Master Bronze), which carried Coed Coch Glyndwr on both sides of her pedigree. Fantasia was initially sold by her breeders to the Brightwells in Suffolk but went to Wales where she was ridden by the Evans children. Eirlys epitomised the true Welsh type which appealed to her new owner, and was subsequently shown lightly but with great success (female and overall reserve champion at the Royal Welsh, and supreme native champion at the NPS Annual Show 1994).

As a foundation mare, it was hoped that Eirlys would provide a female line on which to base the stud, but she bred mainly colts and only three fillies – two of which came late in her breeding career at Cwrtycadno. Of her ten colts, Eirlys produced four which had an impact on the show ring, the first being Cwrtycadno Cadfridog foaled 1989 by Sarnau Rheolwr. Without question Cadfridog was Rheolwr's most successful son – he was spotted early by Rheolwr's breeders, Hugh and Jane Edwards, who purchased him as a foal and successfully campaigned him in youngstock classes before his sale to Pat Johns-Powell. Interestingly Cadfridog's younger full brother, Cwrtycadno Ceredig (f. 1991) was one of the country's top performance ponies for the Cubitt Smith family. Cwrtycadno Taliesin (f. 1997 by Eyarth Celebration) was three times Belgian National Champion and reserve International Champion at Aachen in 2003, while Cwrtycadno Meredydd (f. 2004 by Millcroft Royal Lustre) was champion at the 2008

Cwrtycadno Cadfridog on holiday in France in 2012.

Swedish Spring Stallion Show. Among Eirlys' other colts mention must be made of Cwrtycadno Cymro (f. 1995), owned by Ms Lewis' great friend Robert Jones, whose family bred the Bolgoed ponies of which Automation has had a major impact on the Welsh Section B. The image of his champion sire, Eyarth Celebration, Cymro has been a consistent sire at leading studs both at home and abroad. At home, he suited perfectly the next foundation mare to arrive at Cwrtycadno, the former Royal Welsh champion Thornberry Royal Gem.

It was due to the lack of Eirlys fillies that a decision was made in 1996 to buy Royal Gem to put to Cymro and in 2002 she eventually produced the first of four outstanding fillies by him, including the previously mentioned Cwrtycadno Glain. She was followed by Cwrtycadno Perlen (f. 2003), the overall junior reserve champion at the WPCS International Show in Belgium in 2005. She won at the Royal Welsh that year and again in 2012, when she took the senior brood mare class and reserve female championship; in 2013 Perlen was shown ony once when she won the overall Section B championship at the Royal Welsh. Another sister, Cwrtycadno Colomen (f. 2005), was champion at the Welsh National Foal Show in 2005 and reserve champion at The Royal in 2006; retired from showing since then due to an injury, Colomen was leased to Ingrid Delaitre who owns the de l'Aurore Stud in France. She also selected for stud duties Cwrtycadno Rhys, a colt by Cymro out of Rhosyn-y-Mynydd (f. 1999 Carwed Charmer x Talhaearn Eirlys y Pasq), which has been shown successfully in France and Belgium.

Sarnau Rheolwr as a yearling.

Cymro's sire, Sarnau Rheolwr, was bred by Hugh and Jane Edwards, whose Sarnau Stud became famous for top-class children's riding ponies, palominos and part-bred Arabs as well as Welsh. Jane, a doctor's daughter from Formby in Lancashire, had amassed a great deal of experience when working for the Lee-Smiths (show pony producers of Pretty Polly fame) and Miss Ferguson (Rosevean Stud). During a summer visit to the Royal Welsh Show she took a temporary job producing show horses for Gatty Lewis, her future husband's uncle who lived across the valley from his nephew. Hugh Edwards and his family settled at Sarnau Park near Llandysul, Ceredigion and their children showed ridden ponies around the local shows. Edwards had a very good eye for a horse and he quickly established a stud, which also bred Welsh Section Bs.

Sarnau Rheolwr (f. 1985) lacked size but was a good mover and had the most beautiful Welsh head – he was a prize winner himself when shown as a yearling, and as a two- and three-year-old. In 1987 he was reserve champion at both Lampeter and the West Midlands Stallion Show to Eyarth Celebration and the following year he was again a winner at Lampeter as well as at Glanusk. He was by Sarnau Rhodri (f. 1981), whose dam was by Weston Gigli out of Weston Odett; Rheolwr's dam was Sarnau Blodyn (f. 1980 Sarnau Senator x Weston Welsh Flower).

The Sarnau ponies were shown at major shows in England and Wales. Rhodri's sire, Sarnau Showman (f. 970 by Sarnau Eros), was a major winner as a yearling, having won at both the Devon County Show and the Royal Welsh. He was sold to Germany in 1982, where

Bronheulog Royal, a winner of no fewer than ten firsts as a yearling for his breeders.

he had a major impact on the breeding of Section B – indeed he was champion at the International Show held in France in 1982. Showman's mating with Sarnau Juno (f. 1968) produced a series of winners both at home and abroad including Sarnau Sea Jade (f. 1973), a winner at the Royal Welsh as a yearling before becoming a champion in the Netherlands and granddam of the well-known stallion, Steehorst Freelance; Sarnau Senator, winner at the Royal Welsh as a yearling colt and champion under saddle at the Ponies of Britain as a gelding; Sarnau Saunter (f. 1980), a winner of sixteen first prizes as a yearling, including a win at the Royal Welsh before his export to Australia, where he has been an important Section B sire; and Sarnau Saucy Sprite (f. 1978), sold as a foal to Australia, where she was a champion both in hand and under saddle. Rhodri and Senator feature in the breeding of Rheolwr, whose offspring such as Sarnau Shooting Star (f. 1988), Sarnau Storm (f. 1999), Sarnau Swynol (f. 1993) and Sarnau Rhisiart (f. 1995) made an immediate impact, winning at Lampeter, Glanusk and the Royal Welsh in the case of Storm; Rhisiart was kept entire to replace his sire.

A stud to benefit from Sarnau breeding was the Bronheulog Stud belonging to the Andrews family from Welshpool, whose 1992 colt by Rheolwr, Bronheulog Royal, was both a good winner in the ring and, like his sire, a good stock-getter. His dam, Bryndansi Bella (f. 1980 Brockwell Chuckle x Coed Coch Lynette), was bred by Mr and Mrs Harry Fetherstonhaugh, who took over the Coed Coch Estate from Lt Col Williams-Wynn – she was one of several well-bred mares Mrs Andrews had gathered at her small stud, where initially the 12.2hh stallions Eyarth Figaro and Elmead Minstrel stood at stud. Bronheulog Royal won ten firsts when shown as a yearling, and at two he was second and third at Glanusk and the Royal Welsh respectively. His offspring such as Bronheulog Apricot (f. 1998), Chit Chat (f. 2002), Royal Secret (f. 2003) and Armani (f. 2004) inherited his prettiness and consistent show ring performance – Bronheulog She's A Lady, Apricot and Chit Chat all won at Lampeter, while Secret was champion Section B at the Royal Welsh Winter Fair and Armani won at the Great Yorkshire during a very successful season in England before his export to Kathleen Rawls in the United States. In 2012, the gelding Bronheulog Harvey (f. 2002 Barkway Malibu x Hattongate Heide) added to a remarkable ridden career both on the flat and over working pony jumps when he took the championship at Olympia over all native breeds for the Manners family from Kelso.

The Cwrtycadno Stud is located at Plas Ffrwd Fal, a country house near Pumpsaint on the road from Llanwrda to Lampeter. The house has two very different connections with Section Bs. The first is through Paul Wilding-Davies (Fayre Stud), whose great-grandfather, Charles Froodvale Davies, built it in 1867–8. In 1950 his father, Theron Wilding-Davies, founded the Fayre Stud – principally Mountain ponies of Revel and Craven breeding, although family connections with Section Bs go back to Harvey Jones (1882–1941), who farmed near Builth Wells and bred a few ponies. One such pony was Irfon Marvel (f. 1916 Dyoll Starlight x Henallt Black), sold to Dinarth Hall on T B Lewis' sale in 1925 for 50 guineas. She was the grand-dam of Coed Coch Glyndwr and great-grand-dam of Coed Coch Berwynfa. Jones' son Ronald bred a roan filly which was later registered in Section B of the Foundation Stock Register as Llanarth Fortress. She subsequently became the foundation for the noted Flying Saucer line of Welsh cobs bred at the world-renowned Llanarth Stud belonging to Miss Taylor and Miss Saunders-Davies, who had spotted the roan grazing in a field by the main road to Builth.

Theron Wilding-Davies is associated not only with his Mountain ponies but also with the famous Fayre Oaks sale, now held annually on the Royal Welsh showground at Builth Wells. Prior to this it was accommodated at Hereford Council's market in the town centre and, before that, at Wilding-Davies' own farm of Fayre Oaks on the outskirts of Hereford. It was started in 1955 for the benefit of local breeders, who had no obvious outlet for their Welsh ponies and part-breds. Under the auspices of Brightwells, the official auctioneers for the WPCS, it has grown into a unique two-day event incorporating a High Flyer evening sale which attracts the best of the breed in Sections A and B and Part-bred.

While Paul Wilding-Davies continued the family interest in Mountain ponies after his father's death in 1976, it was his marriage in 1986 that spread the interest to Section B though his wife, Hazel, in whose name the majority are registered. She had always had a pony to ride and then purchased Vinesend Rosaline (f. 1978) by the Section A, Cui Becket out of Cynan Cream Rose, a granddaughter of Coed Coch Berwynfa. With an abiding ambition to breed good children's ponies, they have sought stallions with good temperaments, starting with Carolinas Purple Emperor by Solway Master Bronze, to which they sent Rosaline and a Section A mare, Fayre Whistle. In 1986 the latter produced a colt, Fayre Fanfare, and Rosaline a filly, Fayre Crystal

Carolinas Purple Emperor captured in oils by his breeder Carolyn Bachman.

Rose. Two fillies to Rosaline by Weston Spider – Fayre Amber Rose and Coral Rose – quickly followed, thus providing a foundation for the Section Bs within their stud. Other mares to follow were Bengad Amethyst (f. 1980 Bengad Carrie's Boy x Bengad Diamond), Seaholm Dream (f. 1990 Pendock Peacock x Seaholm Dynasty) and Seaholm Echo (f. 1997 Pendock Peacock x Seaholm Daybreak), the last two out of daughters of Coed Coch Dawn. The brood mares enjoyed a good show record at the Royal Welsh – Amethyst was second in 1992 and third in 1999, while Dream was twice second and twice third in the years 1994 to 1997.

Fanfare was initially used on the mares, followed by Meepswood Rockefeller (f. 1988 by the Downland Chevalier son, Cennen Signature Tune), champion at the Shropshire and West Midlands Show in 1999. Rockefeller gave their ponies height and scope, including Fayre Calypso (f. 1995 out of Crystal Rose), which qualified for the Horse of the Year Show as a show hunter pony. Cottage Tobias (Downland Gold Sovereign x Uplands Butterfly) crossed well with the Rockefeller daughters. Foals have been extensively shown in preference to youngstock, basically as a means of handling them while still young. In this respect the stud has a remarkable record – having bred 75 foals (up to 2012), 66 were shown with their dams and these were first in more than half of the classes entered. Two of them won all-age Section B championships, and three won Bronze Medals – Fayre Aladdin (f. 1999) was the winning foal at the Royal Welsh and Fayre Darcy (f. 2005) was male champion at The Royal in 2005, when his dam, Seaholm Dream, was overall reserve.

The second Section B connection with Plas Ffrwd Fal (Cwrtycadno) rests with its location, which is very close to Glanyrannell, the house and estate owned by Meuric Lloyd (Dyoll Stud). The Lloyd family also had connections with another estate in the district, Glansevin, belonging to the Pryse Lloyds. (Caroline Pryse Lloyd was mother of Mrs Kemmis-Betty, the importer of Sahara from Gibraltar in 1914.) Glansevin lends its name to a stud of Section Bs established in 1969 by Michael and Odette Aylmer and their daughter, Sarah, who lived at Langhale House, not far from Norwich. Aylmer decided to use the prefix in memory of his grandmother, 'Monty' Aylmer, who was brought up in Wales, her family having a close association with the Lloyds of Glansevin. Her own Glansevin kennel of Welsh Terriers was of international repute.

Seaholm Dream, Female Champion at The Royal Show 1997.

Of the original foundation mares, Kirby Cane Gopher (f. 1956 Downland Serchog x Downland Grasshopper) and Meadow Flower of Maen Gwynedd had an immediate impact on the newly established stud. Gopher produced a colt, Glansevin Goshawk (f. 1970 by Kirby Cane Pilgrim), which was third at the Royal Welsh in 1973, and two outstanding fillies by Belvoir Zoroaster, Geisha Girl (f. 1971) and Gaiety Girl (f. 1972), which extended the mare base almost immediately. Both mares provided useful colts in 1978, which became stallions – Geisha Girl produced Gadabout (by Minsterley Captain) and Gaiety Girl produced Gay Gordon (by Seaholm Fabian), which later became an outstanding ridden winner. In the meantime, Meadow Flower of Maen Gwynedd, in foal to Chevalier, was purchased from Mrs Cuff for 420 guineas at the 1969 Fayre Oaks Sale. The result the following year was a grey colt, Glansevin Melick, a principal winner for the stud in hand and under saddle – in 1973 he won at the Ponies of Britain Stallion Show at Ascot and seven years later he won the stallion class and was judged male champion at the Royal Welsh.

Sarah Aylmer was the public face of the stud, although her mother took a keen interest; marriage to bloodstock agent Martin Percival in 1981 and the arrival of children curtailed the stud's activities somewhat over the ensuing years. Melick was used extensively and the stud which was enlarged, with more of Mrs Crisp's Kirby Cane breeding when Firefairy (f. 1974) was added as well as Chetvalley Glitter (f. 1984 Kirby Cane Woodcock x Kirby Cane Generous), the latter bred by Mrs Percival, who lived in Suffolk. To Glansevin Gadabout, Firefairy produced Glansevin Firelight in 1981, which in turn produced Filigree (f. 1990) by Pennwood Malaga. Glitter was sent to Eyarth Rio and a filly, Glansevin Garland, was foaled in 1995. With the success of the North Wales stallions running high, Garland was sent to Carwed Charmer; the successful cross was Glansevin Graffiti (f. 2000), purchased by well-known Dutch breeders Harold and Yvette Zoet (Ysselvliedt Stud), who later leased him to the Hilin Stud which produced him for the show ring in 2007. He was overall ridden Welsh champion at Glanusk as well as stallion class winner and reserve male champion at the Royal Welsh. While in Wales, Glansevin Filigree was sent to him and the result was Glansevin Flavia, a winner as a yearling in 2008. Glansevin Garland was leased by her breeders to Hilin, where she bred nine foals, including the Royal Welsh winners Hilin Garland and Hilin Tattoo; she returned to East Anglia in 2010 with a filly foal at foot, Hilin Guinevere by Eyarth Tayma.

With their family grown up and more time to concentrate on the stud, the Percivals began building up the Glansevin name once more – Glansevin Filigree put to Hilin Carnedd gave them a bay filly, Glansevin Finola, and Garland produced an Eyarth Tayma colt, Glansevin Galileo (f. 2011), which was sold at the Fayre Oaks Sale in 2011. Sadly Sarah Percival died in 2012, but her husband's lifelong passion for breeding animals has meant that the stud continues, along with a flock of pedigree Norfolk Horn sheep and Dexter cattle.

The Aylmers, like many breeders in East Anglia, were great friends of Mrs Crisp, whose visit to Miss Brodrick in 1948 and her venture into breeding Welsh ponies – Section B in particular – had such far reaching effects across the world. The Kirby Cane bloodlines were also very prevalent closer to home and many of the emerging studs sought her ponies as a foundation for their studs. Among

Glansevin Graffiti competes under saddle ridden by Yvette Zoet at Houten, Netherlands.

them Donald Cooke, who ran a caravan park at the holiday town of Great Yarmouth, made use of several local breeders to source foundation mares for his Bureside Stud. Three of them were bred by Mrs Crisp – Kirby Cane Pensive, Gleam and Spellbound. From Mrs Gowing he selected Shimpling Moon Frolic and from Mrs Aylmer he purchased one of her own foundation mares, Merddyn Prydferth, whose foal by Glansevin Melick produced him the first of several well-known stallions, Bureside Bandit (f. 1976). Kirby Cane Sundog (f. 1971 Paith Astronaut x Kirby Cane Spellbound), his son, Bureside Mandate (f. 1992 and a dun like his sire) and Pennwood Malaga were familiar exhibits at shows in the East of England.

Paith Astronaut (f. 1968), a palomino by Downland Chevalier x Criban (R) Moonbeam, was an influential stallion in this area – he stood at Bury Hall near Thetford in Norfolk, where Mrs Angela Broadhead and her daughter Susan (McInnes Skinner) established their Ernford Stud in 1965. They began with a few mountain ponies but, like their great friend Mrs Crisp, soon turned to the Section B because it better suited their land and they wanted to breed good ponies for children to ride and hunt. (While Mrs Crisp and her family hunted with the Waveney Harriers, the McInnes Skinners hunted with the Dunston Harriers and West Norfolk Foxhounds, one of the oldest hunts in England.) Mrs Broadhead first purchased the beautiful show mare, Ready Token Camelia (f. 1963) by Coed Coch Berwynfa bred by Mrs Hope, although Camelia was later sold to Mrs Gadsden for her Bengad Stud in Gloucestershire, better known for its Mountain ponies but also the home to a number of winning Section Bs. Mrs Broadhead's purchases from Mrs Crisp – Kirby Cane Go-Lightly, Bangle and

Witchball – proved most influential on her stud. To Paith Astronaut, Witchball produced Ernford Witchcraft (f. 1972), which in turn produced a stallion for the stud named Ernford Wildfire (f. 1976) by the Chevalier son, Rotherwood Commander. It must have seemed a golden period for the stud as, in 1971, Bangle produced a filly, Ernford Brocade by Kirby Cane Ptarmigan, which became an important brood mare. Among her best-known progeny was Ernford Benjamin (f. 1983 by Sunbridge Alicante) – purchased by Gill Simpson for her Wortley Stud, he was shown with success in hand and sired some top-class ridden ponies at Wortley including

Above: Beaumont Bowbell, foundation mare at the Ernford Stud.

Left: Ernford Bellboy, a prolific winner under saddle including qualifying for Olympia on three occasions.

the full brothers and sisters out of Desarbre Polyanthus, Wortley Cavalier (f. 1991), The Duchess (f. 1993), Wild Jasmine (f. 1994) and Wizard (f. 1995).

Polyanthus, bred by Roger and Jane Card from Yorkshire, was sired by Tetworth Nijinski, as was Desarbre Shoestring – which became an important influence in the Netherlands for the Hoekhorst Stud – and the Cards' 1984 Royal Welsh champion, Desarbre Folksong. Nijinski was out of Tetworth Czarina, as was Tetworth Mikado, a stallion owned by Bill Stamp and stabled in a pub yard at Stradishall near Newmarket. Mikado sired one of the most famous of all the Ernford Section Bs, Ernford Bellboy (f. 1989), whose dam, Beaumont Bowbell, became the most important of all the mares brought into the Ernford Stud. By Solway Master Bronze out of a mare by Abercrychan Spectator, she was unsold as a yearling at the 1979 Fayre Oaks Sale, but Mrs Broadhead managed to buy her privately. Her first foal, Ernford Bo-Peep (f. 1982), a filly by Wildfire, became one of many famous offspring out of Bowbell, which had fourteen foals for her owners over her stay at Ernford.

Ernford Bellboy, a distinctive chestnut roan like his sire, had a great record under saddle and at Olympia – from six appearances he was best of breed on no fewer than three occasions. He passed his riding qualities on to his offspring, including the striking palomino stallion Northlight Galliano (f. 1997), bred by Jane Waterhouse – Galliano qualified for the Olympia final three times like his sire, and also won his class at the Horse of the Year Show in 2004. The same venue provided another celebration in 2012 for Bellboy when his daughter, the chestnut mare Stambrook Wedding Belle (f. 2001), claimed the Show Hunter Pony of the Year title ridden by Eleanor Fairley and was reserve champion at the Royal International Horse Show the same year. Belle is out of Nantcol Candette, a daughter of the 1981 Royal Welsh winner Nantcol Kadet (f. 1980). Kadet was by Cawdor Cadet, also the sire of the 1986 Royal Welsh female champion, Nantcol Katrin (f. 1979). All three were bred by Mrs Gwen Cook, whose small height Nantcol Section Bs have held a remarkable record in the mini plaited classes at the Horse of the Year Show, demonstrating, like Wedding Belle, the value of Section Bs in open as well as breed classes.

The Stambrook Stud was founded on both Glansevin and Kirby Cane bloodlines, specialising in performance ponies of which Stambrook Phoebe (f. 1975 Glansevin Melick x Glansevin Pirouette by Chirk Caradoc) was a winner both side saddle and astride, including winning at the Royal Welsh in 1979. Her full sister, Stambrook Portia (f. 1976), was a famous winner in Scotland. Glansevin Gadabout became the resident stallion at Stambrook during the 1980s. Crossing Section B with the riding pony has also been successful for the stud, just as it has been for another pair of East Anglian breeders, Jill and Angela Grummitt, whose Fenbourne Stud of Section Bs in Cambridgeshire was formed with the show mare, Cusop Bonus (f. 1963 Coed Coch Pawl x Coed Coch Bugeiles). They used Ernford Bellboy on the Bonus female line to produce the useful stallion Fenbourne Bellhop (f. 1989), winner of the three–year-old colt class at the Royal Welsh in 2003.

Mrs and Mrs John Greenleaf from Colchester were breeders who didn't follow the Kirby Cane bloodlines, although they did favour those of a near neighbour, Lady Seton, whose Baylaurel ponies had become well established in Section B of the Stud Book. Greenleaf purchased from her the foundation mare Cawdor Heather, bred in West Wales by W J Thomas, who had become a member of the WPCS in 1947 and whose prefix comes from the Cawdor Estate situated near his hometown of Llandeilo. Greenleaf also purchased for stud duties the colt Baylaurel Plover, a grandson of Heather by Baylaurel Mark Anthony (by Tanlan Julius Caesar) out of Cawdor Pride, whose sire, Wingrove Foxtrot, was the first stallion to stand at stud at Brookhall along with Brockwell Berwyn. Later Cawdor Pride joined the stud along with Mrs Archdale's Breccles Filigree and Breccles Windflower.

As the stud developed, several influential mares joined it. In 1980 Downland Damask was purchased from theButterworths at Fayre Oaks, where two years later they bought the Section A mare, Weston Pearly Necklace, the dam of their new resident stallion, Weston Neptune (acquired by David Lawrence for 3,500 guineas at the 1978 Weston Sale). Baledon Bronze Camilla (f. 1972 Solway Master Bronze x Gredington Blodyn) and Weston Lily Langtree (f. 1979 Weston Charmer x Weston Moll Flanders) became principal breeders of Brookhall ponies, which won extensively in the East Anglian show rings and easily found markets both at home and abroad.

I have mentioned the influence of the Cawdor prefix in both the Baylaurel and Brookhall studs. Both Pride and Heather were by stallions whose prefix became more associated at the time with Welsh cobs than Section B, but by the 21st century, show results paint a different picture. I refer to the mountain pony stallions Llanarth Mayday, sire of Heather, and Llanarth Nightlight, sire of Pride's grand dam, Towy Liesbeth. The Llanarth Stud, founded by the remarkable 'Ladies of Llanarth' to which they were often referred, was situated in the heart of 'Cob' country at the village of Llanarth near Aberaeron, south of Aberystwyth. Miss Pauline Taylor, Miss Barbara Saunders-Davies and Miss Enid Lewis shared a love of music, in which they were all trained, and a love of the Welsh breeds which included cattle, ponies, cobs, pigs and Corgis. Their main interest lay with the breeding of cobs, something for which they would become extremely well-known – in fact, many would credit them with taking the ridden Welsh cob out of Wales, thus exposing it to greater publicity in England and further afield. However they also bred mountain ponies in a small way and Section Bs which at the time (the 1950s) encompassed the pony we now know as the Section C. They also had a special interest in palomino colour – an example of this was Llanarth Sentinel (f. 1957) mentioned below, first registered as Section B and later transferred to Section D.

A decision was made in 1961 to disperse the Section A and B herd, which was offered at the Gredington Sale held that September. The foreword to the catalogue read:

Owing to the impending retirement of Miss Saunders-Davies, it has been felt advisable for the Llanarth Stud to concentrate mainly on the Welsh Cob and Section C Pony. It has, therefore, been decided to take this opportunity of disposing of the Stud's Section A and Section B stock.

Many of the best of these are homebred and represent the fruits of 16 years' breeding policy in which the free true Welsh action, with a good shoulder, has always been of paramount importance. All these ponies move well and some are of exceptional brilliance. Here is a chance too, for breeders to obtain first-class stock not related to the popular and outstandingly well-known Show-winning strains.

The service stallion, Llanarth Sentinel (Section B), was unbeaten in the Show ring last year, and was 1st and Reserve Champion Palomino at the Ponies of Britain. He has grown over-height, but, mated with these Section A mares, should produce the larger Section B ponies which are so scarce.

The farm was transferred in 1979 to the University of Wales, the cobs subsequently sold, and the Llanarth prefix transferred to Len Bigley, who had managed and shown the cobs latterly with such great success. He and his family have maintained the stud's name within the 'cob' world but, some 50 years after the Section A and B herd was dispersed in 1961, it was a Llanarth-prefixed Welsh Section B that was judged overall section champion at the Royal Welsh in 2012.

It was not until the mid-1980s that the Section B side of the Llanarth Stud started to gather momentum, although a few Section B mares had been purchased by Len and Ann Bigley following their marriage in 1975. Besides his work in Wales for Miss Taylor and Miss Saunders-Davies, Bigley had been brought up with horses through his uncle, a farmer living at Bishop's Castle, Shropshire and a neighbour of Mr and Mrs Borthwick, who encouraged Bigley when he was a boy. Ann's mother came from Wales –indeed her grandfather and uncle bred the famous Hewid cobs. While they both had equine interests beyond cobs, it was Bigley's experience of the famous Llanarth cobs which took him to the very top as a producer and showman, qualities which he would bring to the section B in due course and which he would share with his children, Simon and Catryn, who have inherited their father's love of the show ring.

A visit to the Carolinas Stud in 1981 introduced Bigley to a young colt which impressed him greatly. This was Carolinas Purple Emperor (f. 1980), a son of Solway Master Bronze and Eden Blue Bunting (f. 1966) by Gorsty Dark Shadow (full brother to Gorsty Firefly) out of Talfan Snow Bunting by Ceulan Gondolier. Like many of the successful ponies bred at Weston, she was full of the best Coed Coch breeding, particularly Glyndwr several generations back. Blue Bunting had been purchased cheaply at the Fayre Oaks sale in 1977 by Mrs Carolyn Bachman, wife of an American, Tom Bachman, and daughter of Major General Goff Hamilton, who had served in India, and the writer, M

Carolinas Mistlethrush, winner as a yearling before his export to Australia.

M Kaye, whose novel *The Far Pavilions* was a best-seller and subsequently made into a film. The family lived in Sussex with sufficient land to sustain a small stud of both Section Bs and riding ponies, but moved to a farm at Bentley in Hampshire. Initially, Mrs Bachman bought Downland Cavalcade to put on her mares of Coed Coch Berwynfa breeding such as Shawbury Ballerina, which was in foal to Downland Mohawk. The product of that mating was Carolinas Moccasin (f. 1979), a colt that was bought as a foal by Vanessa de Quincey for her Erimus Stud in Sussex and shown with success (she later sold him to Mrs Nina Clayton in Yorkshire). However Mrs Bachman changed tack, buying the very best of Downland breeding to cross with Solway Master Bronze and his son, Purple Emperor. Her mares included the Downland-bred Dresden (f. 1971 Downland Chevalier x Downland Dignity), Edelweiss (f. 1979 Downland Mohawk x Downland Eglantine), Philomel (f. 1984 Downland Woodsmoke x Downland Edelweiss) and Mirth (f. 1985 Downland Mohawk x Mere Fire Myth).

The outcome of that visit saw Carolinas Purple Emperor moving to Llanarth on lease for stud duties and showing. He proved a huge success at both, particularly the latter for as a two-yearold he was reserve champion at the Ponies of Britain Stallion Show at Ascot, and the following year won at the Royal Welsh and National Pony Society Shows and qualified for the Lloyds Bank Final at the Horse of the Year Show. The die was cast for a series of wins for the stud over the next five years under Bigley's production, which would take it to the top at major shows round Britain. Carolinas Swallowtail (f. 1982) was reserve champion youngstock at the Royal Welsh in 1983; Carolinas Red Rose (f. 1980) was novice brood mare winner at the Royal Welsh in 1984; Carolinas Holly Blue (f. 1984) won at The Royal, East of England and Shropshire and West Midlands in 1985; and Carolinas Red Fox (f. 1985), under the ownership of the American breeder, Mrs Nancy Jane Reed, won at the Ponies of Britain Show in 1986 before embarking on a very successful career in the United States. Carolinas Dancing in the Dark (f. 1986) was champion at The Royal in 1987 and Carolinas Queen of Spain (f. 1986) was female champion at Lampeter the same year. Downland Edelweiss beat the stallion Broadlands Coronet (f. 1983 by Varndell Right Royal x Sarnau Moccasin) for the overall championship at the Royal Welsh in 1988, the year after her colt foal, Purple Rain, took the overall foal championship at the National Pony Society Show at Malvern.

Based in the south of England, the stud was well placed for buyers from continental Europe, many of whom liked the combination of Master Bronze and Downland breeding. As a result, Carolinas ponies sold readily to an overseas market and found their way to several European studs as well as those in the United States and Australia. After a relatively short history, the last foals were registered with the Carolinas prefix in 1997, but not before other studs benefited – Carolinas China Rose (f. 1982) went to Cottrell and Shawbury Ballerina and Carolinas Painted Lady found a home at Llanarth. It was Painted Lady that eventually gave the Bigleys their ultimate success, as she had a filly by Ardgrange Difreg named Llanarth Dulcie (f. 1988), which was sold to the Parsons from Worcester before returning as a brood mare at Llanarth.

Among the many producers and showmen in Britain, none can compare with Bigley for the success he has enjoyed over 40 years at the very top. (The Welsh Cob

Llanarth Flying Comet was male champion at the Royal Welsh in 1972, shown by Bigley.) He has shown all sections of the Welsh Stud Book as well as Part-breds across the length and breadth of Britain to qualify for the in-hand finals at the Horse of the Year Show, and his record with Section B stallions knows no equal. As a result, he has also had available stallions which have helped define the type of pony bred at Llanarth typified by their depth, substance, free movement and unquestionable Welsh type. Colour has never been an issue in the breeding policy – the ponies are invariably grey. Contrary to popular misconceptions, this has been achieved by an amalgam of bloodlines including both Downland and Coed Coch.

The first of his own stallions was a chance find while he was on a judging trip to Denmark – his attention was taken by a good-moving little grey stallion, Twyford Signal (f. 1976), which was competing in performance classes. He was owned by John Kristensen, whose road into Section Bs came from his admiration of his neighbour's stallion, Coed Coch Endor. Foaled in 1970, Endor provided a double cross to Coed Coch Berwynfa and in due course Kristensen bought him to develop his Section B interest. Wishing to purchase a Section B mare to put to Endor, Mrs Mountain sold Kristensen a Rhyd-y-Felin Selwyn mare she had bred called Twyford Sparkle with a colt foal at foot, Signal, by Pendock Peregrine. In due course, when Sparkle was put to Endor, she produced a colt, Mollegaards Spartacus (f. 1986), which carried four crosses of Tanybwlch Berwyn in his pedigree within four generations. Twyford Signal also carried much Berwyn on both sides of his pedigree, so it was no surprise that he crossed well with Lydstep Prairie Flower by the Downland Chevalier son, Downland Gondolier. It was while Prairie Flower was being shown by the Bigleys for Mrs and Mrs Jackson that she produced an outstanding colt to Signal in 1991 named Douthwaite Signwriter, which was quickly snapped up by Kristensen, whose Mollegaards Stud was by now well established.

The three stallions mentioned enjoyed amazing success under the Bigleys' production, which by now

Mollegaards Spartacus, Royal Welsh Champion 1992, 1993, 1994.

Cwrtycadno Cymro, prolific breeder for several studs.

had become a family affair. In addition to numerous major show championships, the charismatic Twyford Signal competed four times at the Horse of the Year Show when shown in hand, and twice at Olympia under saddle. Mollegaards Spartacus did exactly the same, although he had the honour of taking both the in hand supreme and overall ridden championships in 1992 at the 9th International Welsh Show at Peterborough. Spartacus arguably had the most impressive record of the three when he equalled Criban Victor's record of three consecutive overall championships at the Royal Welsh (from 1958 to 1960) when he triumphed from 1992 to 1994. Few would argue with the statement that the task Spartacus faced in terms of quality and numbers of his opposition far outweighed that of the great Victor. Douthwaite Signwriter also competed at the Horse of the Years Show on four different occasions.

Twyford Signal, Mollegaards Spartacus and Douthwaite Signwriter have played their part in the success of the ponies bred at Llanarth, starting with Signal, whose daughter Brenob Finch (f. 1990 out of Cusop Birdsnest) was overall reserve champion at the Royal Welsh for the Bigleys in 1999. To Whatehall Cornpoppy, Signal produced Llanarth Welsh Poppy (f. 1993), a good winner including the championship at the Three Counties Show for Hay-on-Wye butcher Chris Gibbons, whose Rhydspence Stud gained top awards in the hands of Bigley. This included his imported stallion, Mollegaards Mr Swing King, by Spartacus and out of Twyford Sparkle – he took the Pony Supreme and Section B championship at The Royal Show in 2002. Another of Signal's daughters was Scrafton Angelina, which was Royal Show champion in 2006 on her first show ring appearance.

Among the early winners to appear by Mollegaards Spartacus during his five-year stay in Britain was the chestnut colt Llanarth Byron (f. 1993). His dam, Elmead Lockets, was all Mountain pony other than a quarter Downland through her sire, Aston Love Knot by Downland Romance. Byron had an excellent youngstock career, taking the championship at Northleach in 1995 and twice second at the Royal Welsh, in 1995 and 1996 – he was later exported to the United States. Coalbourne Cameo (f. 1992) out of Downland Raphael (by Downland Mohawk) was also by Spartacus; unhealthily obese, she was rescued by Ann Bigley and put to Douthwaite Signwriter, who had now taken over stud duties as the stud reached the new millennium.

The filly was Llanarth Camilla (f. 2001), whose youngstock championship at the Royal Welsh in 2003 put down a marker for things to come as she won extensively later as a brood mare. In 2008 her championship at Lincoln led to qualification for the In Hand Final at the Horse of the Year Show and by 2011 she had won at the WPCS International Show. Camilla's son by Cwrtycadno Cymro, Llanarth Da Vinci (f. 2006), has followed in his dam's footsteps with championship wins across the country, including the South of England in 2010 when he was reserve for the Cuddy In Hand qualifier. Other Llanarth-bred winners by Signwriter have included Sotheby (f. 2003) and Scout (f. 2004) out of Horsegate Minuet, Opale (f. 2001) out of Downland Finch and, most famously, Delilah (f. 2007) out of Llanarth Dulcie. Delilah exceeded all expectations from the time when, as a yearling, she was supreme at the Royal Welsh Winter Fair under Dr Wynne Davies. As a two-year-old she was shown twice, coming first at the Royal Welsh and youngstock champion at the International Show in the Netherlands, and at three, she was champion at Lampeter. Shown by Catryn Bigley, she swept all before her at the 2012 Royal Welsh Show when she took the overall Section B championship as well as Cuddy qualification for the Horse of the Year Show. This grey mare gave her breeders their first homebred champion in Section Bs at Builth Wells.

Delilah's success at the Royal Welsh was not the only cause for the Bigleys' celebrations

Llanarth Da Vinci, Champion Section B Ponies UK Summer Championships 2010.

Llanarth Delilah, Royal Welsh Champion and Cuddy Qualifier Royal Welsh 2012.

in 2012 as their current show stallion Boreton White Prince (f. 2002 by Betton Demetri) qualified for the Cuddy Finals for the second year running. Time will tell whether or not his offspring fair as well as those by his predecessors at Llanarth. Spartacus, like the other stallions standing at Llanarth, became a popular choice for other breeders and winners by him began to appear in the 1990s, such as the previously mentioned Thornberry Demelza. The services of Spartacus were also taken up by one of the oldest breeders of Welsh Section Bs, Mrs Evelyn Charlton, who started by registering Foundation Stock with her Kiltinane prefix in 1956 while living at Plas Bellin, near Mold in Clwyd. A great friend of Miss Brodrick, her early breeding policy was based on Coed Coch bloodlines. Mares of this breeding moved with her to the Horseley Stud in Somerset in 1980, where she and her husband also bred top-class racehorses – she was owner of the famous National Hunt chaser Wayward King. Mrs Charlton chose a mare of pure Coed Coch breeding for Mollegaards Spartacus. Kiltinane Daphne (f. 1986), by Coed Coch Nimrod (Cusop Banknote x Coed Coch Nina) out of Coed Coch Donna (Coed Coch Targed x Coed Coch Dawn), produced a filly in 1995 named Kiltinane Dawn, which was third at the Royal Welsh 1998 and was then sold to Mrs Gilchrist-Fisher and to become dam of her Lemonshill Dawn Chorus, by Lemonshill Top Note.

Mrs Angela Poles was keen to add Spartacus to an existing good blood line which she had developed with her husband and well-known veterinary surgeon, Sam Poles, when they lived near Peterborough. The move of the Horsegate Stud to Herefordshire meant that it was close to the Bigleys, who had settled at Quakers Farm at Michaelchurch on the Welsh border. The Horsegate ponies were full of Weston, Criban and Pendock breeding, hardy and full of Welsh character – some would say 'old-fashioned' – and for this they were in keen demand as performance ponies and outcross breeding stock for the Downland and Rotherwood stallions popular at the time.

The foundation mares at Horsegate largely came from Llewellyn Richards – Criban Mayfair, Louisa, Lucy, Red Flash and Brocade were all early producers, as were the Felbrigg mares, Ribby, Scamp and Flutter. The Felbrigg ponies were bred by Mrs Phillips, whose husband bred Thoroughbred horses at their Old Buckenham Stud in Norfolk. Like others in the county, her mares were based on Miss Brodrick's breeding and were all direct descendants of Coed Coch Bettrys, Coed Coch Pomgranad and the stallion Coed Coch Pawl, which she had purchased from Eckley. The Coed Coch bloodlines enhanced the breeding at Horsegate, as did another early foundation mare, the dun Keston Kerry Gold (f. 1969) by Springbourne Golden Flute out of Royal Occasion's dam, Clyphada Periwinkle. Kerry Gold produced a colt by Weston Spider in 1974, Horsegate Ambassador, which became a big winner for the Poles and a major influence in the offspring he produced. Weston Spider (f. 1966 by Chirk Crogan) was a chestnut roan like his sire out of Brockwell Penelope, an FS mare by Brockwell Cobweb out of Coed Coch Pendefiges. When the stud was disbanded Ambassador was retired to the Lindisfarne Stud in Northumberland. The other stallion used at Horsegate which enhanced the 'old' breeding through Criban Victor was Pendock Peregrine (f. 1973), a son of Pendock Plunder and Pendock Prudence.

Horsegate Spark (f. 1993) by Mollegaards Spartacus out of the Lydstep Blondie daughter Horsegate Carol, was sold to the Thornberry Stud at the dispersal of the Polaris Stud. His half-sister by Horsegate Ambassador, Horsegate Sapphire (f. 1985), became a foundation mare along with her full sister Horsegate Gigi for Mrs and Mrs Fillingham's Stockham Stud, initially based at Dulverton in Somerset but latterly situated near Grantham in Lincolnshire. Sapphire came to them in foal to Twyford Signal in 1989, producing the roan filly Stockham Skylark, the first of a long series of winners from this line. The Fillinghams used local stallions but also stood at stud Ardgrange Difreg, leased in 1992–3 from the Carters at Millcroft, and Rotherwood Secret Agent (f. 1991 Ardgrange Dihafel x Rotherwood Penny Royale), which they bought as a yearling from Mrs Mansfield-Parnell. After producing many good winners at Stockham, he was sold to the Glenmore Stud in Australia.

The Stockham stud prides itself in performance ponies, to which its breeding was dedicated. Sapphire once more provided them with an outstanding winner when Stockham Stonechat (f. 1992), by Brockwell Prince Charming, started the 1998 season by winning as a novice working hunter pony at the British Show Pony Championships. He went on from strength to strength in a distinguished working hunter pony career when he was judged twice reserve champion at the Horse of the Year Show, a winner at the Royal International Horse Show and champion of champions at the British Show Pony Society Summer Championships.

Arguably the most famous of his daughters out of Sapphire was the 2001 Royal Welsh winner Stockham Contessa (f. 1998 by Rotherwood Secret Agent), whose full brother, Gold Finger, was second there as a three-year-old the previous year.

Another daughter of Secret Agent, Stockham Domino (f. 1994), this time out of Stockham Skylark, was twice winner at the Royal Welsh and a good winner under saddle before becoming a major force at the Davies family's Cadlanvalley Stud at St Dogmaels on the Ceredigion coast in south-west Wales. The stud gets its name from the picturesque valley of Cwm Cadlan at Penderyn near Aberdare, where Howell Davies farmed on the southern edge of the Brecon Beacons and used ponies for shepherding. His wife Yvonne rode ponies for pleasure as a child and within a year of getting married in 1975 they had visited the Fayre Oaks Sale and purchased their first Section B, Nefydd Llain o'r Hydref (f. 1968), a mare by Langford Phoebus (Solway Master Bronze x Downland Autumn) which was in foal to Brockwell Chuckle. The outcome was a chestnut filly, Honey Bee, the first registered with their Cadlanvalley prefix. Next came the bay Chirk Caradoc mare, Chirk Dilys (f. 1972), which produced a filly, Cadlanvalley Cygnet by Keston Royal Occasion, which gave them their first Royal Welsh victory when she was shown for them with her dam in 1979 by Edwin Prosser.

Mr and Mrs Davies had a preference for ponies with bone, substance and true Welsh character, so they selected the foundation mares Brockwell Joyful (f. 1965 by Brockwell Berwyn), Weston Louisa (f. 1978 by Weston Charmer) and Gunthwaite Sunkissed (f. 1982 by Radmont Tarquin). For stallions, they started with Abercrychan Speculator but sold him out of preference for Georgian Sinbad (f. 1970), a son of Solway Master Bronze and Downland Symbol by Downland Chevalier.

Weston Louisa, foundation mare at Cadlanvalley Stud.

Boston Bonaparte (f. 1979 Varndell Right Royal x Contessa of Kirkhamgate by Radmont Tarquin) was purchased as a two-year-old and by 2006 he had topped the WPCS In Hand Sire rating competition. The 1987 Royal Welsh male champion Millcroft Copper Lustre joined the stud but sadly died prematurely, so Eyarth Troy (f. 1999 Carwed Charmer x Weston Twiggy) and Newtonhill Naughty Boy Charlie (f. 1999 Heaton Romeo x Laithehill Brocade) followed. The senior stallions found new homes to make room for youngsters – Bonaparte was sold to the Netherlands, Troy to France and Charlie to the Menai Stud. Troy's son, Russetwood Elation, and grandson, Cadlanvalley Buzby, were retained.

The black mare Weston Louisa produced some of the stud's biggest show successes and three females out of her were retained – Cadlanvalley Confetti (twice winner of the Royal Welsh) by Linksbury Celebration, Elouise by Millcroft Copper Lustre and Painted Lady by Linksbury Jester. In recent years Millcroft Cappuccino, Duntarvie Venetia and Hilin Serena have been added to the stud with success, but it has been Stockham Domino that has provided the family with its greatest successes to date. To Eyarth Troy she produced Cadlanvalley Sandpiper (f. 2006), a cream gelding which has qualified for the ridden finals at the Horse of the Year Show for the Scott family of Aberdeen every year since a four-year-old and

Cadlanvalley Sandpiper, runner-up at Olympia in 2012, ridden here by Catherine Scott.

in 2012 was second at the Olympia Ridden Native Finals. To Russetwood Elation, Domino has produced two outstanding fillies. The first, Cadlanvalley Georgette (f. 2007), was a Bronze, Silver and Gold medal winner as a yearling, the last captured at the Royal Welsh when she was judged youngstock champion in 2008. Her full sister Cadlanvalley Georgia (f. 2008) not only went one better but swept all before her at the 2011 Royal Welsh when she took the overall Section B championship as well as standing runner-up for the overall Tom and Sprightly Welsh championship and Cuddy In Hand Qualifier.

Having been involved in ridden ponies through both

Cadlanvalley Georgia, Royal Welsh Champion 2011.

their children and grandchildren, Mr and Mrs Davies became conscious as breeders of the importance of the performance qualities in the ponies they breed. This has been endorsed by the success of Cadlanvalley Sandpiper at the Horse of the Year Show and Olympia, as well as that of Cadlanvalley Jackpot (f. 1990 Cusop Steward x Gunthwaite Sunkissed), a Gold Cup working hunter pony victor at the British Show Pony Society Championships. In addition, Cadlanvalley Tomahawk and Totum, both out of Cadlanvalley Louisiana, are performance geldings in France for Section B enthusiast François Lescanne. Increasingly the trend of linking bloodlines in Section B to performance has fallen into line with thinking in continental Europe, and perhaps it is in this direction that future Section B breeders must look.

Chapter XVIII

Reflections on a Century of Welsh Pony Breeding

Four full brothers and sisters bred by Gill Simpson competing in the Section B ridden class at HOYS 2001. (l to r) Wortley The Duchess, Wortley Cavalier, Wortley Wizard and Wortley Wild Jasmine.

Welsh Pony enthusiast Theo ten Brinke from Holland enjoying a day in the snow with his homebred Section B stallion Luciano

When Sir Pryse Pryse, Bt. registered his homebred mare named Arab, registration no 4 in Volume I of the newly-formed Welsh Stud Book in 1902 (interestingly Dyoll Starlight also shared this number), one wonders whether or not he was at the forefront of breeding the Welsh Section B despite the fact that she was registered in Section A. Her details simply record that she was bay, born in 1892, by an unnamed Arab and out of Peggy by an unnamed Arab stallion. Her ancestry was certainly a recipe later recognised by breeders in the Foundation Register as Section B. Her links with the desert were much greater than those with the mountains of Wales, despite their proximity to her home. Some 120 years on, the question is purely academic – but what matters is that during the late Victorian era, as befitted the time, breeders with an interest in the Welsh breeds took the trouble to record what they already had so that the breeding programmes of others might benefit in the years to come.

The early chapters of this account have shown that establishing a stud book was not easy and was driven largely by well-intentioned breeders and landowners who acted not only as enthusiasts but also as great benefactors of the Welsh Pony and Cob Society. Until 1919, a year after World War I had ended, several breeders had registered sizeable numbers of ponies and cobs in the Welsh Stud Book. William S Miller of Forest Lodge had registered 488 mares (including 33 Section B of cob-type and ten Section C Welsh cobs). He also registered 67 stallions including Forest Morning Star, a chestnut with no breeding but which was recorded as having won 36 first prizes. None of the stallions were by Dyoll Starlight and only two were by his sons Greylight and Bleddfa Shooting Star. J Marshall Dugdale (Llwyn prefix) was another of the early supporters to register large numbers of his ponies and cobs, of which 36 were born in the 1890s and the rest born before 1910. He registered 195 mares (of which 17 were Section B, 17 Section C and eight Section D) and 40 stallions – Llwyn Cock Robin was registered as no 6 in Section A of Volume I. By contrast, Meuric Lloyd (Dyoll Stud) registered only 56 mares (all Section A but only 20 bred by him, and of these 19 were by Dyoll Starlight and the other by his son, Bleddfa Shooting Star); he also registered 16 stallions including Dyoll Starlight, 11 of his sons and two of his grandsons. Mrs D H Greene, another principal breeder of the time, registered 188 mares (nine Section B, two Section C and one Section D), of which 72 were bred by her; she registered 35 stallions, all Section A and mainly by Dyoll Starlight or one of his sons.

Much of the very early breeding details were based on memory. A glance of the early Stud Books belonging to that great historian of the breed E S Davies (father of Dr

Wynne Davies) reveals the corrections which he made to the breeding, colour or markings of individuals recorded therein with which he was personally acquainted. Nevertheless it was a start, and little did it matter – for, as we know now, within six generations the genetic inheritance from individuals becomes so diluted that it almost becomes meaningless. The introduction of 'foreign' blood to the indigenous ponies of Wales in the 18th and 19th centuries was not done out of thoughtlessness or malice towards the breed but for practical and financial purposes by people of the time who either used them or bred them to sell to others. Pedigree breeding as we know it simply did not exist – nor did the breeding of ponies for pleasure. It was not until the beginning of the 20th century that Britain witnessed a warming to the use of ponies for recreation, hunting in particular, and for showing, something that had only previously been enjoyed by children of the privileged élite.

It is at this point that the story of the Welsh Section B begins to take shape, as people like Mrs Inge, with a love of hunting and Welsh mountain ponies from her beloved Tan-y-Bwlch Estate, recognised the need to breed a riding pony with all the characteristics of the Welsh ponies she knew and loved but which were larger in height. It coincided with the men of the Polo Pony Society, such as John Hill from Church Stretton, who recognised the worth of native ponies to cross with Arabs, Barbs and small Thoroughbreds in order to supply ponies for their sport. The development of the Section B also coincided with the ambitions of the social élite, who saw the show ring as a vehicle into the very heart of Britain's establishment at the highest level – one only has to observe the list of Presidents of the shows held at Islington and Olympia, which closely resemble pages out of *Debrett's* or *Who's Who*.

The type of Mountain pony often described as the 'Starlight' strain, regardless of its actual origins (of which there has been much debate), was exactly the type of Welsh pony required to ensure the quality, prettiness and riding characteristics needed in the new generation of larger ponies in the Welsh Stud Book. So too were the type bred by breeders such as Howell Richards and his sons from Criban, whose personal preference leant towards Mountain ponies with good shoulders, long fronts, quality bone and free riding action. In the long term, we have seen that the two strains of ponies complemented one another as the small Thoroughbred was used to upgrade the latter and the Arab to improve the former.

The demand for the Welsh Mountain pony was always changing, as Captain Howson, Secretary of the Royal Welsh Agricultural Society, noted in a report published in 1931. He commented:

It will probably be conceded by all who have followed the fortunes of the Welsh Mountain pony for any length of time that the outlook for the breed has changed almost completely within the last ten or a dozen years. There was a day when there were many outlets for surplus and inferior stock. Ponies were required in large numbers to perform runabout harness jobs in town and country, to draw the vehicles of small traders, and to carry out haulage work in the pits; and, often enough, strength and bone were of more importance in the eyes of the purchasers than was appearance.

The outlets have now virtually all disappeared and, just as altered circumstances in other spheres of endeavour compel reorganisation, in order that the changed conditions may be met, so is it now necessary for the breeder of Welsh Mountain ponies to ponder over his position and to lay new plans.

In describing the qualities desired in the Welsh Mountain pony, he used terms such as 'deep, well-placed, riding shoulders', 'long reachy fronts', 'neat, gamey little heads' and 'free and sweeping action' – all characteristics that can now be recognised in the Section B. The WPCS, of which Howson was also Secretary, was itself making a significant contribution to the development of the Section B. Its type was also defined by commentators such as A L Williams and, more importantly, the heavier shepherding pony with which the riding type of Section B had shared a section, was valued in its own right and given a new classification of its own – the Section C, the Welsh Pony of Cob Type. This was a major breakthrough for the Welsh Pony, Section B.

There was a demand for such ponies, as observed by Dr Wynne Davies in his report of the WPCS Performance Award Ceremony in 1977 at which the Society's Chairman, Lord Kenyon, introduced the guest speaker, the Duke of Beaufort. He wrote:

> [Lord Kenyon] described in an amusing anecdote how, just before the First World War when he was rather disillusioned by many hunting falls, it was riding a Breconshire-bred Welsh pony which reinstated his confidence and interest in riding. After this successful experience with Welsh ponies, about 5 or 6 were purchase annually for Cub-hunting and with these duties completed, several found a ready market as Polo Ponies; indeed, some of them became quite famous.

Ironically, it was a Breconshire-bred pony – Criban Victor – which became the mainstay of Lord Kenyon's Gredington Stud many years later.

The discussion on the future of the Welsh pony was not restricted to Wales, as shown in an exchange of views published in the popular 'Riding' magazine in October 1944. In the feature entitled 'The Horseman's Brains Trust', Miss Burnaby-Atkins of Stamford asks, 'Will the Brains Trust say whether in their opinion there is any future in the breeding of the Welsh Mountain Pony?'

Significantly it was the leading showman of hacks and hunters of the day, Count Robert Orssich, who replied:

> Most certainly. They are, I think, the most beautiful of all our indigenous pony breeds, for they are hardy, sound, most tractable, and eminently suitable, especially if crossed with Arab blood, as children's ponies. The ones I have seen lately seem a little too coarse or common, getting away from the real type; therefore, I think some infusion of Arab blood would be of use.

Of course he reflected a current trend in the breeding of children's riding ponies, which had already been embraced by the likes of Mrs Inge and Tom Jones Evans. Their use of the Arab offspring Tanybwlch Berwyn and Craven Cyrus, respectively, would be acknowledged as the means to upgrade the height of the Welsh pony while maintaining many of its characteristics. Orssich used the term 'real type' without describing what it was, and it was the perception of type that plagued the Section B through a significant part of its history as breeders made use of the Foundation Stock Register in order to bulk up numbers to meet a growing demand for this expanding section of the Welsh Stud Book.

Two women, Miss Brodrick in North Wales and Mrs Cuff in the south, while tackling the task of breeding the Section B from different directions, shared a common goal which was to breed a pony of correct conformation and movement which was suitable for a child to ride, be it in the show ring, out hunting or for pleasure. They both had their followers, breeders who would expand the gene pool, and eventually the merging of the Downland and Coed Coch bloodlines would shape the future of the breed. The interpretation of how bloodlines interacted was perfected by the likes of Jack Edwards at Weston and expanded upon more recently by Jess Parry at Eyarth, who would be the first to admit that success has only come about by the efforts of others, often small breeders with a vision of their own.

The early promotion of the Section B must be attributed to Miss Brodrick and people like the Bullen family and Mrs Heyburn, who exposed the showing public to beautifully-produced Welsh ponies such as Coed Coch Pryderi and Coed Coch Powys in open riding pony classes. Howson warned breeders against changing the type of the Mountain pony to accommodate the riding classes – little did he realise that they never would, and that initially it would be the Section B that reached the top in 12.2hh classes. Ponies such as Chirk Seren Bach, Senlac Fleurette and Firby Fleur de Lys held their own in any company right up to the 1970s, but as the child's riding pony demanded more Thoroughbred quality, they too would be sidelined – although not for long, as the working hunter and show hunter pony classification found a new role for the Welsh Section B.

The role of the Section B as a crossing sire for breeding the quality show pony found considerable success as far back as 1954, when 'Horse and Hound' reported,

Among the widely-known pony mares sent to Welsh riding type stallion Reeves Golden Lustre is Mr Deptford's champion pony Pretty Polly and the same owner's well-remembered Firefly. Mr F E Blythe is sending his 12.2hh winner Glide On. Golden Lustre is standing at the Reeves Stud, Penn, Bucks with Mr & Mrs Gordon Gilbert.

More recently, stallions such as Chirk Caradoc and Keston Royal Occasion and the Part-breds Fairley Rembrandt and Sandbourne Royal Ensign found new heights as show pony sires.

The show ring itself witnessed great changes in the breeding classes. Over a 60-year period of the Royal Welsh Show – the Mecca for all Welsh pony and cob enthusiasts – numbers shown in Section B classes have risen from 16 in 1959 to 322 in 2009, and the number of classes on offer has increased from three to 12 over a similar period. For the first time in the history of the Section B, there was an overseas judge to place the classes at the 2009 Royal Welsh – American judge Mrs Gail Thomson attracted a record entry for the breed. A sign of the times and a reflection of the interest in the Section B in continental Europe, Mrs Thomson's overall champion that day was the stallion Liezelshof Macho, shown by Dutchman Peter de Rade for the owner/breeder, Danny Daelemans from Belgium. The Dutch-bred stallion Shamrock Mr Oliver was male champion in 1995; furthermore, history was made in 2012 when Stougjeshoeve Escudo, a yearling colt bred in the Netherlands by Geert Verbaas, was judged youngstock

Colne Heiress, winner of the Brightwells Championship for the best Welsh pony or cob shown at the Horse of the Year Show during the Welsh Pony and Cob Society's centenary year in 2001.

Section B champion at the Show. It proves that the Welsh breeds, Section B included, no longer belong only to their homeland but have become universally adopted by countries across the globe.

At home, there has also been a proliferation of classes for ponies in the performance field, including the Champion Native Ridden Pony of the Year, staged for the first time at Olympia in 1978, when the championship was won by the Section B, Criffell Casper. Today classes have embraced the full height range of the Section B, ranging from the small lead rein to the full height pony, which is capable of winning an open working hunter pony competition. Examples of all of these have been recorded at the Horse of the Year Show, which has led the charge of all showing disciplines. The Welsh Pony, Section B, has benefited from ridden classes both on the flat and over jumps – Tetworth Catelpa (f. 1983) was winner of the first Mountain and Moorland Working Hunter Pony of the Year in 1993; the full-height Blainslie Wedding Breakfast was judged Working Hunter Pony of the Year in 1997; the

Crystal Vision Supreme, Pony of the Year at the Horse of the Year Show 2011

lead rein, Colne Heiress, was winner of the Brightwells Championship for the best Welsh pony or cob shown at the Horse of the Year Show during the WPCS's centenary year in 2001; Wortley Celebration was judged Show Hunter Pony of the Year in 2010 and then reserve mini native ridden champion the following year, when Crystal Vision was judged Supreme Pony of the Year.

The success of ridden Section Bs explains why such high prices have been achieved for them in recent years. There is nothing new in the fact that some enthusiasts breed Section Bs exclusively for the show ring and others are prepared to buy ponies specifically for that purpose. It is most fortunate that the sale ring has provided a vehicle for both and without question the Fayre Oaks Sale, started in 1955, has been a godsend for breeders and buyers of Section B alike. Brightwells and its predecessor, Russell, Baldwin and Bright, have been the official auctioneers at Fayre Oaks since then and the service that they have provided has been exceptional by any standards. Few breeds can boast a sale facility such as the one the company provides every year, and good prices have been recorded since the 1960s. In a report published in the WPCS Journal, Dr Davies wrote:

> The 1967 Fayre Oaks Sale was a very different affair from the first Sale of 1955. It is Section B ponies that have been realizing the record prices for the past two or three years but whilst their figures this year were still very good by any standards, their prices seem to have levelled off and become more realistic and in line with the smaller Mountain ponies.

Wortley Celebration, Show Hunter Pony Champion at the Horse of the Year Show 2010.

It was at the 1973 Fayre Oaks Sale that 740 guineas was paid for a service to Downland Chevalier when stallion nominations were auctioned in aid of the Animal Health Trust. Other nominations for well-known stallions were a fraction of the price.

Due to ever-increasing costs, the expensive nature of showing ponies at the very top, whether in hand or ridden, indicates that it may become a very exclusive pastime in the years ahead. There has also been a huge increase in the number of professionals involved in competition, with the attendant notion that one of them has to be involved in order to win at the top – although there are many examples which prove otherwise. It could be that British breeders of the Section B may have to turn to the example set by their continental neighbours, who have shunned the tradition of the showing ponies and cobs in ridden classes in favour of the other performance disciplines such as dressage, showjumping, eventing or driving, all of which are perfectly suited to the Welsh Pony. Various schemes have encouraged the breeding aspect of performance – the WPCS has run both a performance show and a performance points competition; recently Grandstand Media has introduced sire and stud competitions based on animals qualified for the Horse of the Year Show.

Just as the early breeders of the Section B focussed on the breeding of Welsh ponies suitable for children to ride, perhaps the modern breeder will have to take a further step and focus on breeding a Welsh pony suitable for children to compete within the performance world. Just as the Downland Mohawk son, Downland Folklore, is held in high esteem as a performance sire on the Continent, perhaps the day will come when a similar name appears on the lips of competitors in Britain. It may take time and it may take a change in mindset based on the traditional view of hind leg conformation and movement, which does not fit comfortably with the European performance model. But the determination of breeders of the Section B has already proved remarkable, if not revolutionary, and shown that where there was a will, there was a way – and a very successful one at that.

Scottish Team Dressage member Waxwing Star Thyme.

Waxwing Prince Hal, winner Section B Horse of the Year Show 2011.

Bibliography

Welsh Ponies & Cobs 1980 (JR Allen) Dr E W Davies MBE FRAgS
The Welsh Pony 2006 (JR Allen) Dr E W Davies MBE FRAgS
Sixty Years of Royal Welsh Champions 2009 (JR Allen) Dr E W Davies MBE FRAgS
One Hundred Glorious Years 2001 (The Welsh Pony & Cob Society) Dr E W Davies MBE FRAgS
The Early Years – A collection of articles from the early Welsh Stud Books Vols 1-15 2002 (The Welsh Pony & Cob Society)
British Native Ponies and Their Crosses 1971 (Nelson) Phyllis Hinton
Handling Horses, first printed 1943 (Hurst & Blackett) Lt-Col Patrick Stewart
The Most of the Bubble – Reminiscences 1993 (Deadline Press, Washington) Kathleen Aldridge
A Lifetime with Ponies 1945 (Abbey Press Hexham) R Charlton
A Lifetime with Horses 1962 (Nicholas Kaye) R S Summerhays
Desert Heritage 1980 (AH & AW REED) P Upton
Out of the Desert 2010 (Medina Publishing) P Upton
The Welsh Pony, Described in Two Letters to a Friend (Forgotten Books) C A Stone
Live Stock Journal Almanac 1900 (Vinton)
Live Stock Journal Almanac 1901 (Vinton)
Welsh Pony & Cob Society Annual Journals
Welsh Pony & Cob Society Stud Books
Welsh Pony & Cob Society of Australia Annual Journals
The Scottish & Northern Welsh Pony & Cob Association Annual Journals
National Pony Society Stud Books
Riding Magazine
Horse & Hound
Country Life
Sale Catalogues: Brightwells; Russell, Baldwin & Bright; D L Jones (Brecon); Bruton, Knowles & Co (Glos); Jones & Sons (Wrexham);
Show Catalogues: Royal Welsh Agricultural Society; Royal Agricultural Society of England; National Pony Society Spring Show Islington

Year	Sec A	Sec B	Sec C
1894	*Not applicable*	*Not applicable*	*Not applicable*
1902	*Welsh Mountain Pony not exceeding 12.2hh*	*Welsh Pony not exceeding 13.2hh*	*Cob from 13.2hh to 14.2hh*
1906	*No change*	*No change*	*No change*
1908	*Part 1 Welsh Mountain Pony not exceeding 12hh* *Part 2 Welsh Pony over 12hh but not exceeding 12.2hh*	*Welsh Pony not exceeding 13.2hh*	*Cob from 13.2hh to 14.2hh*
1914	*Produce of registered sire and dam no requirement for inspection before registration*	*Produce of registered sire and dam no requirement for inspection before registration*	*Produce of registered sire and dam no requirement for inspection before registration*
1921	*New registration rule applies*	*New registration rule applies*	*New registration rule applies*
1925	*New registration rules introduced in a move to close Stud Book* *Register opened for mares with non-Welsh ancestry after inspection. They would be referred to as Foundation Stock (FS)*	*New registration rules introduced in a move to close Stud Book* *Register opened for mares with non-Welsh ancestry after inspection. They would be referred to as FS*	*New registration rules introduced in a move to close Stud Book* *Register opened for mares with non-Welsh ancestry after inspection. They would be referred to as FS*
1927	*Register closed to further FS entry*	*Register closed to further FS entry*	*Register closed to further FS entry*
1930	*Closed to registered sire and registered dam* *New Foundation StockAppendix introduced to boost numbers*	*Closed to registered sire and registered dam* *New Foundation StockAppendix introduced to boost numbers*	*Closed to registered sire and registered dam* *New Foundation StockAppendix introduced to boost numbers*
1931	*Ponies of mountain type, undocked and not exceeding 12hh*	*Ponies of riding type not exceeding 13.2hh*	*Cobs and ponies of Cob-type exceeding 12hh (i.e.inclusive of all heights)*
1935	*No change*	*No change*	*No change*
1949	*Ponies of mountain type, undocked and not exceeding 12hh*	*To be completely reconstituted with extensive list of conditions*	*Welsh ponies of Cob type exceeding 12hh but not exceeding 13.2hh*
1950	*Ponies of mountain type, undocked and not exceeding 12hh*	*Ponies of riding type not exceeding 13.2hh*	*Welsh ponies of Cob type exceeding 12hh but not exceeding 13.2hh*
1951	*No change*	*No change*	*No change*
1951	*No change*	*No change*	*No change*
1959	*New entries restricted to entries within this section*	*Entries will be closed to entriesfrom Sec C and D*	*Lower height limit of section C to be removed*
1960	*Register closed to FS. No transfers from B due to height* *Overseas register opened*	*Register closed to FS.No more C or D blood allowed* *Overseas register opened*	*Register closed to FS* *Overseas register opened*
1994	*International Register opened*	*International Register Opened*	*International Register Opened*
1997	*No change*	*No change*	*No change*

Sec D	Sec E	Welsh Part-bred	Comments
Not applicable	*Not applicable*	*Not applicable*	*First Welsh ponies registered in Vol 1 of PPS Stud Book*
Cob from 14.2hh to 15.2hh	*Not applicable*	*Not applicable*	*Initial entrants foundation mares and stallions inspected before registration. Stallions required vet's cert. of soundness. Produce with fully registered parents were eligible for entry.*
Cob from 14.2hh	*Not applicable*	*Not applicable*	*Upper height limit abolished for cobs*
Cob from 14.2hh to 15.2hh	*Not applicable*	*Not applicable*	*Part 1 not docked and not hogged* *Part 2 docking and hogging allowed but not necessary*
Produce of registered sire and dam no requirement for inspection before registration	*Not applicable*	*Not applicable*	
New registration rule applies	*Not applicable*	*Not applicable*	*No pony or cob eligible for entry in Hackney or any other stud book will be eligible for registration*
New registration rules introduced in a move to close Stud Book	*Not applicable*	*Not applicable*	*Registration requires: 1) by a registered sire 2) out of a registered mare or FS mare or mare by a registered sire*
Register opened for mares with non-Welsh ancestry after inspection. They would be referred to as FS	*Not applicable*	*Not applicable*	
Register closed to further FS entry	*Not applicable*	*Not applicable*	*Foundation Stock registration stopped. Original rules no longer apply.*
Closed to registered sire and registered dam *New Foundation Stock Appendix introduced to boost numbers*	*Not applicable*	*Not applicable*	*After three crosses on the top line using fully registered sires in the Welsh Stud Book, offspring would be eligible for full registration: FS>FS1>FS11>*
Abolished	*Not applicable*	*Not applicable*	
Reintroduced for geldings	*Not applicable*	*Not applicable*	
For Cobs exceeding 13.2hh	*Geldings from all sections*	*Not applicable*	
For Cobs exceeding 13.2hh	*Geldings from all sections*	*Not applicable*	*1948 regulations for section B rescinded due to lack of impact they were achieving*
No change	*No change*	*Part-bred Register started ponies must have 50% known registered Welsh blood. Height up to 13.2hh*	*After 2 crosses on top line using fully registered sires in Welsh Stud Book, offspring eligible for full registration: P>P1>P2*
No change	*No change*	*Welsh Riding Pony Stud Book established*	*Very few entries in this section of the Welsh Stud Book so closed in 1960*
No change	*No change*	*No change*	
Register closed to F.S. *Overseas register opened*	*No change*	*Ponies must have min. 25% known registered Welsh blood. No height limit*	*FS Appendix closed to FS mares*
International Register opened	*No change*	*No change*	
No change	*No change*	*Part-bred Register changed ponies must have min. 12.5% known registered Welsh blood*	

Index of Horses and Ponies

R

S

Index of People and Places